An Intuitive Guide to Compensating Switching Power Supplies

APPLY CHRISTOPHE BASSO's RECIPES ON LOOP CONTROL

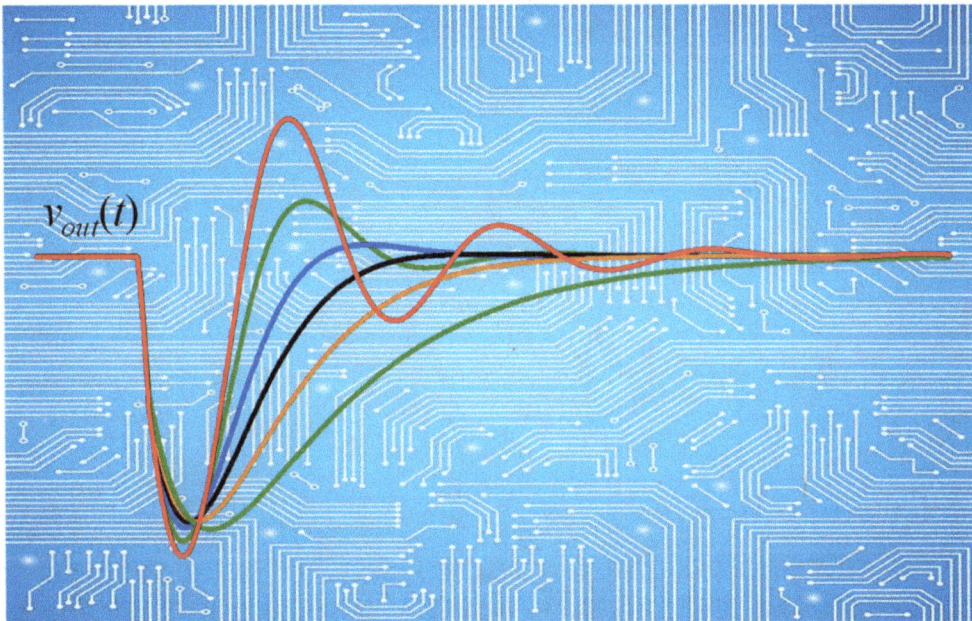

$v_{out}(t)$

Christophe Basso

An Intuitive Guide to Compensating Switching Power Supplies

Other books by Christophe Basso:

The Fast Track to Determining Transfer Functions of Linear Circuits: The Student Guide (2023)
Transfer Functions of Switching Converters: Fast Analytical Techniques at Work with Small-Signal Analysis (2021)
Linear Circuit Transfer Functions: An Introduction to Fast Analytical Techniques (2016)
Switch-Mode Power Supplies, Second Edition: SPICE Simulations and Practical Designs (2014)
Designing Control Loops for Linear and Switching Power Supplies: A Tutorial Guide (2012)
Switch-Mode Power Supplies: Spice Simulations and Practical Designs (2008)
Switch-Mode Power Supply SPICE Cookbook (1996)

Print ISBN 978-1-960405-37-1
eBook ISBN 978-1-960405-38-8

Cover Design by Guy D. Corp, www.GrafixCorp.com

Faraday Press
1000 West Apache Trail—Suite 126
Apache Junction, AZ 85120 USA

Acknowledgements

I WOULD LIKE to warmly thank the following people who kindly reviewed this book for identifying typos but also for correcting my English: Michael Schutten (United States), Iain Mosely (United Kingdom), Alain Laprade (United States), Eleazar Falco (Germany), Tomáš Gubek (Czech Republic), Patrick Wang (Taiwan), Joël Turchi (France), Maarten Laurijssen (The Netherlands) and Nicola Rosano (Italy).

Thank you to my sweet wife, Anne, who knows what it takes to be with a technical writer—hours confronting equations or crafting electrical diagrams—especially when this book is the tenth of a list started in 1996: there is more to come *mon Cœur*!

Merci also to Ken Coffman with Stairway Press, who spent time shaping my manuscript into the book you hold.

Preface

AMONG THE SUBJECTS I cover when teaching seminars, designing the control loop often comes back as the most difficult issue encountered by engineers. The theory is usually well understood but the implementation of what is described in textbooks is lacking in practical details, often unique to the specific control scheme. Many converters are stabilized by trial-and-error then shipped without the assurance that the product will remain stable during its operating lifetime. A badly compensated loop can result in a poor start-up sequence with voltage overshoots latching off the adapter, undershoots hanging the downstream logic circuits, operating noise when components start degrading and so on. All these issues can lead to a line down situation or worse, a product recall.

Literature abounds on the subject of loop control, and I wanted this book to be different. First off, I concisely wrote the theoretical part, concentrating on important parameters such as crossover frequency and phase margin. After reading this introductory section, you will no longer arbitrarily pick these parameters and understand how they set the transient response once the loop is closed. Second, it is important to understand the way compensators are built, how the active element affects the design and the expected ac response. For instance, a device like the TL431 can make your design life complicated if you ignore its constraints. This book really goes straight to the point but for those of you wanting deeper fundamentals, books and articles referenced in the end will offer the option to learn more on this important subject.

Finally, as I want this book to be design-oriented, most of the examples I used for illustration will run on Elements, the free demonstration version of SIMPLIS®. All models are available from my web page and ready to run. Each circuit is associated with an automated calculation macro, determining components values for the selected crossover frequency and phase margin. Then, run an ac or transient simulations to see immediate effects.

I hope you'll like the style and pace of this new *practical* manual and, as usual, please send comments to my email address cbasso@orange.fr.

Happy compensation to you all!

— Christophe Basso, September 2024

About the Author

CHRISTOPHE BASSO WORKS as a business development manager for Future Electronics in France. In his role, he provides technical assistance to customers developing power switching converters in Europe. Before this position, he was a Technical Fellow with *onsemi* in Toulouse, France. He led an application team dedicated to developing new offline PWM controller's specifications. Christophe has originated numerous integrated circuits among which the NCP120X series has set new standards for low standby power converters

Christophe has released several books on power conversion and simulation. His last title was published with Faraday Press and is entitled *The Fast Track to Determining Transfer Functions of Linear Circuits*. In this work, he shows how fast analytical circuits techniques, or FACTs, can help determine transfer functions of linear circuits in the fastest and most efficient way.

Christophe has over 25 years of power supply industry experience. He holds 25 patents on power conversion and often participates in conferences and trade magazines including How2Power.com. Prior to joining *onsemi* in 1999, Christophe was an application engineer at Motorola Semiconductor in Toulouse. Before 1997, he worked for 10 years at the European Synchrotron Radiation Facility (ESRF) in Grenoble, France. He holds a *diplôme universitaire de technologie* from the university of Montpellier (France) and a MSEE from the Institut National Polytechnique de Toulouse (France). He is an IEEE Senior Member.

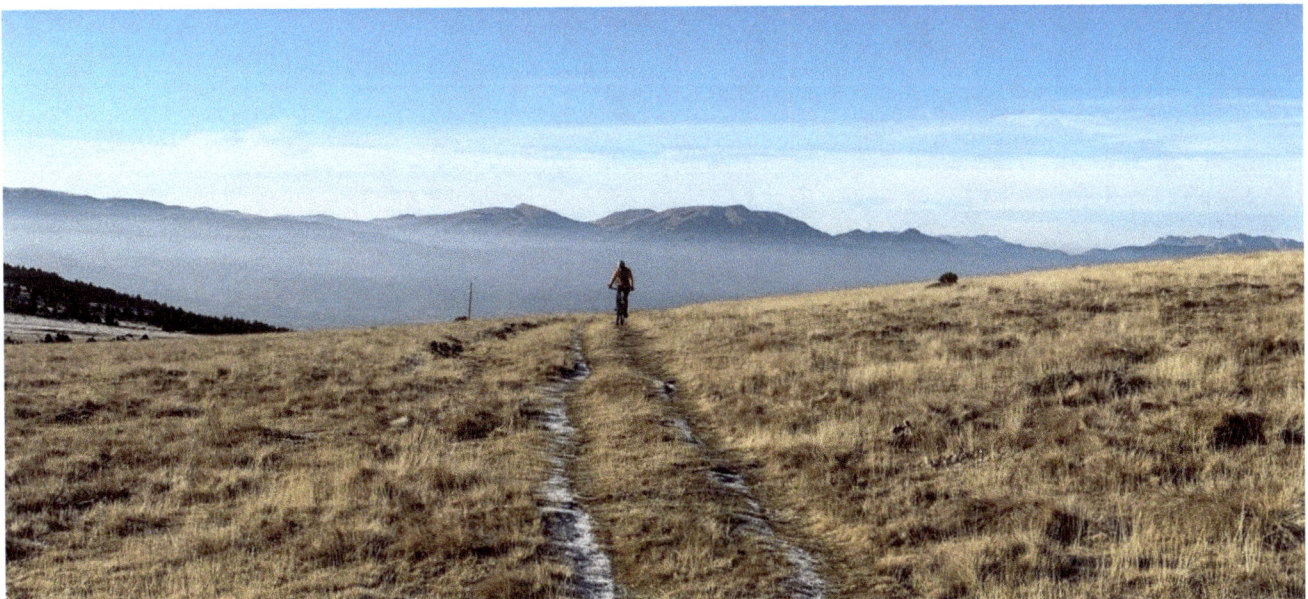

When he is not writing, Christophe finds inspiration in mountain bike outings.

Table of Content

- Introduction
- A few Pages on Theory   2
- Buck Converter in Voltage-Mode Control   76
- Buck Converter in Current-Mode Control   84
- Buck Converter in Constant On-Time (COT)   88
- Forward Converter in Voltage-Mode Control   91
- Forward Converter in Current-Mode Control   94
- Forward Converter Active Clamp Voltage-Mode Control   97
- Full-Bridge Converter in Current-Mode Control   99
- Phase-Shifted Full-Bridge in Voltage-Mode Control   101
- Boost Converter in Voltage-Mode Control   104
- Boost Converter in Current-Mode Control   107
- Power Factor Corrector in Borderline Conduction Mode   110
- Power Factor Corrector in Continuous Conduction Mode   113
- Buck-Boost Converter in Voltage-Mode Control   116
- Buck-Boost Converter in Current-Mode Control   119
- Flyback Converter in Voltage-Mode Control   122
- Flyback Converter in Current-Mode Control   125
- Flyback Converter in QR Current-Mode Control   129
- Flyback Converter with a UC384x Controller   137
- Single-Stage Flyback Power Factor Corrector   139
- Coupled-Inductor SEPIC in Current-Mode Control   141
- LLC Converter in Direct-Frequency Control   144
- LLC Converter in Bang-Bang Charge-Control   148
- LLC Converter in Current-Mode Control   151
- Practical Experiments   154
- References and books   157

The Sections on Theory

- An Open-Loop System   4
- Feedback to the Rescue   7
- Building an Oscillator   10
- Keep Away from Oscillations   13
- Transient Response   14
- Selecting Phase Margin   15
- The Crossover Frequency   17
- Generating the Duty Ratio   19
- How to Force Crossover   22
- Shaping the Loop Response   23
- Poles and Zeros   25
- The Three Compensators   26
- Compensating with type 1   28
- A Type 2 with an Op-Amp   31
- A Type 3 with an Op-Amp   35
- The PID Block   40
- The Optocoupler   44
- Type 1 with a TL431   47
- Type 2 with a TL431   50
- Type 3 with a TL431   54
- Type 1 Design with an OTA   59
- Type 2 Design with an OTA   60
- Digital Compensators   62
- Stabilizing Switching Converters   68
- Monte Carlo   71
- Front-End Filter Interactions   72

Introduction

LOOP CONTROL is a vast territory featuring many trails difficult to explore. Fortunately, being a good engineer in this playground does not imply you have to venture in all directions and discover everything. Unlike multi-variable control systems, like a car or an aircraft, the power converters covered in this book feature two or three *state variables*, making modeling and simulation a simpler exercise. Furthermore, these converters deliver a fixed value – the output voltage V_{out} or the current I_{out} – but need to fight perturbations such as the input voltage V_{in} or the output current. For that purpose, voltage (or current) regulation is ensured by a circuit permanently comparing V_{out}, or a fraction of it, with a stable reference voltage V_{ref}. This circuit, *the compensator*, takes action on the *control variable* if a deviation exists between the two values. It is the object of this book to offer guidance on how to design the compensator to obtain the desired dynamic performance, once the converter works in closed-loop conditions.

The control variable common to many switching structures is the duty ratio D. It can either be *directly* generated from the error voltage (voltage-mode control or VM), or *indirectly* obtained after setting the inductor peak current (current-mode control or CM). It is also possible to implement constant on-time (COT) or fixed off-time (FOT) control; or change the switching frequency F_{sw}. In all cases, for a given bias point, the duty ratio D will be identical whether the converter uses VM, CM or COT. As an example, assume a 5-V VM buck converter supplied by a 12-V rail. The duty ratio in this case, neglecting ohmic drops, is $5/12$ or 41.7%. Changing the scheme to CM or FOT for the same bias point, the duty ratio won't change. It is the way this duty ratio D is built that affects the converter dynamics.

The parallel exists with a resonant converter like an LLC: the error voltage can either directly control the switching frequency or set the resonant peak current, *indirectly* controlling the switching frequency. For the same operating point, the switching frequency remains the same but the power stage dynamics will drastically differ, depending on the control scheme.

When stabilizing a converter, it is important to understand how the wanted dynamic response correlates with the compensation strategy. It all starts with the *control-to-output* transfer function: if the control variable D is ac-modulated by a sinusoidal *stimulus* – say from 10 Hz to 100 kHz –, how does this signal propagate inside the circuitry to generate a *response* at the output V_{out}? The ratio $V_{out}(s)/D(s)$ is the transfer function we need to study the compensation network.

An Open-Loop System

AN OPEN-LOOP SYSTEM transforms a control signal, the *stimulus*, into an action, the *response*, following a specific relationship linking both signals. The *transfer function* is designated by H. In this device, the control input u is set independently from the output y as shown below:

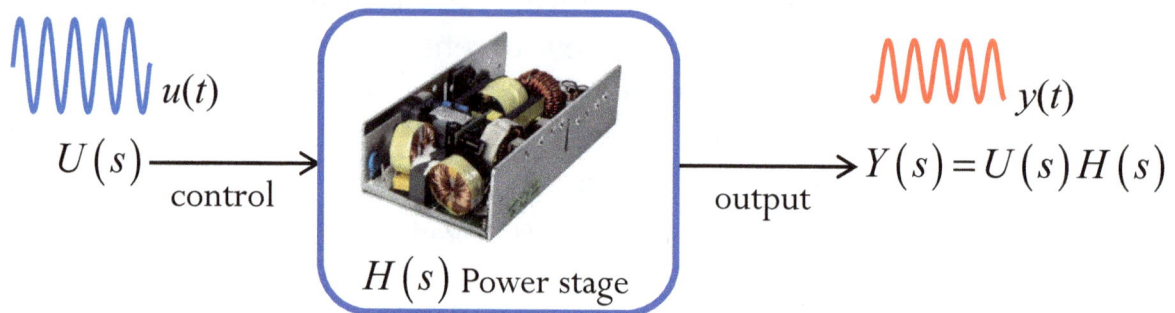

$u(t)$

$U(s)$ ——→ control

$H(s)$ Power stage

output ——→ $Y(s) = U(s)H(s)$

$y(t)$

The transfer function, a mathematical relationship expressed in the Laplace-domain, describes the change in magnitude and phase undergone by the stimulus when propagating through the power stage. By power stage, I imply the conversion stage of boost or buck converters for instance: if I apply a 1-kHz sinusoidal waveform at the control input, what are the amplitude and the phase of this signal observed on the output? This *control-to-output* transfer function is the *cornerstone* of any compensation exercise and you must obtain it before considering a stabilization strategy. How can you determine it?

- Analytical analysis: large-signal models are used to describe the *average* behavior of the converter. These models are then linearized to form a small-signal circuit from which the transfer function is extracted. I personally like the PWM switch model, derived by Dr. Vatché Vorpérian in 1990 [1, 2]. In my opinion, for the 2-switch converters studied in this book, it supplants state-space average (SSA) modeling by its amazing simplicity and ease of implementation. Why having an expression versus a plot is important? Because the numerator and denominator will reveal the ac behavior of the circuit by highlighting the position of poles and zeroes. These contributors shape the response but move in relationship with components values and their stray elements such as the equivalent series resistance (ESR) for a capacitor. Knowing who the offenders are, is of paramount interest to neutralize their variability.

The Transfer Function

- Ac simulations: you build your circuit with a dedicated *averaged* model such as the *PWM switch* previously mentioned. Averaged models simulate fast as they do not consider the switching component and lend themselves well for obtaining a transient response or the control-to-output transfer function you want.

- Transient simulations: some converters, like LLCs for instance, cannot be modeled with an averaged circuit simply because the switching component and its harmonics convey energy. Therefore, you cannot ignore them for analysis as you do in an averaged model. Piece-wise linear (PWL) simulators such as SIMPLIS® or PSIM® are thus perfectly suited for extracting the ac small-signal response from the switching circuit, letting you verify the correct operating point and obtain the needed transfer function in a few seconds. I know LTspice® also offers the possibility to extract the ac response from a switching circuit, but, in my opinion, it is less convenient than SIMPLIS® which has been purposely designed for this approach.

- Build a prototype: if you don't have access to a simulator, you can always resort to a bench measurement and extract the control-to-output Bode plot with a frequency-response analyzer (FRA). There are plenty of affordable types to choose from nowadays and you will have to measure the loop anyway for validating your compensation strategy. However, it takes time to build the prototype and components may not be immediately available, further delaying the experiments. Better to begin with a computer first and then verify results in the lab.

- ❖ I recommend combining all these methods: analytical analysis, simulations and final measurements on the bench. If parasitics have been well extracted and reflected in the simulation setup, simulations are likely to be well matched by the bench measurements. This final step is mandatory to verify that your approach was correct from the start.

Modeling by Blocks

WE CAN LOOK at a switching converter operated in open-loop conditions via a simplified graphic illustration. In this drawing, you see the reference voltage – a stable and precise dc source – driving the power stage whose output, let's assume a voltage level V_{out}, is affected by the input source V_{in} and also current I_{out}. We could add other variables like temperature or pressure depending on the operating conditions. All these contributors are modeled as *perturbations* and affect the output voltage:

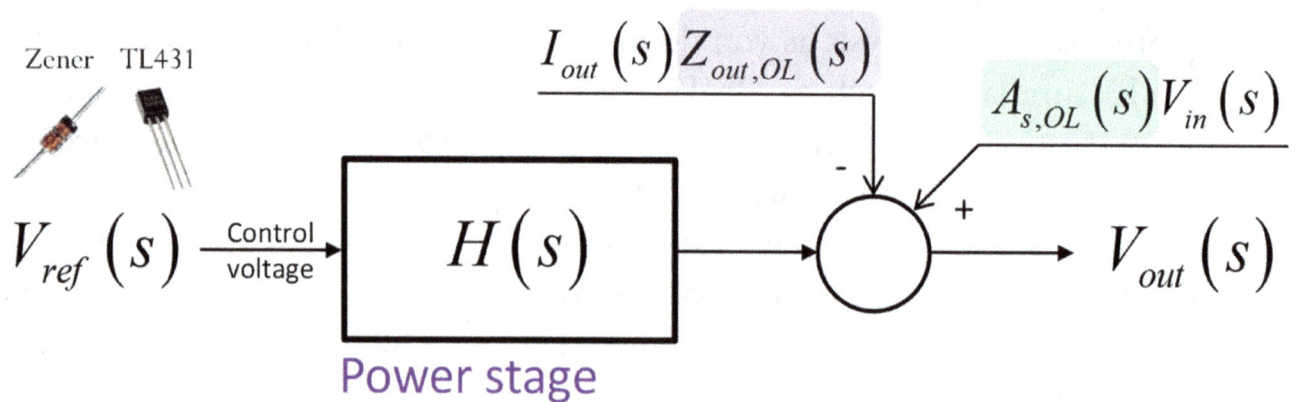

Zener   TL431

$$V_{ref}(s) \xrightarrow{\text{Control voltage}} \boxed{H(s)} \xrightarrow{} \bigotimes \longrightarrow V_{out}(s)$$

$$I_{out}(s)Z_{out,OL}(s)$$

$$A_{s,OL}(s)V_{in}(s)$$

Power stage

You can easily express the output voltage by following the flow from left to right:

$$V_{out}(s) = \underbrace{V_{ref}(s)H(s)}_{\text{Power stage}} - \underbrace{I_{out}(s)Z_{out,OL}(s)}_{\substack{\text{Open-loop output impedance} \\ \text{contribution}}} + \underbrace{V_{in}(s)A_{s,OL}(s)}_{\substack{\text{Open-loop input} \\ \text{voltage contribution}}}$$

In this open-loop equation – hence the *OL* subscripts – you can immediately see how the output impedance affects the output by dropping voltage proportionally to the delivered current. Similarly, if the input source changes, V_{out} will invariably be affected by the ability of the power stage to reject these perturbations as it can. This is called *audio-susceptibility* or *power supply rejection ratio* (PSRR). Please note that all this discussion assumes a linear system and the terms appearing in the expression are *small-signal* ones.

Feedback to the Rescue

CONSIDERING OUR GOAL to build a stable and precise system, we need a way to counteract or limit the effects of these perturbations. A known way to do it, is to build a *control system*. In such a scheme, a portion of the regulated variable (V_{out} but it could also be I_{out}) is permanently observed and compared with our reference voltage V_{ref}. Any deviation between the two will produce an error signal forcing the control system to take action so as to minimize the deviation. The error voltage is usually noted ε (Epsilon) and shaped through a compensator to drive the power stage. Therefore, the control variable which was V_{ref} in our previous block is now affected by a *feedback* from the variable to be regulated:

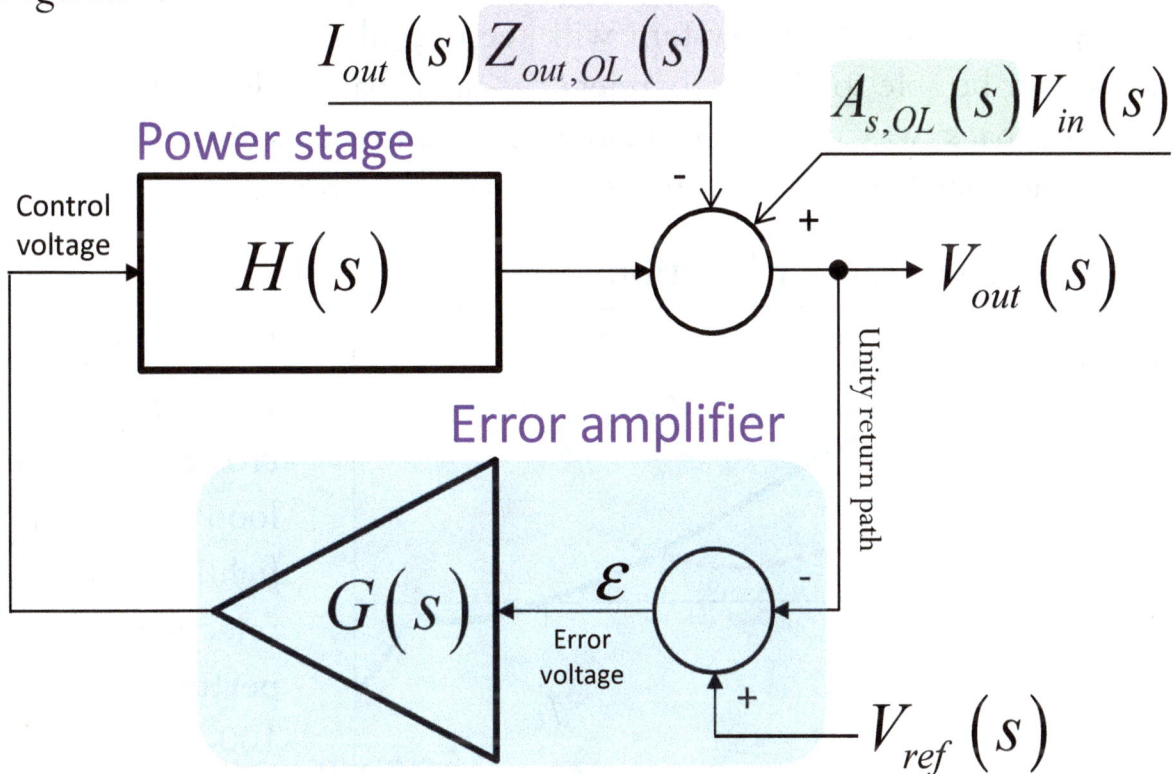

$$I_{out}(s)Z_{out,OL}(s)$$

$$A_{s,OL}(s)V_{in}(s)$$

Power stage

Control voltage

$$H(s)$$

$-$

$+$

$$V_{out}(s)$$

Unity return path

Error amplifier

$$G(s)$$

ε

Error voltage

$-$

$+$

$$V_{ref}(s)$$

With this new circuit, it becomes possible to write the closed-loop expression now determining the output voltage:

$$V_{out}(s) = V_{ref}(s)\frac{T(s)}{1+T(s)} - I_{out}(s)\frac{Z_{out,OL}(s)}{1+T(s)} + V_{in}(s)\frac{A_{s,OL}(s)}{1+T(s)}$$

Theoretical output

Closed-loop output impedance

Closed-loop input rejection ratio

We need Gain for Regulating

IN THE CLOSED-LOOP expression, the term $T(s)$ represents the *loop-gain* that you obtain by multiplying $H(s)$ – the power stage transfer function – by $G(s)$, the compensator ac response you have to determine: $T(s) = H(s)G(s)$. By combining gain or attenuation, pole(s) and zero(es) in $G(s)$, you will shape the loop gain $T(s)$ magnitude to cross over the 0-dB axis at a selected frequency f_c – the *crossover frequency* at which the loop gain magnitude is unity. Before this value, the control system has gain and is active in performing its regulation work, e.g. rejecting perturbations occurring within this frequency range. Beyond f_c, the loop no longer has gain and the system operates in *ac open-loop*: remember this adage, *no gain, no feedback*! If a perturbation occurs at a frequency higher than f_c, the system will keep regulating the dc value but it will no longer be able to efficiently reject the disturbance, doing what it can as if its ac feedback was gone. The below graph illustrates this principle and highlights the zone before and after the loop gain has dropped to 1:

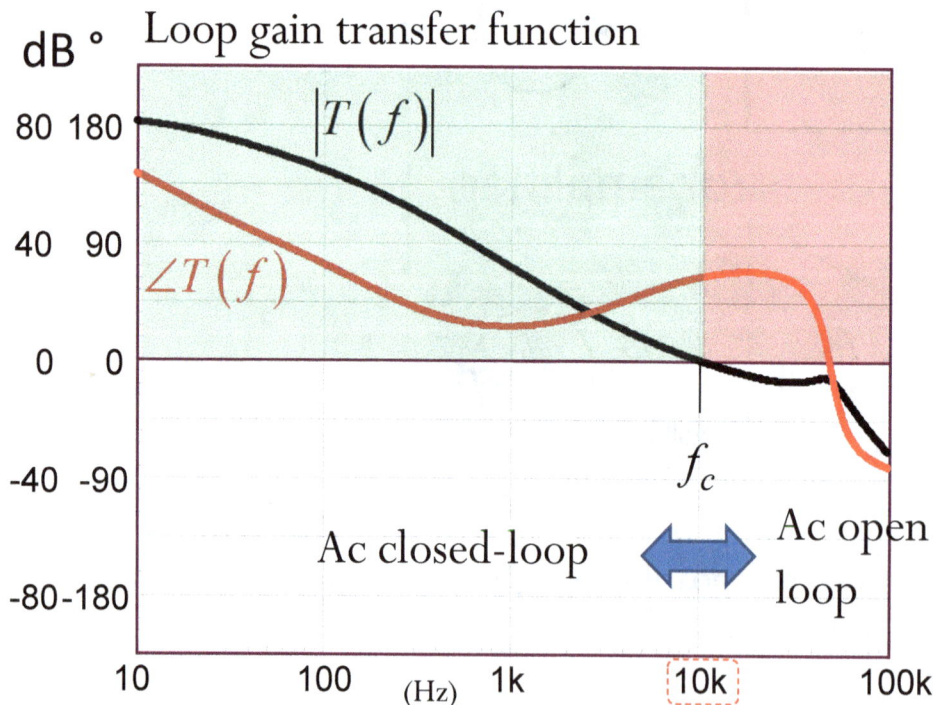

Having a high gain well before crossover, implies a loop vigorously fighting and rejecting incoming perturbations, like a 100-Hz ripple from a full-wave rectifier.

Looking at this graph, does it mean we always need to push f_c close to its maximum theoretical value which is $F_{sw}/2$? Certainly not, as you would build an unnecessarily-sensitive circuit, over-reacting to the environmental noise and highly unstable. So tailor f_c to suit your needs and not more than that.

The Benefits of Feedback

CLOSING THE LOOP offers a series of benefits as shown in our updated equation below where the subscript *CL* designates a closed-loop term:

$$V_{out}(s) = V_{ref}(s)\frac{T(s)}{1+T(s)} - I_{out}(s)Z_{out,CL}(s) + V_{in}(s)A_{s,CL}(s)$$

Each term is affected by the sensitivity function, $S(s)$: $\quad S(s) = \dfrac{1}{1+T(s)}$

If we rewrite this equation in static conditions, for fixed output current and input voltage – their small-signal values are zero, i.e. $\hat{i}_{out} = 0$ and $\hat{v}_{in} = 0$ – we can write:

$$V_{out}(s) = V_{ref}(s)\frac{T(s)}{1+T(s)} \quad \xrightarrow{T_0 \to \infty} \quad \begin{array}{l} \varepsilon \to 0 \\ V_{out}(s) \approx V_{ref}(s) \end{array} \quad \text{No static error}$$

If you close the loop with an operational amplifier (op-amp) featuring a high open-loop gain A_{OL} – then you see how the dc static error is theoretically nulled. Similarly, the perturbations are now reduced in proportion to the loop gain magnitude.

$$Z_{out,CL}(s) = Z_{out,OL}\frac{1}{1+T(s)} \quad \xrightarrow{T_0 \to \infty} \quad R_{0,CL} \to 0\,\Omega \quad \text{No output drop}$$

$$A_{s,CL}(s) = A_{s,OL}\frac{1}{1+T(s)} \quad \xrightarrow{T_0 \to \infty} \quad A_{s,CL} \to 0 \quad \text{Infinite input rejection}$$

You now realize that when $T(s)$ drops in magnitude, the sensitivity function approaches unity and weakens the feedback action. The previous graph showing the system running in ac open loop beyond f_c illustrates the impact of the sensitivity function. As a practical example, if your ac-dc converter is fed by a rectified rail affected by a strong 100-Hz ripple, e.g. you cannot afford a large bulk capacitor, then you will have to make sure the loop offers a significant gain below 200 Hz otherwise the ripple will show up on V_{out}.

Building an Oscillator

THE CROSSOVER FREQUENCY is one parameter to consider when closing the loop. Ensuring stability margins is another important aspect. Let's consider the below closed-loop system in which the compensator lies in the return path:

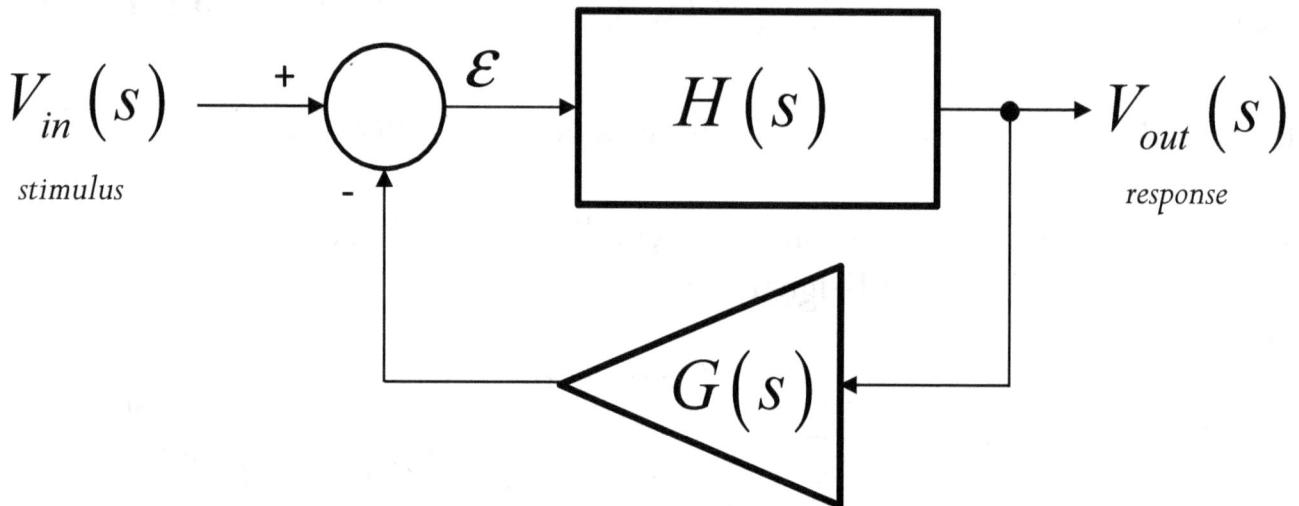

Now assume you would like to build an oscillator, meaning that if energy is brought to the system – like noise at power up or via the stimulus V_{in} – then oscillations will appear and remain at a constant amplitude when the stimulus is gone. What conditions in this circuit would guarantee this operation?

$$\frac{V_{out}(s)}{V_{in}(s)} = \frac{H(s)}{1+H(s)G(s)} = \frac{H(s)}{1+T(s)}$$

Write the closed-loop transfer function linking V_{out} to V_{in}.

To sustain oscillations when the stimulus V_{in} disappears, what are the conditions to be fulfilled?

$$V_{out}(s) = \lim_{V_{in}(s) \to 0} \left[\frac{H(s)}{1+G(s)H(s)} V_{in}(s) \right]$$

To obtain oscillations while the stimulus has gone, the quotient must approach infinity, implying a denominator equal to zero.

Oscillations are obtained when the magnitude of $T(s)$ is unity and V_{out} returns in phase at its control input. It is the *Barkhausen* criteria:

$$1+G(s)H(s)=0 \longrightarrow \begin{array}{l} |G(s)H(s)|=1 \\ \angle G(s)H(s)=-180° \end{array} \longrightarrow \begin{array}{c} Nyquist \\ \overline{+} \\ -1, j0 \end{array}$$

Time-Domain Simulation

TO ILLUSTRATE this theory, the SPICE circuit below is a phase-shifted oscillator. Each RC network contributes a phase shift and affects the return path in magnitude and phase. The gain of the loop is adjusted via resistor R_f. The goal is to plot V_{out}/V_{in} open-loop transfer functions obtained with different R_f and observe the time-domain response when the circuit is excited.

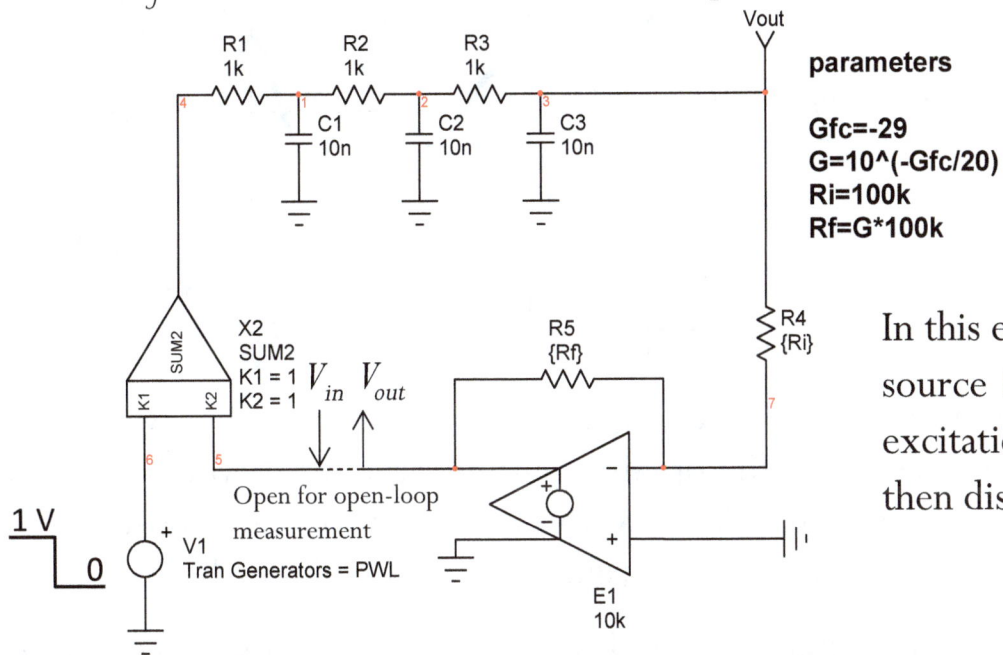

In this example, source V_1 brings an excitation signal and then disappears.

In the first attempt, the loop gain crosses the 0-dB axis at 31 kHz but the loop phase lag is less than 180° at this point.

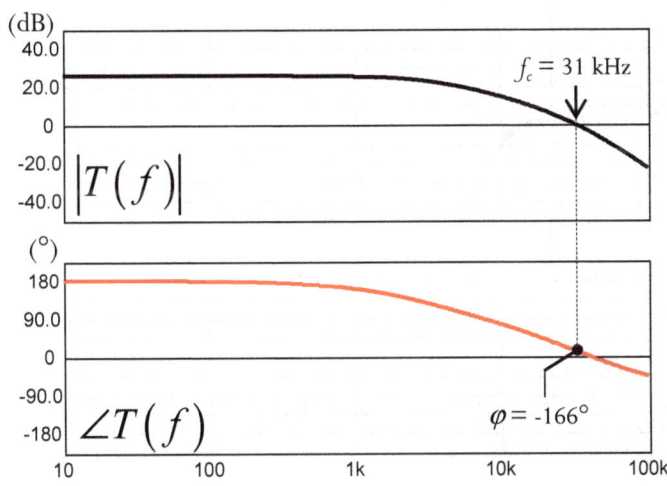

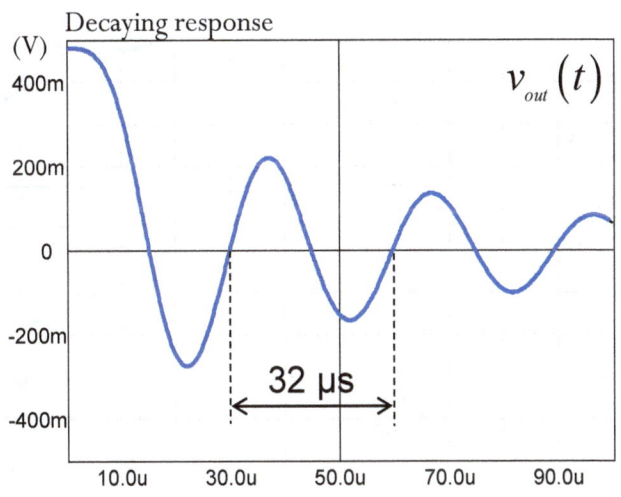

The time-domain response is damped and oscillatory, implying that the denominator of $T(s)$ features negative roots: the poles are stable and located in the left half plane. They are designated as left-half-plane poles or LHPP.

An Unstable System

FOR THE NEXT EXPERIMENT, I will increase the loop gain so that crossover occurs when loop phase is -190° at that frequency which is 50 kHz.

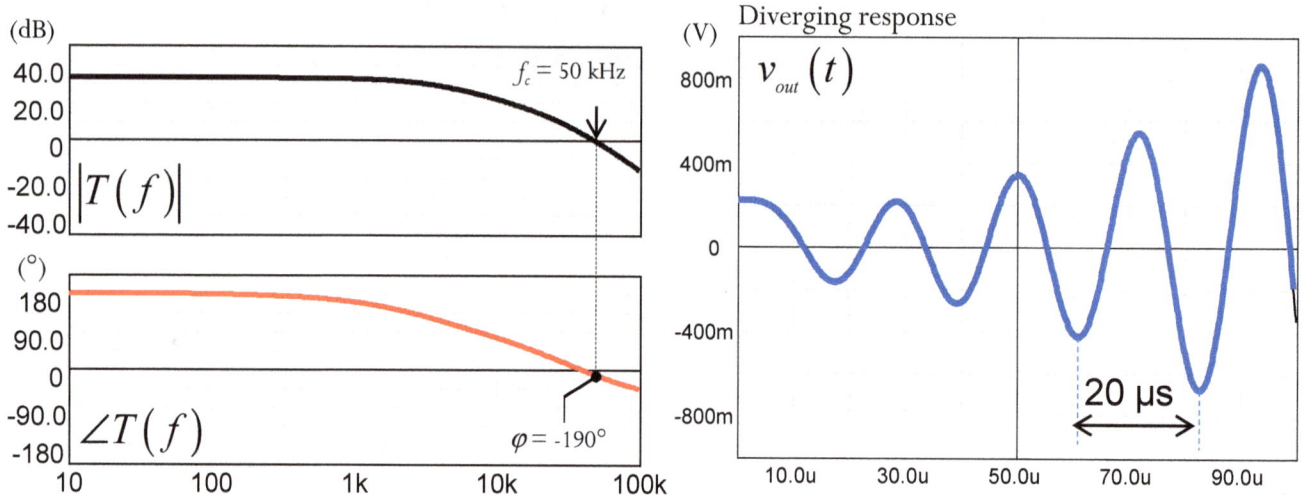

The time-domain waveform is diverging and you have an unstable system. The amplitude will increase until the op-amp output saturates. If you analyze the open-loop gain denominator in this mode, the roots are positive and correspond to *unstable* poles known as right-half-plane poles or RHPP.

For the third and final experiment, R_f is adjusted to cross over exactly at the point where the loop phase lag is 180°.

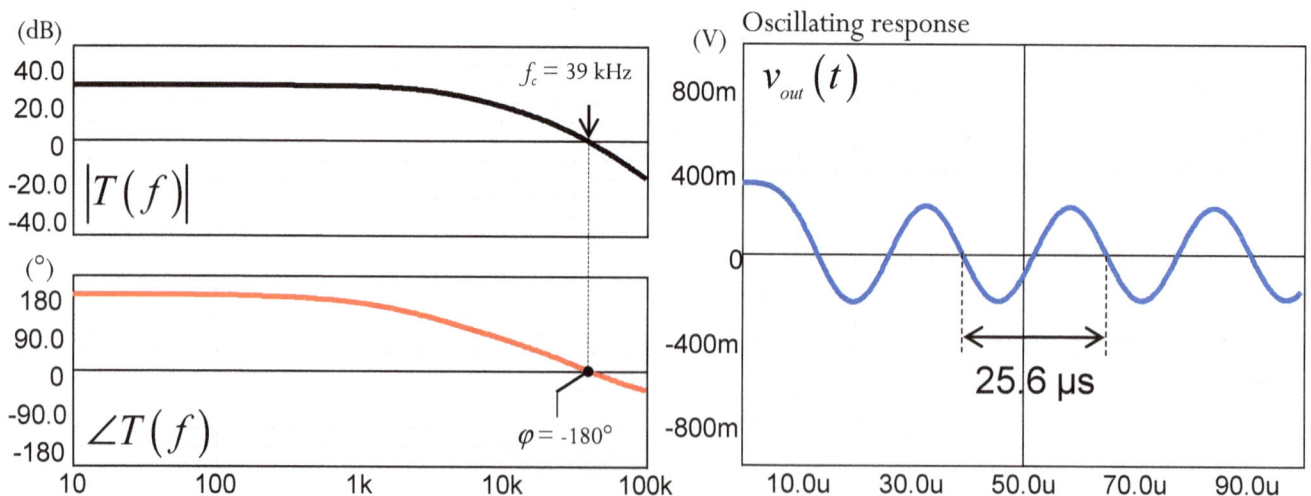

In this last step, oscillations are sustained: the poles are imaginary and complex conjugates. Oscillations remain as long as the gain magnitude is 0 dB.

Keep Away from Oscillations

OUR GOAL, as a power supply designer, is not to build an oscillator but ensure a stable, non-oscillatory, transient response of the converter. From the previous experiments, we can see that oscillations are obtained and self-sustained only when the *Barkhausen* criteria is met: if the stimulus returns with the *exact* same amplitude and in phase at the injection point, then you have conditions for oscillations. Practically speaking, on the loop gain Bode plot, you locate the point where the magnitude is 1 or 0 dB – and that is the *only* point to consider – then measure the loop phase at this point. The distance to -360° or 0°, represents the *phase margin* often noted PM or φ_m. The below graph illustrates this concept with the compensated loop gain of a voltage-mode buck converter:

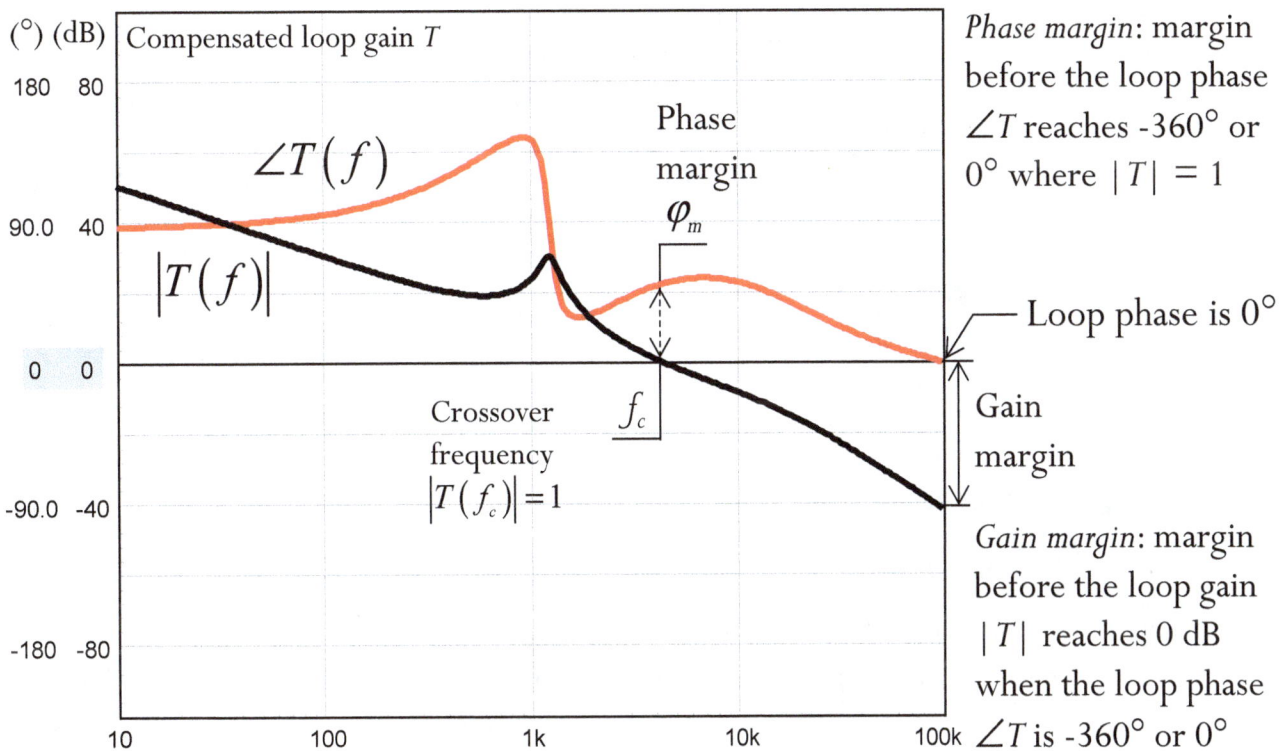

If the above magnitude curve shifts up – because of components tolerance or temperature variations – and forces a new unity gain where the loop phase is -360° or 0°, you would have conditions for oscillations and instability. This safety distance is called the *gain margin,* noted GM. For a system with low variability, a gain margin of 10 to 15 dB is an acceptable value. If the magnitude curve is highly variable, you may want to increase this number.

Transient Response

FROM THE TIME-DOMAIN responses we obtained with the phase-shifted oscillator, we can see how phase margin of the *open-loop* gain affects the response to a step when the system operates in *closed-loop* conditions. To further illustrate this important link, I have simulated a converter whose crossover frequency is fixed to 5 kHz but the compensator is tailored for different phase margins. A load step is applied to its output and transient responses are recorded:

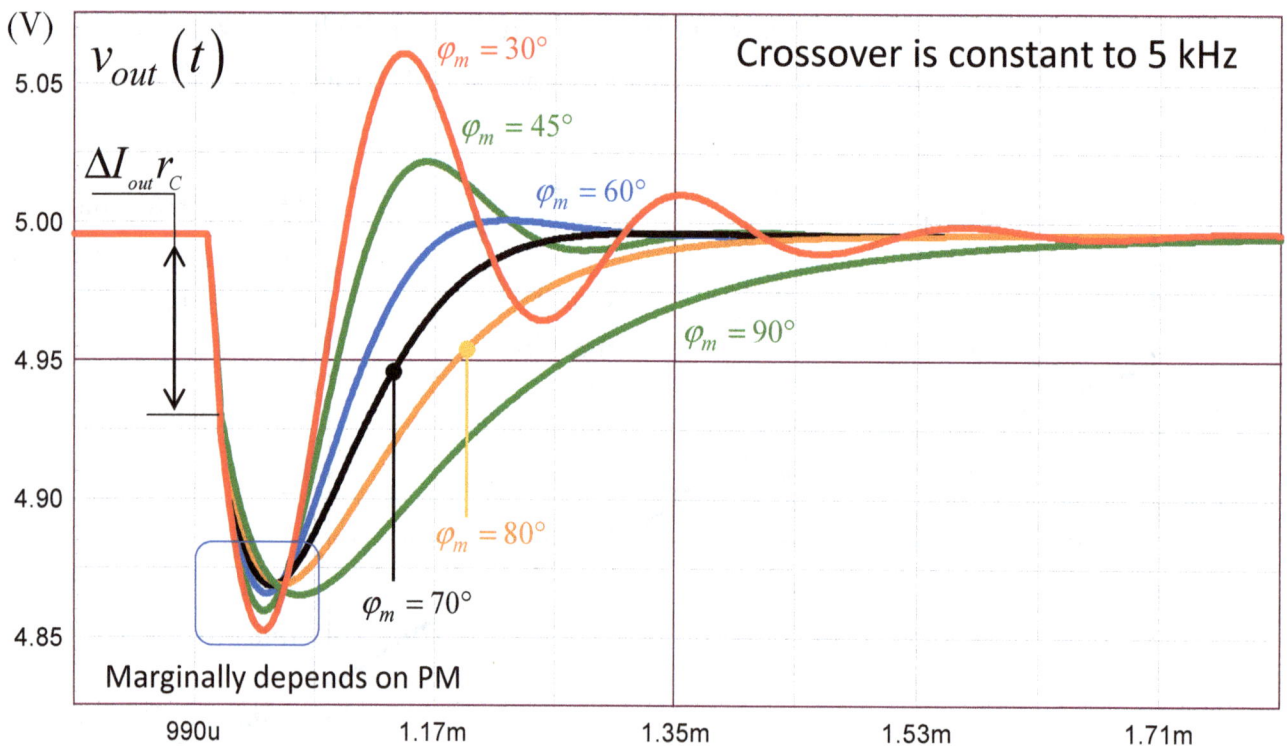

The step response depends on the small-signal output impedance Z_{out}, shaped by the output capacitor equivalent series resistance (ESR) r_C and the effective output capacitance (we neglect the parasitic inductance here). If the phase margin is large, approaching 90°, the recovery time is long but there is no overshoot. When phase margin reduces, recovery time improves and the response becomes more nervous or oscillatory. As phase margin reaches 45°, the recovery is quick but overshoot appears. At 30°, this overshoot becomes even larger and may impact the downstream electronics or trip an over-voltage protection (OVP) circuitry.

Selecting a Phase Margin

A USEFUL APPROACH for illustration purposes, is to consider a closed-loop control system as a *simplified* 2nd-order circuit, affected by a resonant frequency ω_0 and a quality factor Q_c. By observing the loop gain around crossover and by excluding other contributors (zeroes or poles in a higher-order system), it is possible to link the **closed-loop** quality factor Q_c with the **open-loop** phase margin φ_m. The idea behind this concept is to no longer arbitrarily pick a phase margin value but connect it with a desired behavior once the loop is closed.

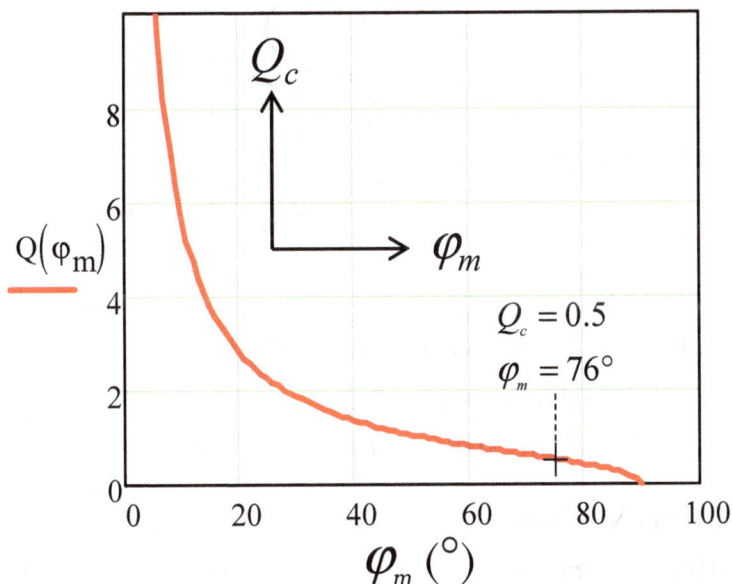

$$Q_c = \frac{\sqrt{\cos(\varphi_m)}}{\sin(\varphi_m)}$$

Q_c Closed-loop quality factor

φ_m Open-loop phase margin

- High values of Q_c will lead to faster, more oscillatory behavior.
- Lower Q_c will make the system response slower but better damped.

(graph shows Q_c vs φ_m ($^\circ$), with $Q_c = 0.5$, $\varphi_m = 76^\circ$)

Based on this graph, the two poles of the 2nd-order circuit are real and coincident – fast recovery with no ringing – for a phase margin of 76°: far from the 45° I learned from academia! So, what phase margin shall we select? There is no straight answer and the value will be adjusted based on the transient response you want:

- If overshoot is a concern, but recovery time is not an issue, select a comfortable phase margin for a rock-solid design, e.g. PM $> 70^\circ$ or even 90° for aerospace projects as an example.

- A quick recovery is important while a small overshoot is tolerated, then reduce phase margin to $50\text{-}60^\circ$ or so, making sure it never approaches 30° in the worst case. Keep in mind that loop phase will move in production but also with converter's age so *margin* is the important term here.

Where is the Limit?

WHEN YOU READ the literature on control systems, authors often refer to a -180° asymptote for evaluating the phase margin. In different documents, they talk about -360° or 0°. Who is correct? Actually, all three values refer to the same amount. It depends on where you open the loop:

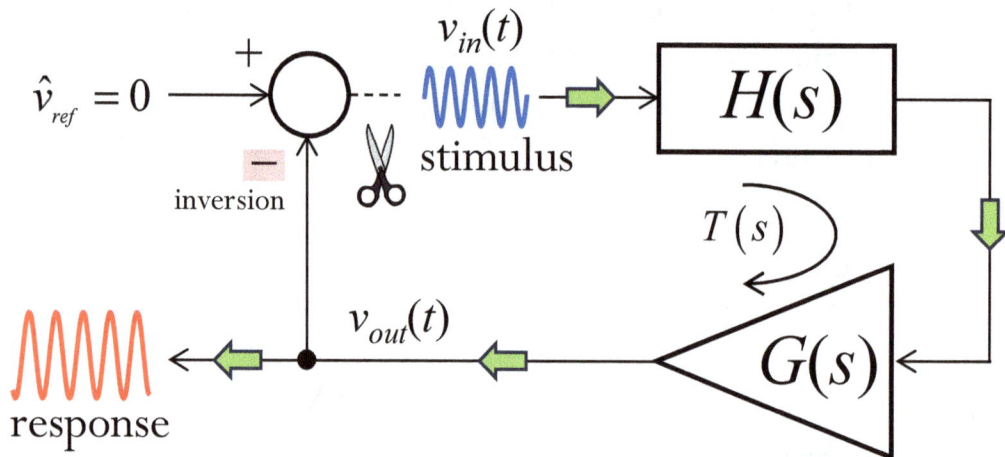

In the above circuit, the response is observed *before* the negative sign and the -180° shift inherent to the control system is excluded from the analysis. The limit for assessing the loop phase $\angle T$ is therefore -180° since another -180° will be added by the inverting operation. The sum of both is -360° or 0°.

In the bench, the operation is different. You open the loop with an injection transformer and consider the entire return path, naturally including inversion:

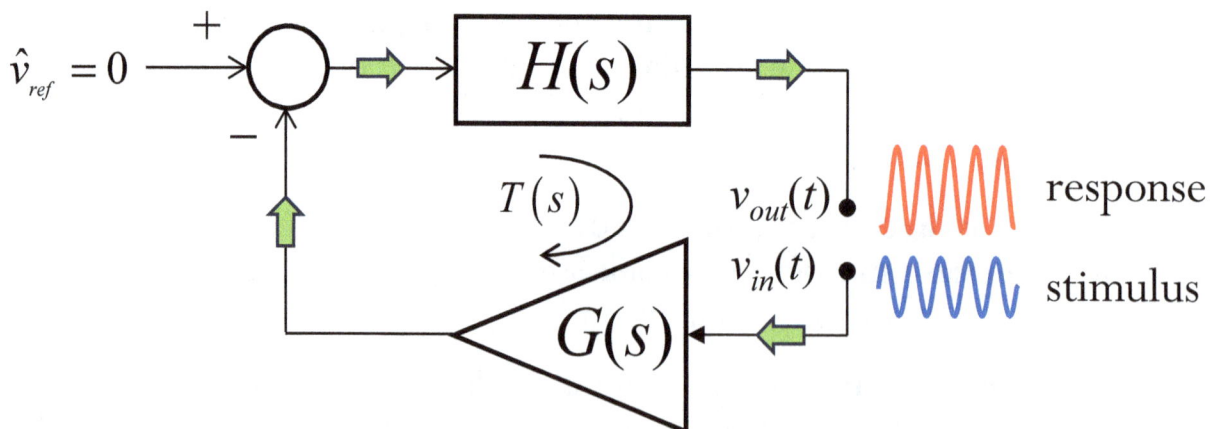

In this setup, you want to ensure the injected stimulus does not return in phase at the frequency where the loop gain $|T|$ is unity: the limit to consider is thus -360° or 0° when measuring with the FRA.

The Crossover Frequency

SELECTING A CROSSOVER frequency cannot be done arbitrarily based on the switching frequency. Assume a simple open-loop buck converter having a fixed duty ratio in continuous conduction mode (CCM). With a step-load, you will observe a damped oscillatory V_{out} response tuned at f_0, the resonant frequency of the LC filter. If you want to get rid of these oscillations once the converter operates in closed loop, you will have to tailor the compensator G so that this unwanted behavior is attenuated by the loop. Thus, the system must exhibit a sufficient amount of loop gain at f_0 to fight the oscillations and deliver a stable response. Since we know a control system needs gain to operate, you realize that the crossover frequency f_c, in this particular case, must be selected to be positioned at least at 3-5 times f_0: the system needs gain in this frequency band to react and damp the converter's output impedance.

With converters like boost, buck-boost or flyback also operated in CCM voltage-mode control, f_0 moves with the duty ratio D, making the selection slightly more difficult. On top of that, the right-half-plane zero (RHPZ) severely affects the power stage phase and you must absolutely slow down the loop reaction time by forcing a crossover frequency below 20% of the lowest RHP zero position (which depends on V_{in} and I_{out}): the RHPZ sets an upper limit for choosing the crossover frequency for these converters. If you want to increase the RHPZ value, selecting a smaller inductance will make the converter more reactive. As discontinuous conduction mode (DCM) is entered, the RHPZ is pushed to higher frequencies and is less of an obstacle to increase f_c in this mode.

In current-mode control, the peaking inherent to the LC filter is gone but subharmonic poles located at $F_{sw}/2$ are present. For a buck converter, the *theoretical* limit for f_c is now half the switching frequency. Of course, you don't want such a wide bandwidth unless you have to meet very stringent transient response requirements while minimizing the output capacitance. I recommend you choose f_c to meet your transient specifications and no farther. For boost and buck-boost converters, unfortunately, the RHPZ remains at a similar position in current-mode control and imposes the same upper limit as in voltage mode.

Pick the Right Value

IN THE PREVIOUS page, I have explained why it is important to understand the limits for selecting a crossover frequency. Considering resonance in the control path or the presence of a RHPZ, the table below provides some guidelines on where to position the crossover frequency for the three basic switching cells operated in voltage- or current-mode control:

Topology	Voltage Mode	Current Mode
Buck	$3 \cdot f_0 < f_c < \dfrac{F_{sw}}{2}$ *	$f_c < \dfrac{F_{sw}}{2}$ *
Boost	$3 \cdot f_0 < f_c < 0.2 \cdot f_{RHPZ}$	$f_c < 0.2 \cdot f_{RHPZ}$
Buck-boost	$3 \cdot f_0 < f_c < 0.2 \cdot f_{RHPZ}$	$f_c < 0.2 \cdot f_{RHPZ}$

Continuous conduction mode   * *Theoretical* upper limit

A control system opposes output current perturbations by minimizing the output impedance of the power stage. It is possible to show that the voltage undershoot ΔV_{out} brought by a current step ΔI_{out} can be *approximated* as [3]

$$\Delta V_{out} \approx \frac{\Delta I_{out}}{2\pi f_c C_{out}} \quad \xrightarrow{\text{Extract } f_c} \quad f_c = \frac{\Delta I_{out}}{\Delta V_{out}} \frac{1}{2\pi C_{out}}$$

It is valid if we have an ESR r_C of negligible value with respect to the output capacitor impedance at crossover: $r_C << Z_{Cout} @ f_c$. It is a simplified formula, of course, but its goal is *not* to precisely predict a voltage undershoot. Rather, it illustrates the link between the transient output deviation and the crossover frequency f_c. I therefore recommend to determine the output capacitor based on ripple specification, rms current, nominal voltage and so on. Then choose what crossover frequency would be needed to match your transient requirements. You now have the tools to choose phase margin and crossover: no guess work!

Generating the Duty Ratio

IN THE SIMPLIFIED ILUSTRATIONS, the power stage is designated as $H(s)$ and this transfer function represents the control-to-output ac response from the stimulus input – the feedback pin of the controller – to the output voltage of the buck or flyback converter for instance. Practically speaking, the error voltage delivered by the compensator needs to be transformed into a duty ratio D which is the control variable of the switching stage. For that purpose, we build a pulse-width modulator or a PWM block:

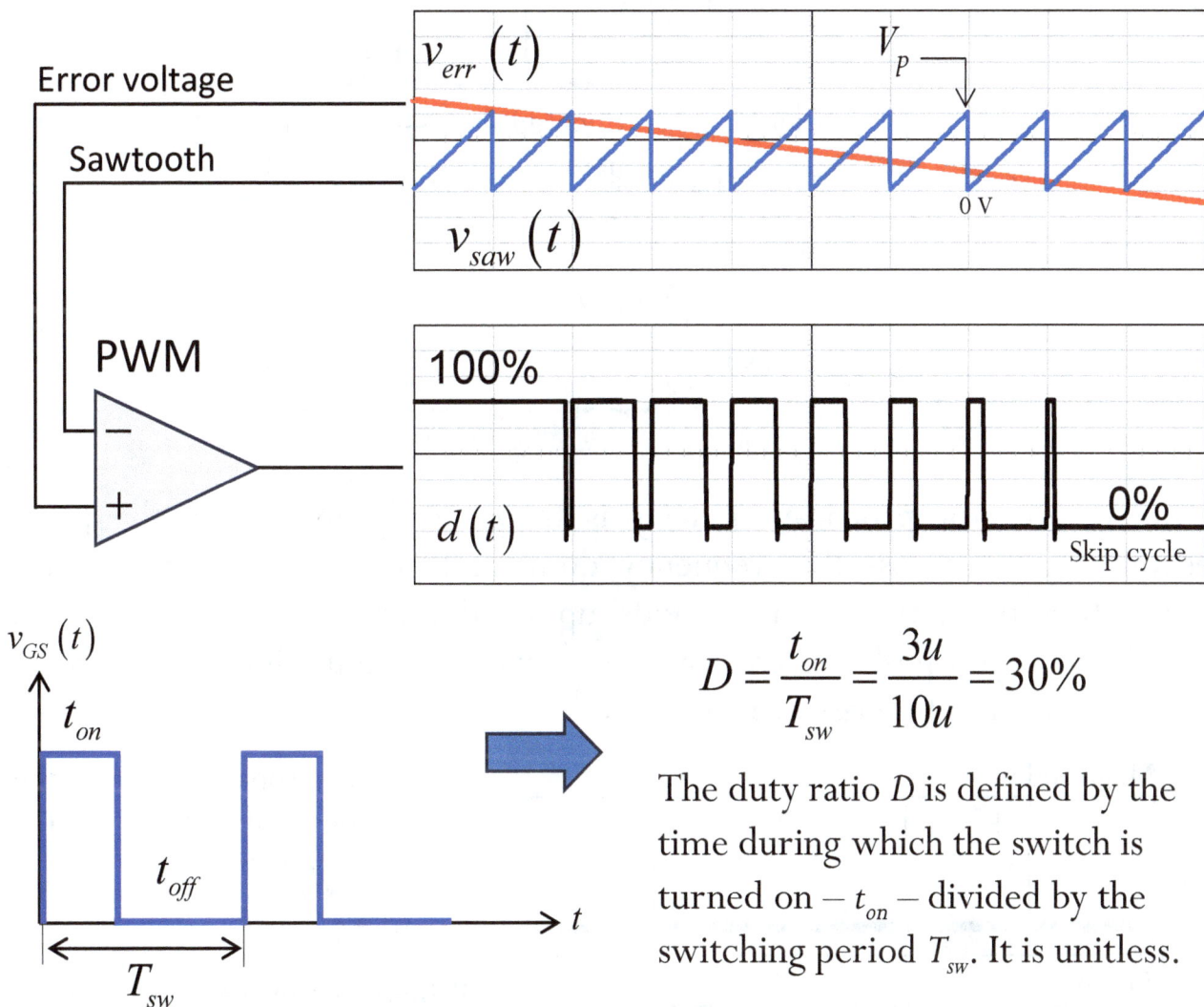

Error voltage

Sawtooth

$v_{err}(t)$

V_p

$v_{saw}(t)$

0 V

PWM

100%

0%

Skip cycle

$d(t)$

$v_{GS}(t)$

t_{on}

t_{off}

T_{sw}

t

$$D = \frac{t_{on}}{T_{sw}} = \frac{3u}{10u} = 30\%$$

The duty ratio D is defined by the time during which the switch is turned on – t_{on} – divided by the switching period T_{sw}. It is unitless.

A PWM block is made of a fast comparator which receives a ramp characterized by a frequency and a peak value V_p. By comparing the dc error voltage V_{err} with the ramp, toggling events occur and can last from 0% to 100% of the switching period. The duty ratio D is a discrete value, updated cycle by cycle.

Frequency Response of the PWM Block

THE CIRCUIT BUILT with a comparator and a sawtooth generator is designated as a *naturally sampled* modulator. It processes a continuous-time signal, the error voltage $v_{err}(t)$. A so-called *uniformly sampled* modulator receives a discrete-time signal $V_{err}[k]$ and is found in digital control loops. A typical analog modulator is shown below:

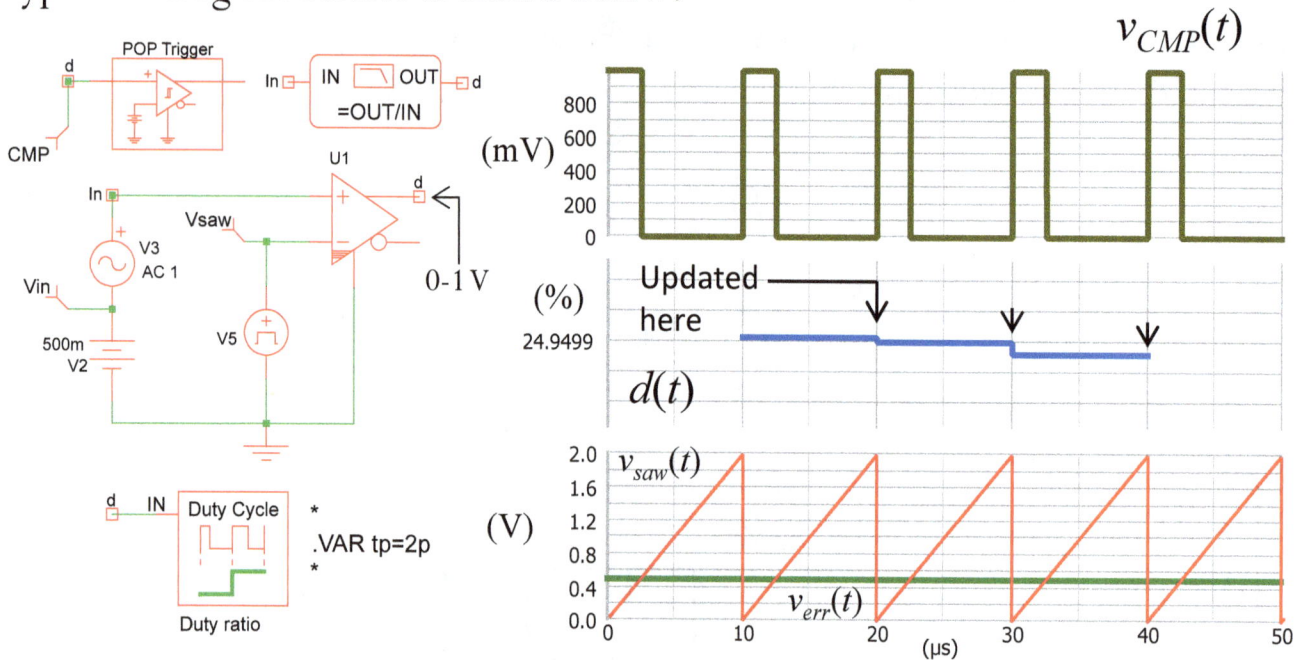

If the comparator propagation delay is theoretically zero, then the phase response is flat across the frequency domain. If you now consider some propagation time, then a delay adds up to the transfer function of this modulator. If you push the crossover frequency, you must include this delay which contributes to phase erosion.

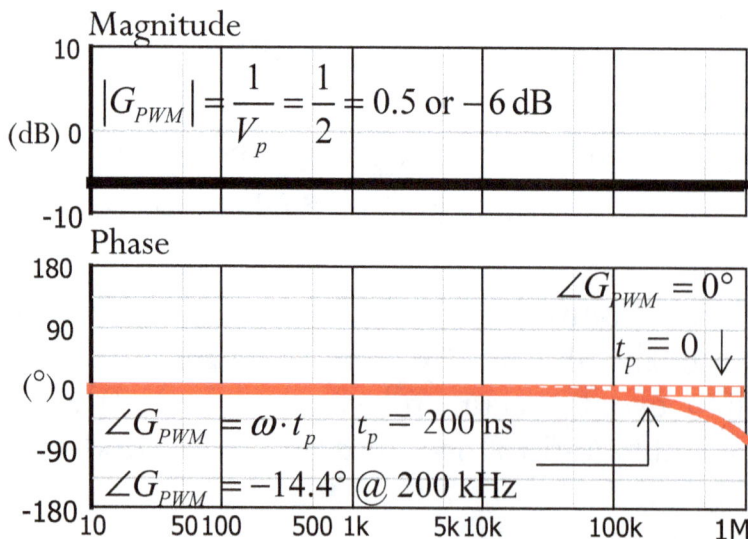

For a zero propagation delay, the magnitude and phase of the modulator are given by:

$$|G_{PWM}| = \frac{1}{V_p} \qquad \angle G_{PWM} = 0°$$

If there is a propagation delay t_p, it becomes:

$$G_{PWM}(s) = \frac{1}{V_p} e^{-s \cdot t_p}$$

$$|G_{PWM}| = \frac{1}{V_p} \qquad \angle G_{PWM}(\omega) = \omega \cdot t_p$$

A Typical Transient Response

TO UNDERSTAND THE RESPONSE of the control system, I took the example of a 5-V voltage-mode buck converter supplied from a 12-V dc input. The crossover is set to ≈ 7 kHz with a phase margin of 54° and a 330-µF output capacitor. If there is an output step-load, this is what you observe:

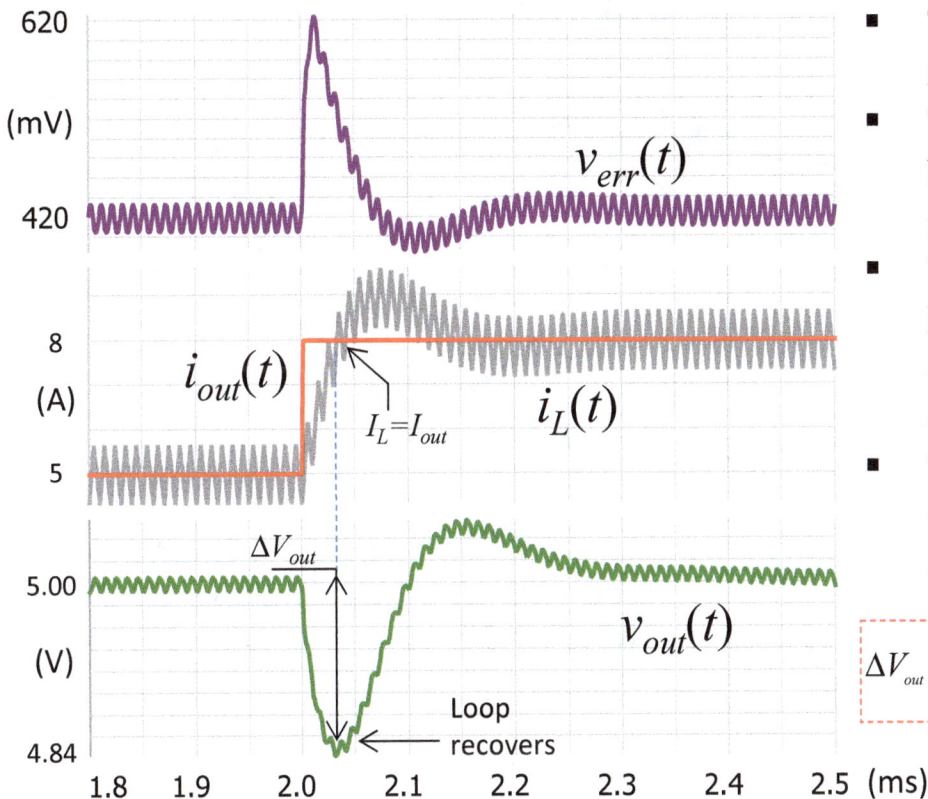

- The output current goes from 5 to 8 A in 1 µs
- The output voltage drops by $r_C \Delta I_{out}$ and continues falling ($r_C = 10$ mΩ)
- The error voltage goes high and forces more current in the inductor cycle-by-cycle
- The loop recovers when the inductor current equals the output current.

$$\Delta V_{out} \approx \frac{3}{2\pi \times 7k \times 330u} \approx 207 \text{ mV}$$

If I now explore different crossover frequencies with a constant phase margin, the transient response changes as illustrated below:

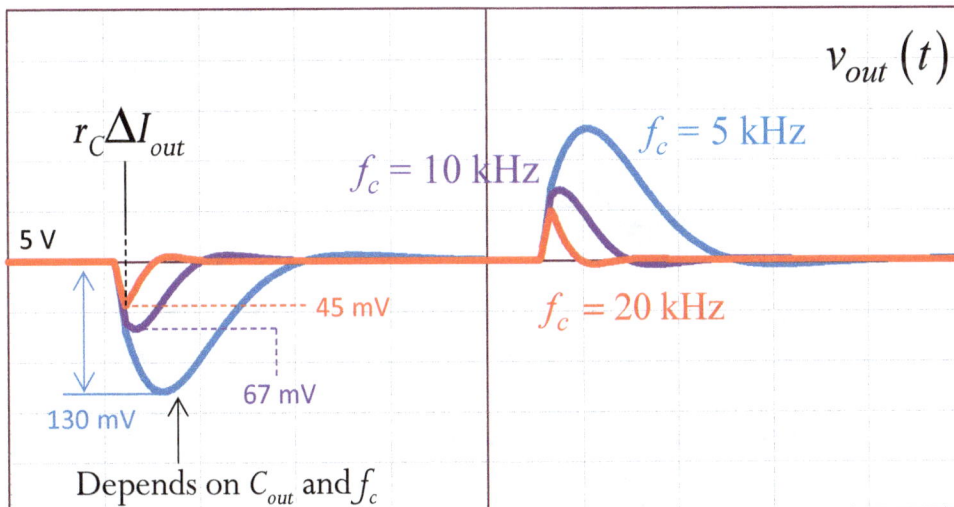

- The capacitance contribution reduces as f_c increases
- At some point, the drop is fixed by r_C: no need to push f_c farther

How to Force Crossover?

YOU HAVE SELECTED a 10-kHz crossover frequency and would like to shape the loop response accordingly. First off, you *must* have the control-to-output transfer function in hand. Assume it looks like the Bode plot below:

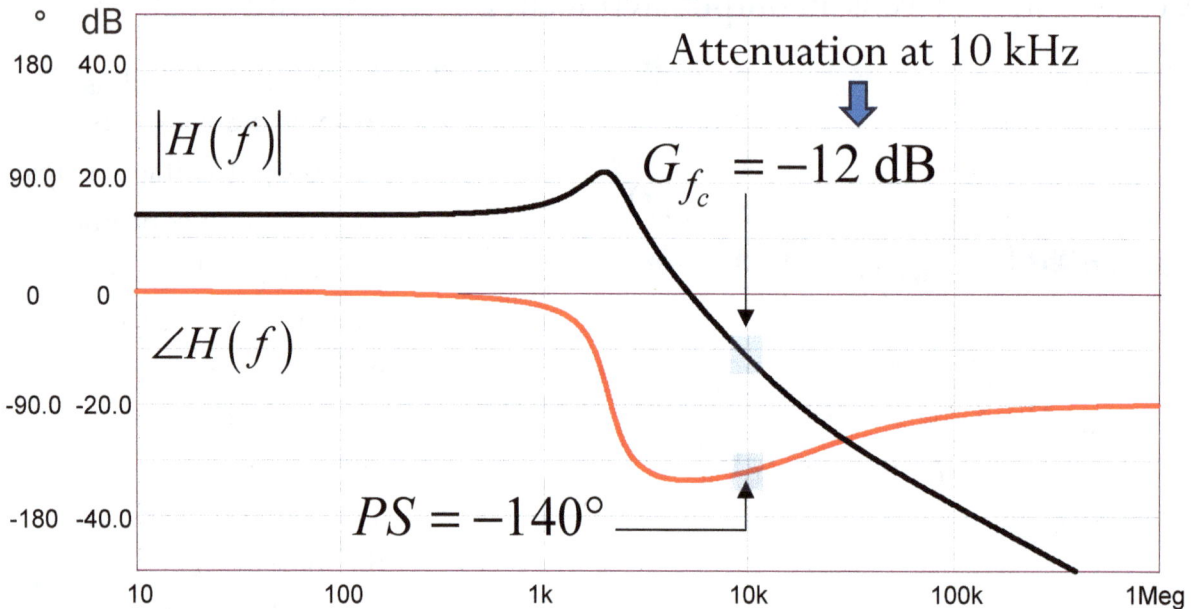

You read the Bode plot and extract the power stage magnitude G_{f_c} with the phase value at the considered crossover, 10 kHz: G_{f_c} = -12 dB, PS = -140°. To force the loop gain to be 0 dB at 10 kHz, we simply shift the curve by 12 dB:

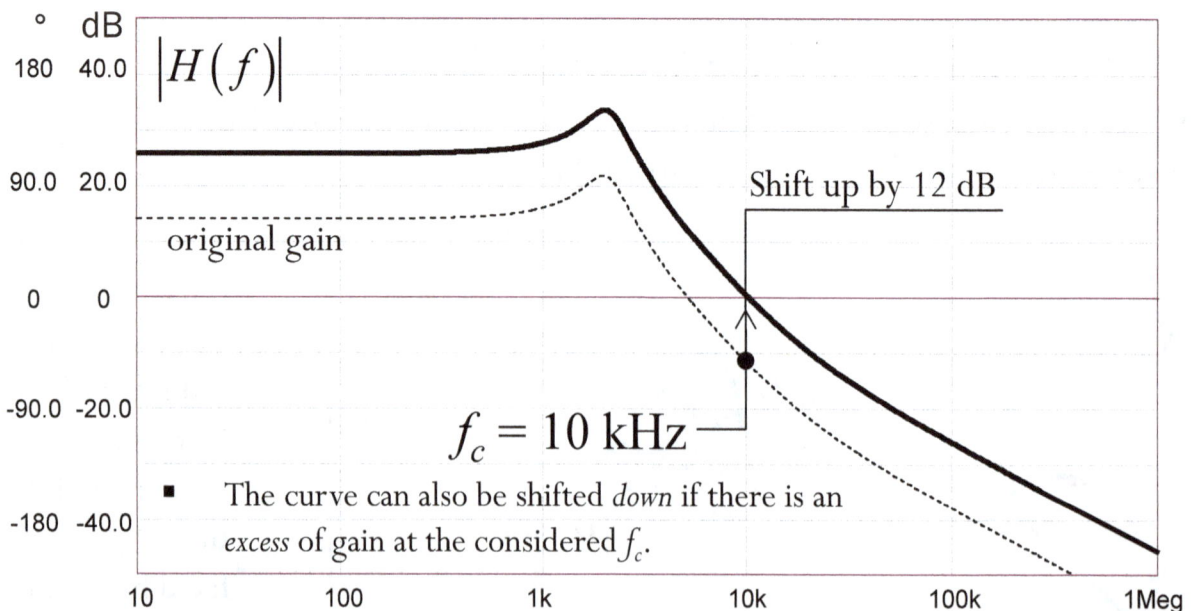

The compensator G — which is an active filter — will be tuned to offer a gain of 12 dB at 10 kHz: $|T(f_c)|_{dB} = |H(f_c)|_{dB} + |G(f_c)|_{dB}$ = 0 dB at f_c = 10 kHz.

Shaping the Loop Response

WE NOW NEED to build some phase margin before safely closing the loop. An op-amp wired in an inverting configuration provides the necessary gain or attenuation but brings a 180° phase lag:

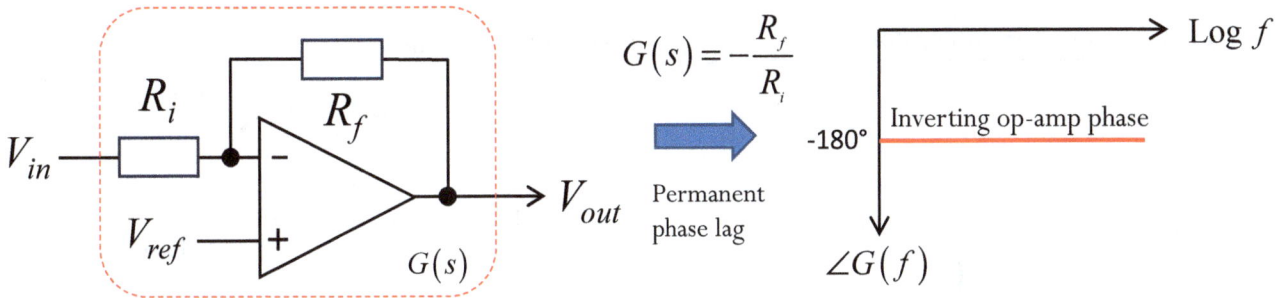

$$G(s) = -\frac{R_f}{R_i}$$

Permanent phase lag

-180° Inverting op-amp phase

$\angle G(f)$

If we want a high gain at dc – to lower the static error in the output – a pole at the origin is needed and we replace R_f by a capacitor. In this case, for $s = 0$, the gain G is A_{OL}. Ac analysis shows that this integrator brings an additional 90° phase lag for a total phase lag of -270° or 90° (same angle):

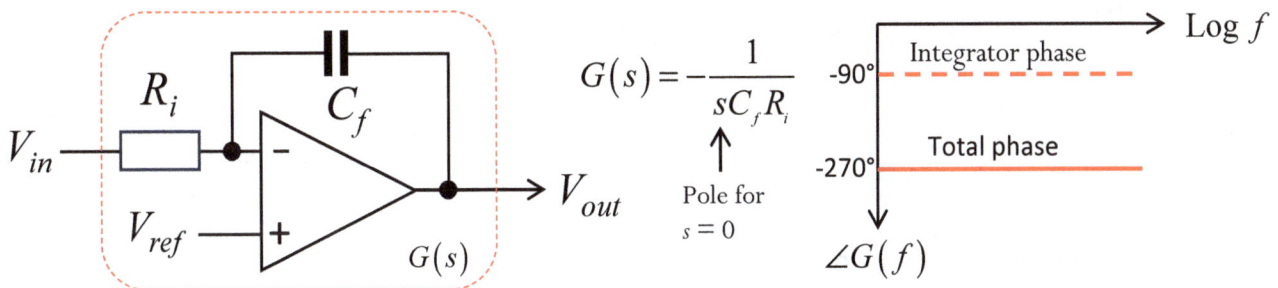

$$G(s) = -\frac{1}{sC_f R_i}$$

Pole for $s = 0$

-90° Integrator phase

-270° Total phase

$\angle G(f)$

When the compensator transfer function $G(s)$ is associated with the power stage ac response $H(s)$, both combine to realize the loop gain $T(s)$:

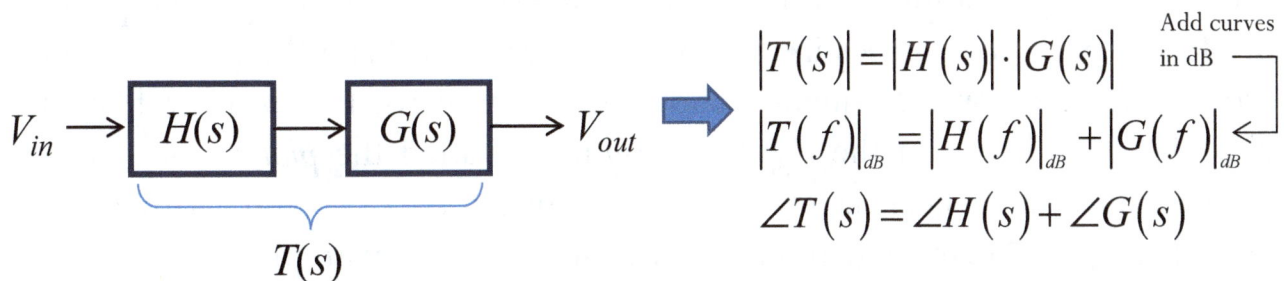

$V_{in} \rightarrow \boxed{H(s)} \rightarrow \boxed{G(s)} \rightarrow V_{out}$

$T(s)$

$$|T(s)| = |H(s)| \cdot |G(s)|$$

Add curves in dB

$$|T(f)|_{dB} = |H(f)|_{dB} + |G(f)|_{dB}$$

$$\angle T(s) = \angle H(s) + \angle G(s)$$

If an integrator is part of our compensation strategy, you can see that adding a phase lag of 270° to the existing lag of the power stage, naturally pushes the loop phase towards the -360° limit: the phase response of the compensator needs to be shaped in the vicinity of the crossover frequency to build margin.

Boosting the Phase

TO ILLUSTRATE THE NEED for shaping the loop phase, assume you cascade the previous power stage with the inverting integrator. As the phase of H and G add up, it is easy to construct the loop gain phase:

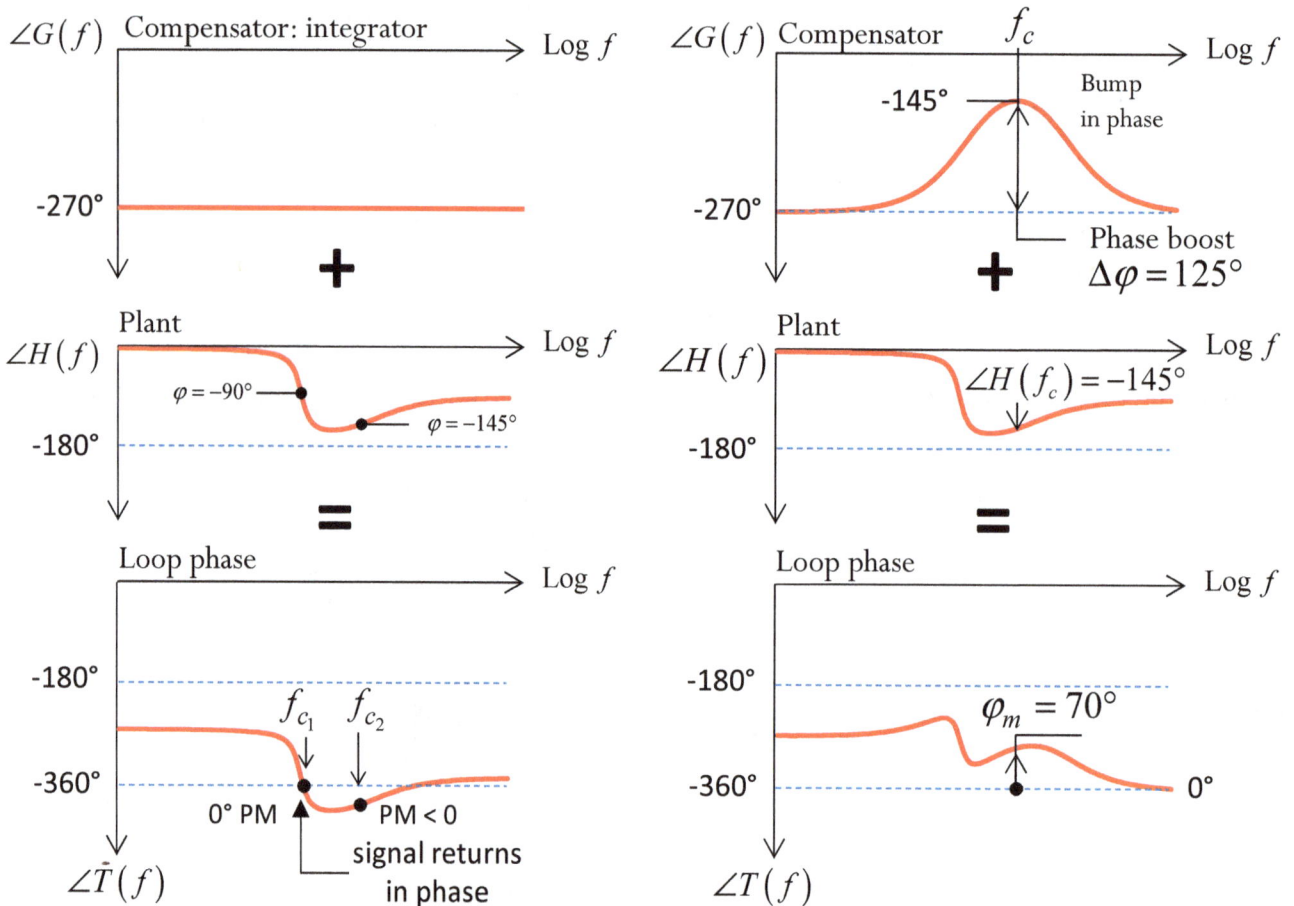

In the left side, you see that if we combine the integrator with the power stage, the phase margin is either $0°$ at f_{c1} or even negative at f_{c2}. The purpose of placing poles and zeroes in the compensator is to shape the phase response around the crossover frequency and create a positive bump away from the $270°$ lag. The bump in the right-side figure is called the *phase boost* and it is usually tailored to peak at crossover. The boost is computed to meet your open-loop phase margin goal like the $70°$ shown as an example:

You add phases and stay away from -360° by φ_m

$$\angle H(f_c) - 270° + boost = -360° + \varphi_m$$

Inverting integrator

The needed boost — The PM you want — The power stage phase at f_c

$$boost = \varphi_m - \angle H(f_c) - 90°$$

Poles and Zeros

ASSUME A 1st-ORDER transfer function combining a pole ω_p and a zero ω_z:

$$G(s) = \frac{1 + \dfrac{s}{\omega_z}}{1 + \dfrac{s}{\omega_p}}$$

$$\angle G(f_c) = \underbrace{\tan^{-1}\left(\frac{f_c}{f_z}\right)}_{\text{Zero phase}} - \underbrace{\tan^{-1}\left(\frac{f_c}{f_p}\right)}_{\text{Pole phase}}$$

Zero phase   Pole phase

The phase peaks at

$$f_c = \sqrt{f_p f_z}$$

(0 to 90°) + (0 to -90°)

Total phase varies from 0 to 90°

Considering the zero and the pole respectively leading and lagging the phase, combining the two gives a way to position the phase boost at crossover and adjust it to the exact needed value:

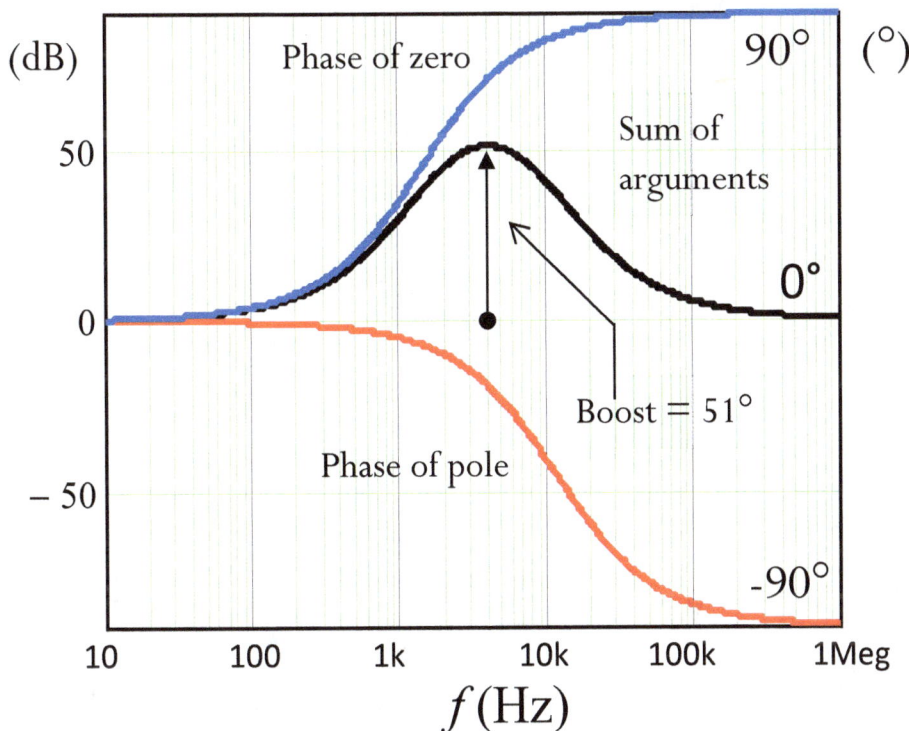

- The zero occurs first and boosts the phase up to 90°.
- The pole then kicks in and brings the phase down to 0°.
- $\omega_z = 1.4$ kHz, $\omega_p = 11.3$ kHz and $f_c = 4$ kHz

The compensation technique uses a pole-zero pair to adjust the phase boost between the two roots. As the zero is placed before the pole, the distance between the two lets you tweak the phase boost between 0 and 90°. And if you add a second pole-zero pair, then you can *theoretically* adjust the phase boost between 0 and 180°. I say in theory because the selected active amplifier (an op-amp or a TL431) will bring its own ac response and sets a limit to what you can obtain. The k factor uses this approach of pole-zero pair.

The Three Compensators

WE HAVE SEEN that placing poles and zeroes offers a means to tailor the frequency response of the compensator G. You force a given crossover frequency but also locally boost the phase response to ensure adequate phase margin. Since a high gain at dc is also a desirable feature for reducing the static error, all three compensators will host a pole at the origin, the integrator, supplemented by a single or double pole-zero pair. The three possible associations are designated by types:

Type 1: a pure integrator, no phase boost, used to set gain or attenuation only.

Type 2: an integrator plus a pole-zero pair. Phase boost up to 90°.

Type 3: an integrator plus a double pole-zero pair. Phase boost up to 180°.

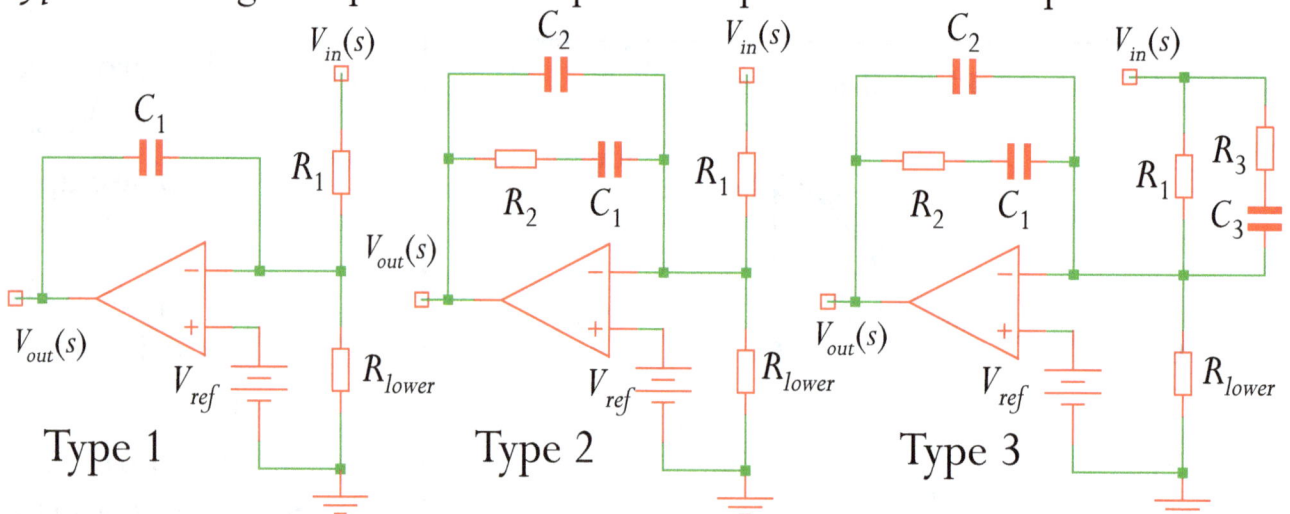

Type 1

$$G(s) = -\frac{1}{\dfrac{s}{\omega_{po}}} \qquad \omega_{po} = \frac{1}{R_1 C_1}$$

Type 2

$$G(s) = -G_0 \frac{1 + \dfrac{\omega_z}{s}}{1 + \dfrac{s}{\omega_p}}$$

$$G_0 = \frac{R_2}{R_1} \frac{C_1}{C_1 + C_2} \approx \frac{R_2}{R_1}$$

$$\omega_z = \frac{1}{R_2 C_1} \qquad \text{If } C_2 << C_1$$

$$\omega_p = \frac{1}{R_2 \dfrac{C_1 C_2}{C_1 + C_2}} \approx \frac{1}{R_2 C_2}$$

Type 3

$$G(s) = -G_0 \frac{\left(1 + \dfrac{\omega_{z_1}}{s}\right)\left(1 + \dfrac{s}{\omega_{z_2}}\right)}{\left(1 + \dfrac{s}{\omega_{p_1}}\right)\left(1 + \dfrac{s}{\omega_{p_2}}\right)}$$

$$G_0 = \frac{R_2}{R_1} \frac{C_1}{C_1 + C_2} \approx \frac{R_2}{R_1} \quad \begin{array}{l} \text{If } C_2 << C_1 \\ R_3 << R_1 \end{array}$$

$$\omega_{z_1} = \frac{1}{R_2 C_1} \qquad \omega_{z_2} = \frac{1}{(R_1 + R_3) C_3} \approx \frac{1}{R_1 C_3}$$

$$\omega_{p_1} = \frac{1}{R_2 \dfrac{C_1 C_2}{C_1 + C_2}} \approx \frac{1}{R_2 C_2} \qquad \omega_{p_2} = \frac{1}{R_3 C_3}$$

- The definition of poles and zeroes can be simplified if C_2 is much smaller than C_1.
- R_1 and R_{lower} set the dc output.
- R_{lower} plays no role in the *ac* response considering the virtual ground at the (-) pin.

Compensators AC Responses

I HAVE PLOTTED the three small-signal responses of the compensators. Please note that the given transfer functions are derived when considering an ideal op-amp, featuring an infinite open-loop gain A_{OL} and no internal poles. In reality, especially in projects where you want to push f_c, you will need to ensure that the op-amp does not distort the expected frequency response by imposing its own asymptotes. I recommend you design the components and simulate the response with an ideal circuit and, once you confirm the converter is stable, plug the non-ideal model to confirm parameters are not significantly altered.

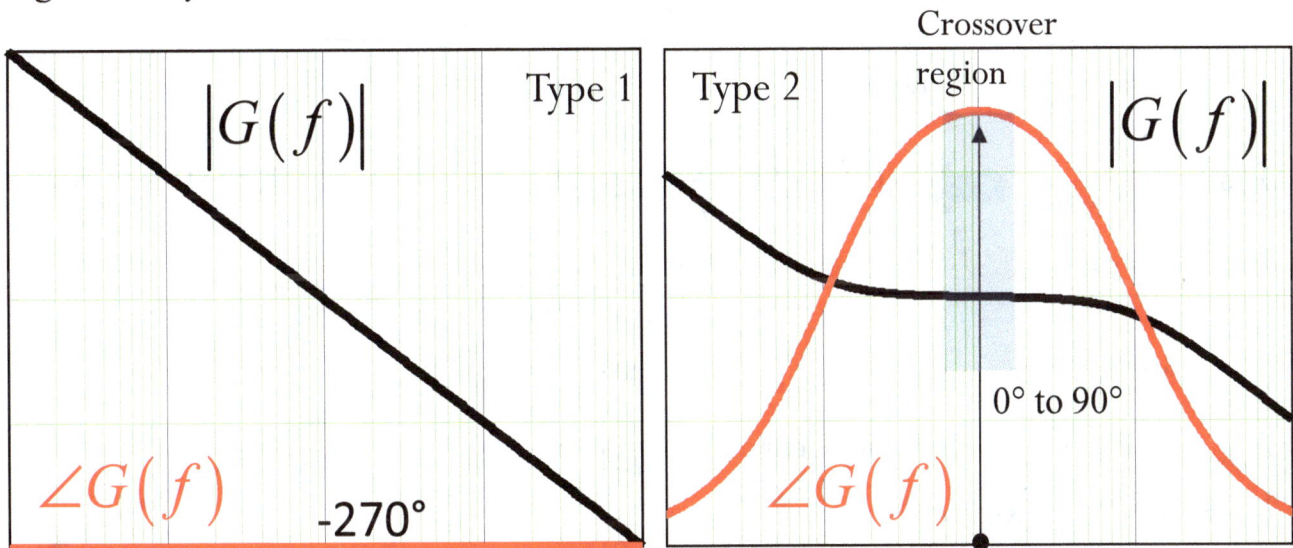

- The compensator types 1, 2 and 3, were forged by Dean Venable in a paper he published in 1983 and entitled *The k Factor: A New Mathematical Tool for Stability Analysis and Synthesis*.[8]
- Food for thought: if you remove capacitor C_2, a type 2 becomes a type 2a which is a PI compensator. If you add an extra pole to a *filtered-*PID, you have a type 3 compensator.

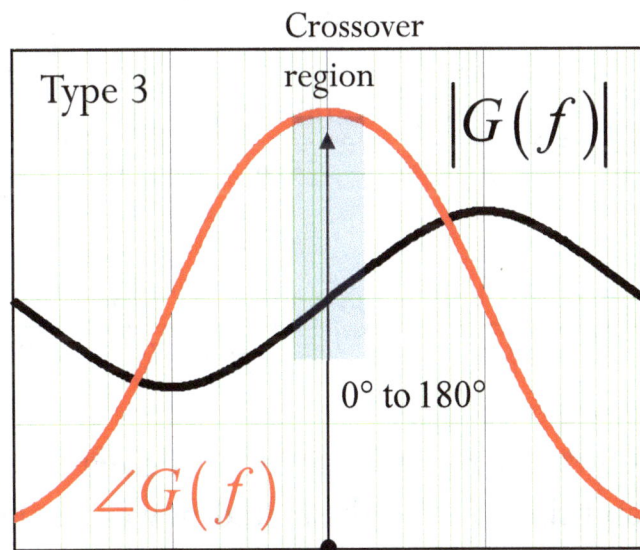

Compensating with a Type 1

THE INTEGRATOR does not boost the phase but lets you build gain or attenuation at the selected crossover frequency f_c. In this example, the power stage of this high-voltage converter exhibits a gain deficiency G_{f_c} of 20 dB at a selected frequency of 10 Hz. Where should we place the 0-dB crossover pole ω_{po} so that the integrator gain be 20 dB at 10 Hz?

$$G_{f_c} = -20 \text{ dB at } f_c = 10 \text{ Hz} \rightarrow G = 10^{-\frac{G_{f_c}}{20}} = 10^{-\frac{-20}{20}} = 10$$

Read from power stage Bode plot

Selected crossover target

Compensator gain so that $|T(f_c)| = 1$

$$G(s) = -\frac{1}{\dfrac{s}{\omega_{po}}} \quad \Longrightarrow \quad |G(f_c)| = \frac{f_{po}}{f_c} \quad \Longrightarrow \quad f_{po} = G \cdot f_c = 10 \times 10 = 100 \text{ Hz}$$

$V_{out} = 400 \text{ V}$ Regulated voltage

$I_{bias} = 250 \, \mu\text{A}$ Divider bridge current

$V_{ref} = 2.5 \text{ V}$ Reference voltage

$$R_{lower} = \frac{V_{ref}}{I_{bias}} = 10 \text{ k}\Omega$$

$$R_1 = \frac{V_{out} - V_{ref}}{I_{bias}} = \frac{400 - 2.5}{250u} = 1.59 \text{ M}\Omega$$

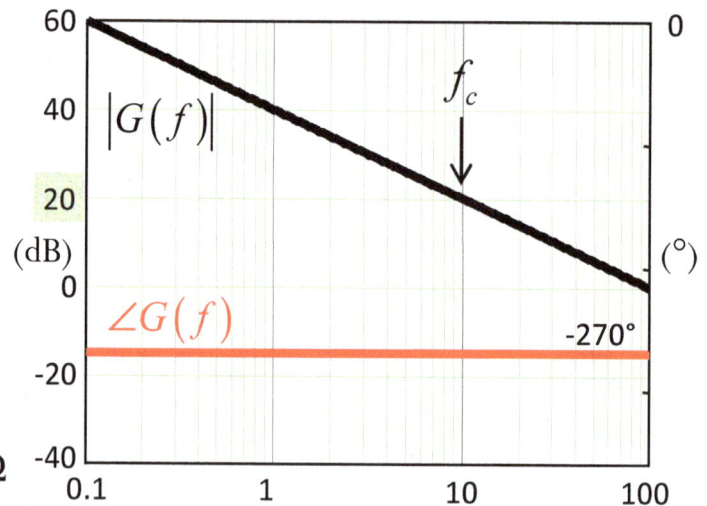

$$\omega_{po} = \frac{1}{R_1 C_1} \rightarrow C_1 = \frac{1}{2\pi f_{po} R_1} = \frac{1}{6.28 \times 100 \times 1.59 Meg} = 1 \text{ nF}$$

Once the resistive divider is determined, considering the regulated voltage V_{out} and the allowed dissipated power in this divider, you can determine capacitor C_1's value for a 20-dB gain at 10 Hz. The capacitor is 1 nF and the plot confirms the calculation.

Simulating the Type 1

ONE IMPORTANT aspect of loop stability analysis is simulation. First off, you can immediately check if the chosen components values meet your design goals in terms of frequency response. Second, once the model is validated by bench measurements, simulations let you verify the loop robustness through Monte Carlo or worst-case analyses. It will also bring confidence that the product you release safely addresses the parasitics naturally varying with age, production or temperature.

The below template shows a type 1 compensator featuring a generic op-amp which, once the ac response is acceptable, can be replaced by the adopted model. When running this type of high dc gain circuit, it is important to keep the op-amp output within a linear range (above ground and away from V_{cc}). It is the role of the L_{OL}/C_{OL} passive elements which force the adequate level on R_1 via E_1 during the bias point simulation (400 V in this example) and keep the op-amp output at an arbitrary 2-3-V level via V_2.

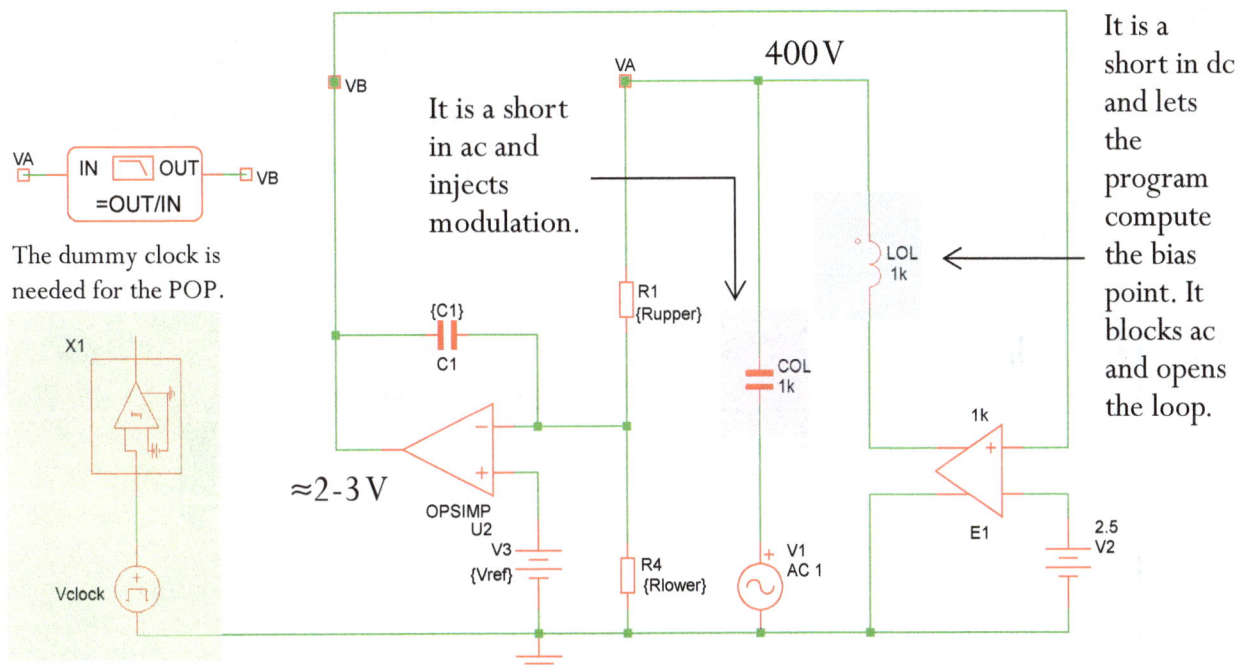

The templates I built in SIMPLIS® use a dedicated macro which automates the components values calculation. You enter the design parameters, simply the gain you have read on the power stage at 10 Hz for this type 1 (there is no phase margin goal) and parts will automatically be calculated.

Results in a few Seconds

THE AUTOMATED TEMPLATE represents a good alternative to a solver or even Excel® as it not only calculates the value but immediately displays the results in dB and degrees. The macro looks like this:

```
*
.VAR Gfc=-20 * magnitude at crossover *
*
* Enter Design Goals Information Here *
*
.VAR fc=10 * targeted crossover *
*
* Enter the Values for Vout and Bridge Bias Current *
*
.VAR Vout=400
.VAR Ibias=250u
.VAR Vref=2.5
.VAR Rlower=Vref/Ibias
.VAR Rupper=(Vout-Vref)/Ibias
*
* Do not edit the below lines *
.VAR G=10^(-Gfc/20)
.VAR fpo=G*fc
.VAR C1=1/(2*pi*fpo*Rupper)
*
```

```
* Choose op amp characteristics *
*
.VAR AOL=90 * open-loop gain in dB *
.VAR POLE=30 * low-frequency pole *
.GLOBALVAR VHIGH=5 * upper output level *
.GLOBALVAR VLOW=100m * lower output level *
*
* Do not edit these lines *
.VAR GAIN=10^(AOL/20)
.GLOBALVAR COL=1/(6.28*(GAIN/100u)*POLE)
.GLOBALVAR ROL=GAIN/100u
*
```

```
{'*'}
{'*'} Rupper = {Rupper}
{'*'} Rlower = {Rlower}
{'*'} C1 = {C1}
{'*'}
```

You can also chose the op-amp characteristics, e.g. its loop gain, low-frequency pole or V_{out} swing.

You enter the data extracted from the power stage Bode plot with the target frequency f_c. Then the resistive divider is calculated based on the regulated V_{out} and the bias current you set. Finally, components values are passed to the simulation engine and you have simulation results in a few seconds:

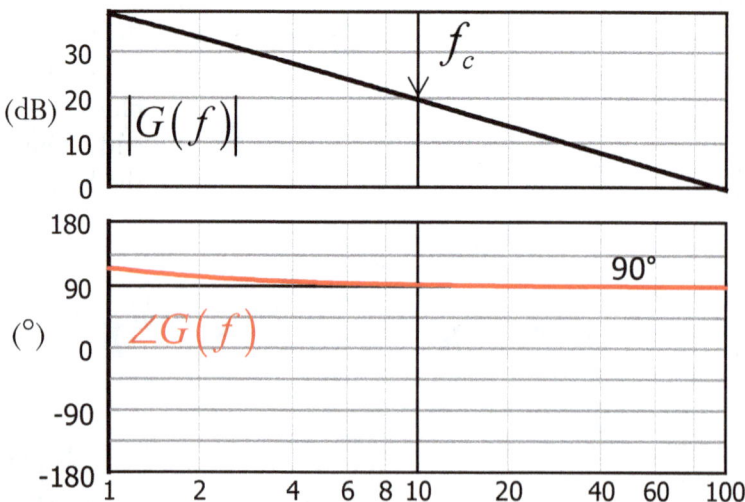

Calculated results show up in:

Simulator>Edit Netlist (after preprocess)

```
* Rupper = 1590000
* Rlower = 10000
* C1 = 1.00097448485469e-09
```

The phase is 90° (which is the same angle as -270°) and the gain is 20 dB at 10 Hz as expected. Components values match what we have previously calculated.

A Type 2 with an Op-Amp - I

IN THIS EXAMPLE, we need to boost the phase at the selected crossover frequency of 1 kHz. The information extracted from the power stage H is the following: $G_{f_c} = |H(1 \text{ kHz})| = -6.8$ dB and $\angle H(1 \text{ kHz}) = -66°$. First, determine the amount of needed phase boost for a 70° phase margin goal:

$$boost = \varphi_m - \angle H(f_c) - 90° = 70 - (-66°) - 90° = 46°$$

We place a pole-zero pair to boost the phase by 46° but also bring a gain at f_c equal to

$$|G(f_c)| = 10^{\frac{G_{f_c}}{20}} = 10^{\frac{-6.8}{20}} \approx 2.19$$

The magnitude and phase of a type 2 compensator are as follows:

Mid-band gain   Inverted zero

$$G(s) = -G_0 \frac{1 + \dfrac{\omega_z}{s}}{1 + \dfrac{s}{\omega_p}}$$

$$|G(f_c)| = G_0 \frac{\sqrt{1 + \left(\dfrac{f_z}{f_c}\right)^2}}{\sqrt{1 + \left(\dfrac{f_c}{f_p}\right)^2}} \quad \text{Magnitude at } f_c$$

$$\angle G(f_c) = \pi - \tan^{-1}\left(\dfrac{f_z}{f_c}\right) - \tan^{-1}\left(\dfrac{f_c}{f_p}\right) \quad \text{Phase at } f_c$$

You can now place the pole-zero pair manually:

$$f_p = \left[\tan(boost) + \sqrt{\tan^2(boost) + 1}\right] f_c \approx 2.48 \text{ kHz}$$

$$f_z = \frac{f_c^2}{f_p} = 404 \text{ Hz}$$

✓ k factor links the pole and the zero position with respect to f_c

Or resort to the k factor which works great in type 2 designs:

$$k = \tan\left(\frac{boost}{2} + \frac{\pi}{4}\right) = 2.48$$

$$\begin{cases} f_p = k \cdot f_c \approx 2.48 \text{ kHz} \\ f_z = \dfrac{f_c}{k} = 404 \text{ Hz} \end{cases}$$

✓ Use manual placement if you want more flexibility

A Type 2 with an Op-Amp - II

NOW THAT WE have the pole and zero in hand, we can determine the components values, using manual placement or the k factor. In these expressions, $R_1 = 2.5 \text{ k}\Omega$ is the upper resistance of the divider:

Manual placement

$$R_2 := \frac{R_1 \cdot f_p \cdot G_0}{f_p - f_z} \cdot \sqrt{\frac{1 + \left(\dfrac{f_c}{f_p}\right)^2}{1 + \left(\dfrac{f_z}{f_c}\right)^2}} = 6.53639 \cdot \text{k}\Omega$$

$$C_1 := \frac{1}{2 \cdot \pi \cdot R_2 \cdot f_z} = 60.26608 \cdot \text{nF}$$

$$C_2 := \frac{C_1}{2 \cdot \pi \cdot f_p \cdot C_1 \cdot R_2 - 1} = 11.75681 \cdot \text{nF}$$

k factor

$$C_{2a} := \frac{1}{2 \cdot \pi \cdot f_c \cdot G_0 \cdot k_1 \cdot R_1} = 11.75681 \cdot \text{nF}$$

$$C_{1a} := C_{2a} \cdot \left(k_1^2 - 1\right) = 60.26608 \cdot \text{nF}$$

$$R_{2a} := \frac{k_1}{2 \cdot \pi \cdot f_c \cdot C_{1a}} = 6.53639 \cdot \text{k}\Omega$$

Once the components values are calculated, you can use a Mathcad® sheet to plot the response and verify the magnitude and phase at f_c:

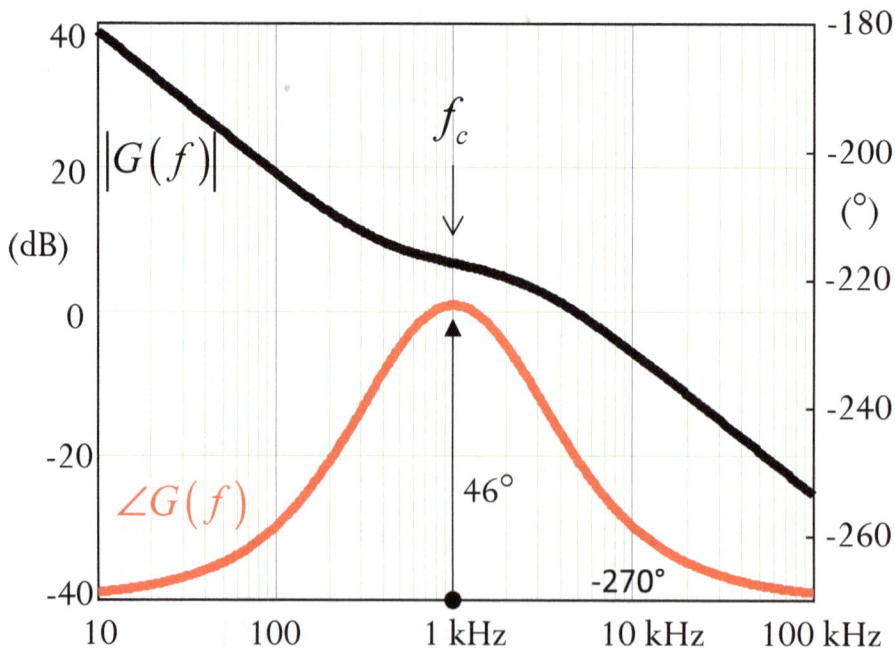

The gain is 6.8 dB at 1 kHz and the boost in phase approaches 50° as expected. We are good to go with this compensator.

Simulating the Type 2 — the Op-Amp

I HAVE BUILT a template for a type 2 compensator which appears below. You will recognize the auto-biasing circuitry which ensures a regulated 5-V rail at node V_A (the output of our converter). The clock source, in the left, is there to "cheat" SIMPLIS® and make it believe this is a switching circuit. As a time-domain simulator, the program requires switching events to determine a periodic operating point (POP) first, then run ac analyses.

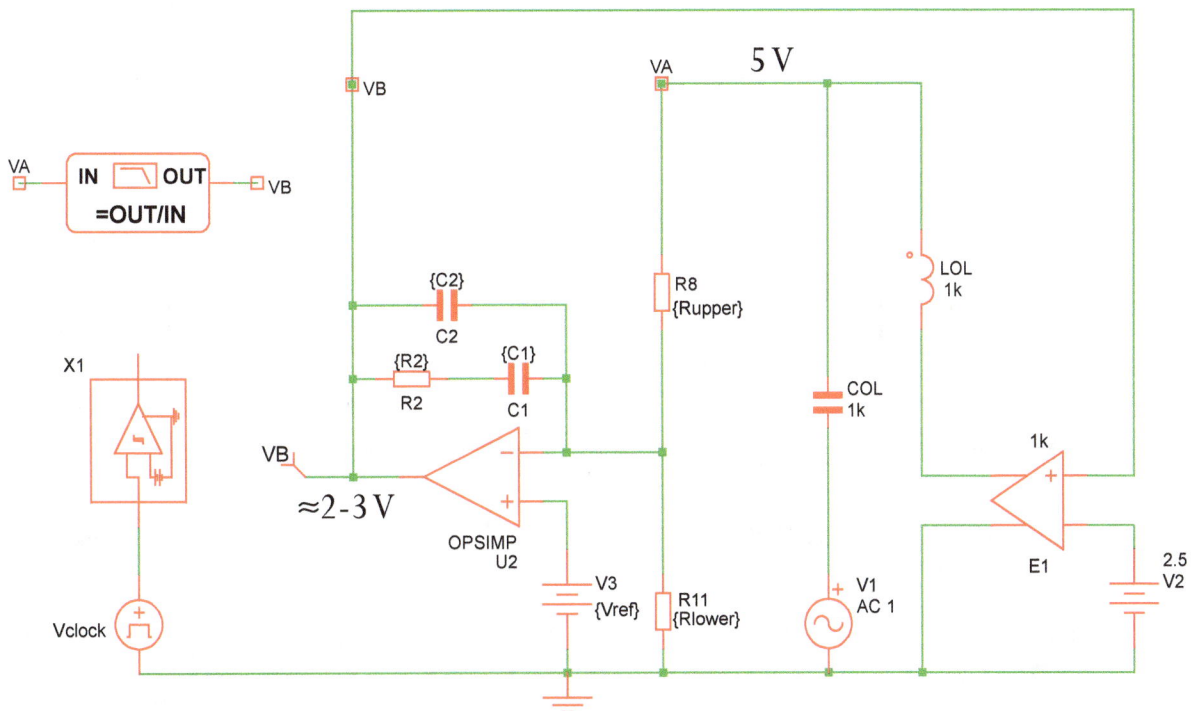

The macro is a bit more complicated than for the type 1 as you now set a phase margin goal of 70°. The resistive divider is determined for the wanted 5-V output with a 2.5-V voltage reference. The pole-zero pair is placed and components values calculated:

```
.VAR Gfc=-6.8 * magnitude at crossover *
.VAR PS=-66 * phase lag at crossover *
*
* Enter Design Goals Information Here *
*
.VAR fc=1k * targetted crossover *
.VAR PM=70 * choose phase margin at crossover *
*
* Enter the Values for Vout and Bridge Bias Current *
*
.VAR Vout=5
.VAR Ibias=1m
.VAR Vref=2.5
.VAR Rlower=Vref/Ibias
.VAR Rupper=(Vout-Vref)/Ibias
*
* Do not edit the below lines *
.VAR boost=PM-PS-90
.VAR G=10^(-Gfc/20)
.VAR fp=(tan(boost*pi/180)+sqrt((tan(boost*pi/180))^2+1))*fc
.VAR fz=fc^2/fp
.VAR a=sqrt((fc^2/fp^2)+1)
.VAR b=sqrt((fz^2/fc^2)+1)
.VAR R2=((a/b)*G*Rupper*fp)/(fp-fz)
.VAR C1=1/(2*pi*R2*fz)
.VAR C2=C1/(C1*R2*2*pi*fp-1)
*
```

```
* Rupper = 2500
* Rlower = 2500
* R2 = 6536.38522423468
* C2 = 1.17568148698654e-08
* C1 = 6.02660788477459e-08
* Boost = 46
* Fz = 404.026225835157
* Fp = 2475.0868534163
```

Type 2 AC Simulation

THE SIMULATED AC response appears below and confirms the phase bump located at 1 kHz as expected:

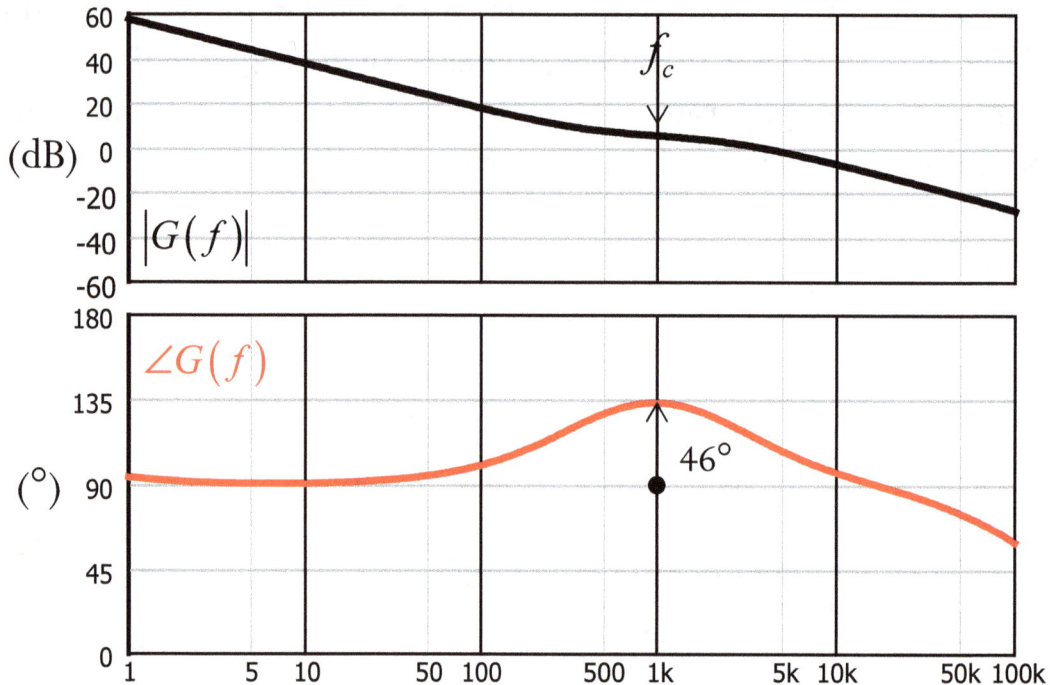

If you push crossover higher, you may want to check the impact of the op-amp GBW product on the overall response. The below generic circuit is easy to modify to obtain any needed open-loop amplification versus frequency curve:

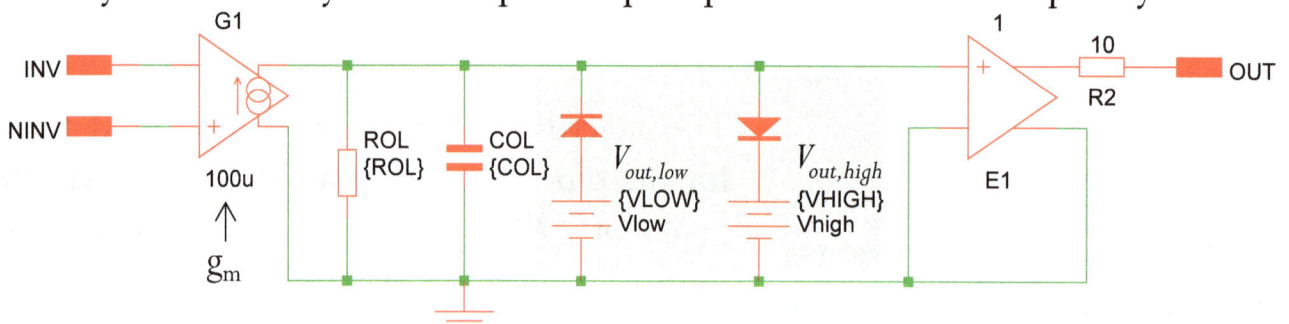

```
* Choose op amp characteristics *
*
.VAR AOL=86 * open-loop gain in dB *
.VAR POLE=8 * low-frequency pole *
.GLOBALVAR VHIGH=5 * upper output level *
.GLOBALVAR VLOW=100m * lower output level *
*
* Do not edit these lines *
.VAR GAIN={10^(AOL/20)}
.GLOBALVAR COL={1/(6.28*(GAIN/100u)*POLE)}
.GLOBALVAR ROL={GAIN/100u}
*
```

By changing some of the variables like AOL or POLE, you tailor the ac response of the op-amp. You can also easily change the output voltage swing and simulate with the simplest structure. Once it works, you may replace the generic model with a more realistic subcircuit.

A Type 3 with an Op-Amp - I

WE NOW NEED to produce a phase boost greater than 90° for stabilizing a second-order system such as a CCM buck converter operated in voltage-mode control. For that purpose, rather than placing a pole at the origin and a single pole-zero pair as with the type 2, we add another pole-zero pair. The *k* factor approach for a type 3 groups the zeroes and the poles (they are coincident) and I often found the resulting compensation strategy questionable in terms of results. It sometimes leads to conditional stability and, for this reason, I developed a way to manually place the poles and the zeroes.

Assume for this example that we target a crossover frequency of 10 kHz. Reading the power stage magnitude at this point returns -20 dB. The phase response is -135° and we want a phase margin of 70°. For my compensation strategy, I want to place a zero at 1 kHz and another one at 3 kHz. I will then place a pole at half the switching frequency ($F_{sw}/2 = 50$ kHz) for building a good gain margin. What matters now is to place the second pole to meet the gain and phase margin goal.

I start with the amount of needed phase boost:

$$boost = \varphi_m - \angle H(f_c) - 90° = 70 - (-135°) - 90° = 115°$$

We want the gain at 10 kHz to compensate the 20-dB attenuation, therefore:

$$|G(f_c)| = 10^{\frac{G_{f_c}}{20}} = 10^{\frac{-20}{20}} = 10$$

The magnitude and phase of a type 3 compensator are as follows:

$$G(s) = -G_0 \frac{\left(1 + \frac{\omega_{z_1}}{s}\right)\left(1 + \frac{s}{\omega_{z_2}}\right)}{\left(1 + \frac{s}{\omega_{p_1}}\right)\left(1 + \frac{s}{\omega_{p_2}}\right)}$$

Gain ↓, Inverted zero ↓

$$|G(f_c)| = G_0 \frac{\sqrt{1 + \left(\frac{f_{z_1}}{f_c}\right)^2}\sqrt{1 + \left(\frac{f_c}{f_{z_2}}\right)^2}}{\sqrt{1 + \left(\frac{f_c}{f_{p_1}}\right)^2}\sqrt{1 + \left(\frac{f_c}{f_{p_2}}\right)^2}} \quad \text{Magnitude at } f_c$$

Phase at f_c

$$\angle G(f_c) = \pi - \tan^{-1}\left(\frac{f_{z_1}}{f_c}\right) + \tan^{-1}\left(\frac{f_c}{f_{z_2}}\right) - \tan^{-1}\left(\frac{f_c}{f_{p_1}}\right) - \tan^{-1}\left(\frac{f_c}{f_{p_2}}\right)$$

A Type 3 with an Op-Amp - II

CONSIDERING TWO ZEROES already positioned with one high-frequency pole, how to position the second pole for adjusting the phase margin? It is usually placed to compensate for the ESR zero brought by the output capacitance but I like to precisely define it:

$$f_{z1} := 1\text{kHz} \qquad f_{z2} := 3\text{kHz} \qquad f_{p2} := 50\text{kHz}$$

$$f_{p1} := \frac{f_c}{\tan\left(\text{atan}\left(\frac{f_c}{f_{z1}}\right) + \text{atan}\left(\frac{f_c}{f_{z2}}\right) - \text{atan}\left(\frac{f_c}{f_{p2}}\right) - \text{boost}\right)} = 16.4599 \cdot \text{kHz}$$

With the two pole-zero pairs in hand, and the reworked magnitude expression, I can extract the value of resistance R_2:

gain at f_c ⟶

$$R_2 := \frac{G_1 \cdot R_1 \cdot f_{p1}}{f_{p1} - f_{z1}} \cdot \frac{\sqrt{1 + \left(\frac{f_c}{f_{p1}}\right)^2} \cdot \sqrt{1 + \left(\frac{f_c}{f_{p2}}\right)^2}}{\sqrt{1 + \left(\frac{f_{z1}}{f_c}\right)^2} \cdot \sqrt{1 + \left(\frac{f_c}{f_{z2}}\right)^2}} = 138\,\text{k}\Omega$$

$$C_1 := \frac{1}{2 \cdot \pi \cdot R_2 \cdot f_{z1}} = 1.2 \cdot \text{nF}$$

$$C_2 := \frac{C_1}{2 \cdot \pi \cdot f_{p1} \cdot C_1 \cdot R_2 - 1} = 74.6\,\text{pF}$$

$$C_3 := \frac{f_{p2} - f_{z2}}{2 \cdot \pi \cdot R_1 \cdot f_{p2} \cdot f_{z2}} = 1.3 \cdot \text{nF}$$

The resistance R_3 comes out naturally:
$$R_3 := \frac{R_1 \cdot f_{z2}}{f_{p2} - f_{z2}} = 2.4\,\text{k}\Omega$$

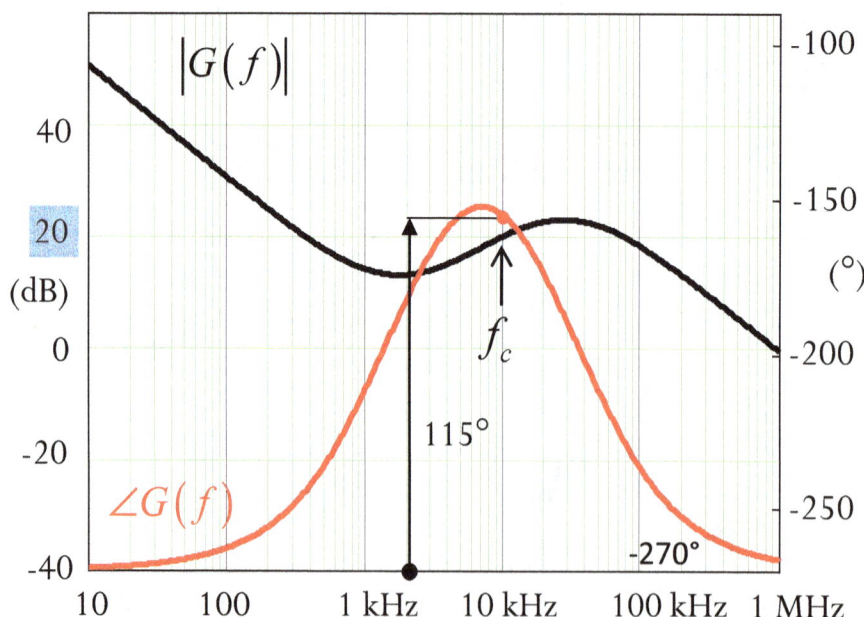

- The value of the upper resistance R_1 was set to 38 kΩ in this example.

- The phase peak is not exactly centered with manual pole-zero placement.

- Engineering judgment is necessary to select zeroes location and avoid negative values.

Simulating the Type 3 – the Op-Amp

A TEMPLATE for a type 3 compensator is shown below and it runs on SIMPLIS®. It could also be easily ported to SIMetrix® or any other SPICE simulator like LTspice®.

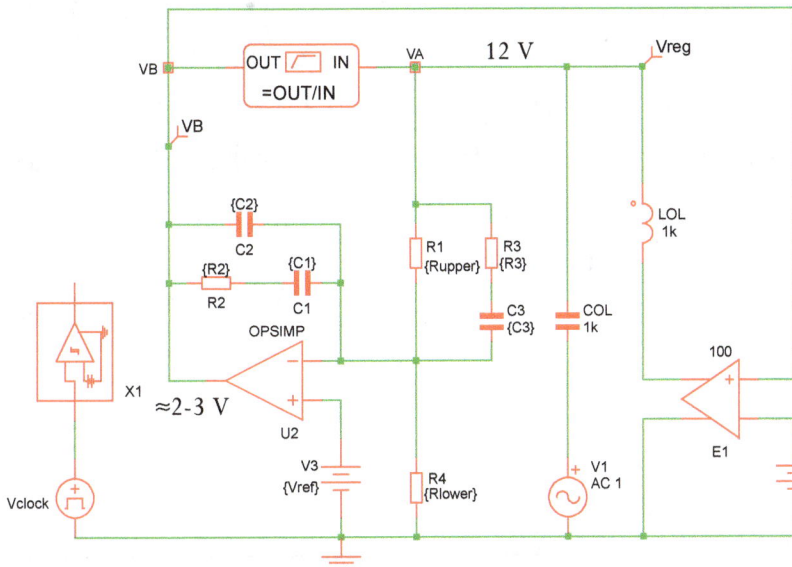

■ The macro calculates similar values as with the Mathcad® sheet:

```
* Rupper = 38000
* Rlower = 10000
* R2 = 138033.936834731
* R3 = 2425.53191489362
* C1 = 1.15301314112668e-09
* C2 = 7.45808819788526e-11
* C3 = 1.31233023251212e-09
* Boost = 115
```

Simulation results confirm our calculations and show how the phase boost is cropped to 100° with an op-amp featuring a 30-Hz low-frequency pole.

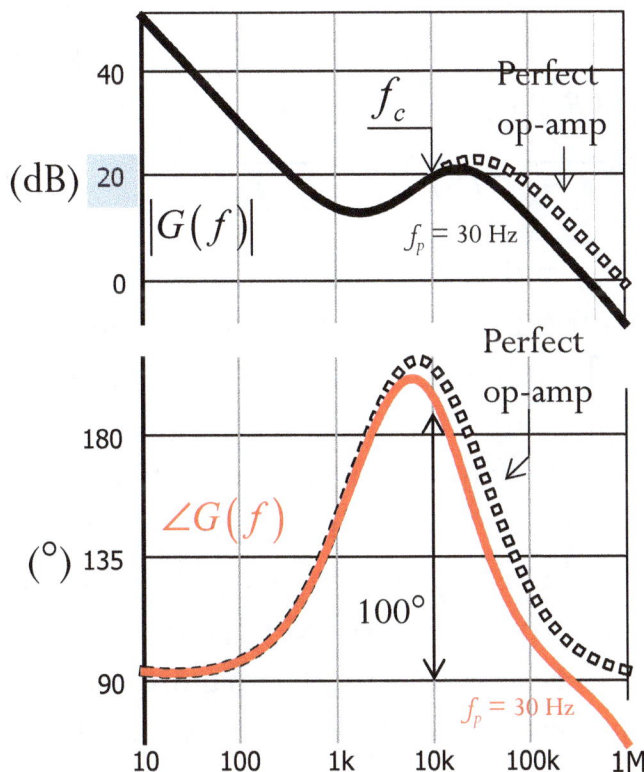

```
.VAR Gfc=-20 * magnitude at crossover *
.VAR PS=-135 * phase lag at crossover *
*
* Enter Design Goals Information Here *
*
.VAR fc=10k * targeted crossover *
.VAR PM=70 * choose phase margin at crossover *
*
* Enter the Values for Vout and Bridge Bias Current *
*
.VAR Vout=12
.VAR Ibias=250u
.VAR Vref=2.5
.VAR Rlower=Vref/Ibias
.VAR Rupper=(Vout-Vref)/Ibias
*
* Capture the double zero position and one of the pole position *
.VAR fz1=1k
.VAR fz2=3k
.VAR fp2=50k ; this pole is usually placed at Fsw/2
*
* Do not edit the below lines *
.VAR boost=PM-PS-90
.VAR G=10^(-Gfc/20)
.VAR fp1=fc/tan(atan(fc/fz1)+atan(fc/fz2)-atan(fc/fp2)-boost*pi/180)
.VAR a=sqrt((fc^2/fp1^2)+1)
.VAR b=sqrt((fc^2/fp2^2)+1)
.VAR c=sqrt((fz1^2/fc^2)+1)
.VAR d=sqrt((fc^2/fz2^2)+1)
.VAR R2=((a*b/(c*d))/(fp1-fz1))*Rupper*G*fp1
.VAR C1=1/(2*pi*fz1*R2)
.VAR C2=C1/(C1*R2*2*pi*fp1-1)
.VAR C3=(fp2-fz2)/(2*pi*Rupper*fp2*fz2)
.VAR R3=Rupper*fz2/(fp2-fz2)
.VAR G0=((R2*C1)/(Rupper*(C1+C2)))*c*d/(a*b) * Gain at fc sanity check *
```

Understanding the Op-Amp Impact

THE OP-AMP RESPONSE is always considered perfect in most analyses and it simplifies the transfer function derivation. When dealing with low to moderate crossover frequencies – a few tens of Hz for a PFC stage or up to a few kHz – this assumption is valid. However, when you start pushing crossover above 10 kHz and you need to boost gain and phase significantly, you will need to ensure that the op-amp own response is not detrimental to the compensator response you want. If it is, then you should select a faster type, with a wider gain-bandwidth product. Below is a classical type 2 compensator in which I have included the op-amp open-loop gain A_{OL}:

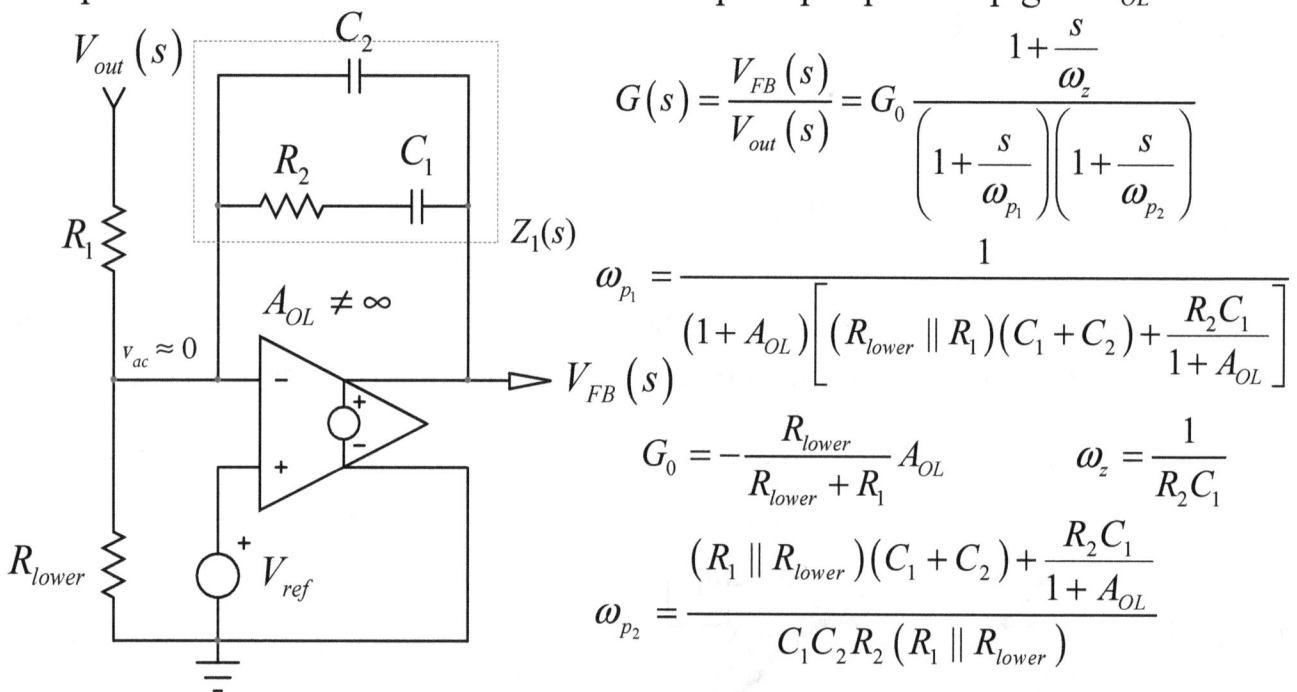

$$G(s) = \frac{V_{FB}(s)}{V_{out}(s)} = G_0 \frac{1 + \dfrac{s}{\omega_z}}{\left(1 + \dfrac{s}{\omega_{p_1}}\right)\left(1 + \dfrac{s}{\omega_{p_2}}\right)}$$

$$\omega_{p_1} = \frac{1}{(1 + A_{OL})\left[(R_{lower} \| R_1)(C_1 + C_2) + \dfrac{R_2 C_1}{1 + A_{OL}}\right]}$$

$$G_0 = -\frac{R_{lower}}{R_{lower} + R_1} A_{OL} \qquad \omega_z = \frac{1}{R_2 C_1}$$

$$\omega_{p_2} = \frac{(R_1 \| R_{lower})(C_1 + C_2) + \dfrac{R_2 C_1}{1 + A_{OL}}}{C_1 C_2 R_2 (R_1 \| R_{lower})}$$

From the above equations, when $s = 0$, A_{OL} and the resistive divider impose a limit on the dc gain, possibly affecting the closed-loop static error. If R_{lower} does not play a role in ac owing to the local feedback of the op-amp – Z_1 imposes a virtual ground at the (-) pin and ≈ 0 V ac across R_{lower} – this local feedback disappears for $s = 0$ and the virtual ground is no longer active: R_{lower} is factored in the gain definition. The given expressions become more complicated if you start including the low- and high-frequency poles present in an internally-compensated op-amp. Once you have plotted your type 2 or 3 response, superimpose the op-amp own ac response over the magnitude plot and make sure there are no overlaps in the boost area. If both curves intersect in this zone, the op-amp will impose its own response: choose a better one.

Effects of the Op-Amp Characteristics

THE BELOW PLOTS compare an ideal op amp with a 50-dB open-loop gain op amp (a TL431 for instance) when the compensator must meet the following targets: f_c = 10 kHz with a 20-dB compensating gain at this frequency and the phase boost must be 65°.

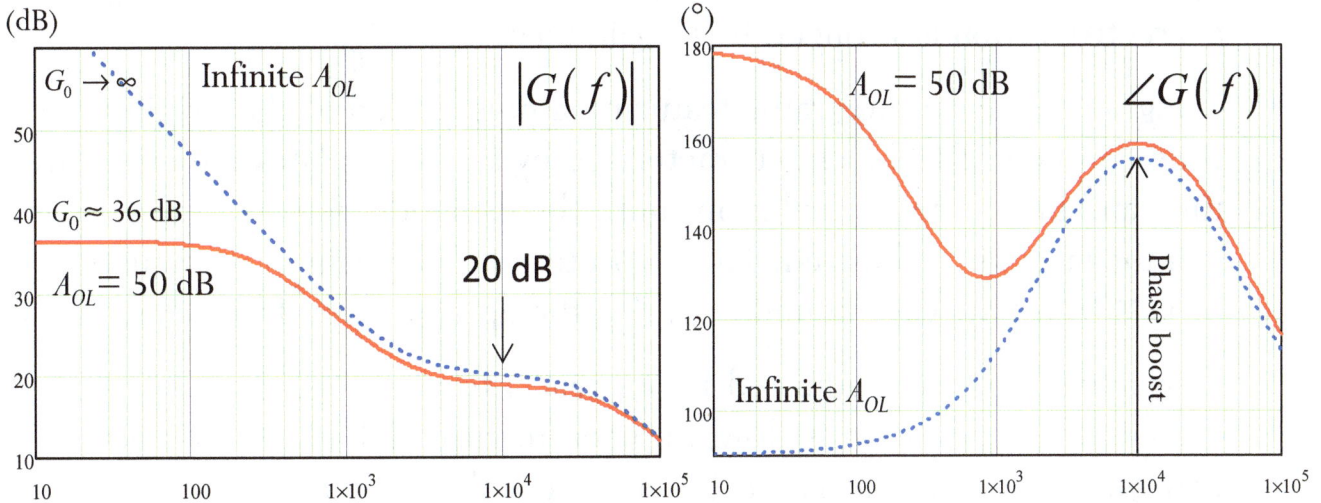

R_1 and R_{lower} are calculated for a 12-V output and a 2.5-V reference voltage. The deviation of the crossover gain and phase boost are negligible. However, the gain is 35 dB at a 120-Hz frequency for the 50-dB A_{OL} while it amounts to 45 dB with the infinite gain. Finally, the quasi-static gain is only 36.4 dB ($\approx$66) for the finite-A_{OL} option while it is infinite with the perfect op amp. What are the impact of these numbers? A lack of gain at twice the mains frequency affects the ability of the control system to reject the rectified ripple. The output variable may be polluted by this component, especially in voltage-mode control. Also, there can be a significant static error in the controlled variable if the plant gain is low. If you now select an op-amp having a higher A_{OL}, 80 dB for instance, the discrepancies disappear and both curves are very close to each other. Pay attention to the open-loop gain value specified in the data-sheet of the selected op-amp and make sure the lowest guaranteed value does not conflict with the compensator response. This statement equally applies to an OTA design in which there is no virtual ground (the resistive divider plays a role in the overall response). The transconductance value g_m naturally appears in the compensator transfer function so carefully check its variability.

The PID Block

A PID COMPENSATOR combines the three functions that are listed below:

- **P**roportional: it sizes the error signal amplitude with the observed deviation. If small, a reduced driving voltage is delivered. On the opposite, if the regulated value deviates significantly from the target, the corrective action is stronger. It is a gain term noted k_p.

- **I**ntegral: this is an integrator featuring a pole placed at the origin with a time constant. This block, characterized by k_i, accumulates the error over time and biases the control input until the deviation between the setpoint and the target is eliminated. The theoretical infinite gain of the equation at dc ($s = 0$) is physically bounded by A_{OL}, the op-amp open-loop gain.

- **D**erivative: this part, adjusted by k_d, reacts to the slope of the observed change. If the deviation is slow, no need to rush and a smooth reaction is enough. Should the deviation be fast, then the reaction shall be swifter. The term k_d has no action at steady-state since the differentiation of a constant value is zero.

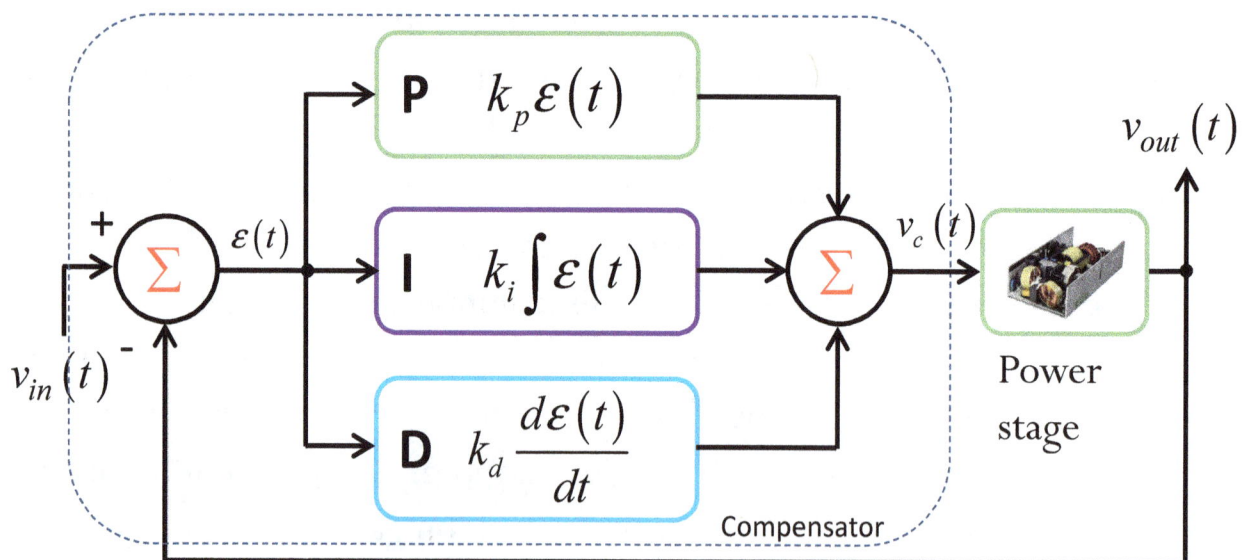

The transfer function linking the response V_c to the stimulus ε is the sum of the three contributors:

$$G_{PID}(s) = \frac{V_c(s)}{\varepsilon(s)} = k_p + \frac{k_i}{s} + sk_d = k_p\left(1 + \frac{1}{s\tau_i} + s\tau_d\right) \text{ with } \tau_i = \frac{k_p}{k_i} \quad \tau_d = \frac{k_d}{k_p}$$

Adding an Extra Pole

IF YOU REWRITE the transfer function of the PID block, you realize that its gain never stops increasing with frequency. Indeed, the Laplace-domain expression reveals a pole at the origin (the integrator) and two zeroes:

$$G_{PID}(s) = G_0 \left(1 + \frac{\omega_{z_1}}{s}\right)\left(1 + \frac{s}{\omega_{z_2}}\right) \qquad \omega_{z_1} = \frac{k_p}{k_d} - \frac{k_p + \sqrt{k_p^2 - 4k_d k_i}}{2k_d}$$

$$\underset{s \to \infty}{\to 0} \qquad \underset{s \to \infty}{\to \infty}$$

$$\omega_{z_2} = \frac{k_p + \sqrt{k_p^2 - 4k_d k_i}}{2k_d} \qquad G_0 = \frac{k_i}{\omega_{z_1}}$$

The plot of this expression clearly shows the gain endlessly growing as f increases. A system running with this configuration will suffer noise immunity problems and, potentially, an impossibility to force a crossover at f_c. We need a pole to roll-off the gain: this is the *filtered* PID in which a second pole is implemented.

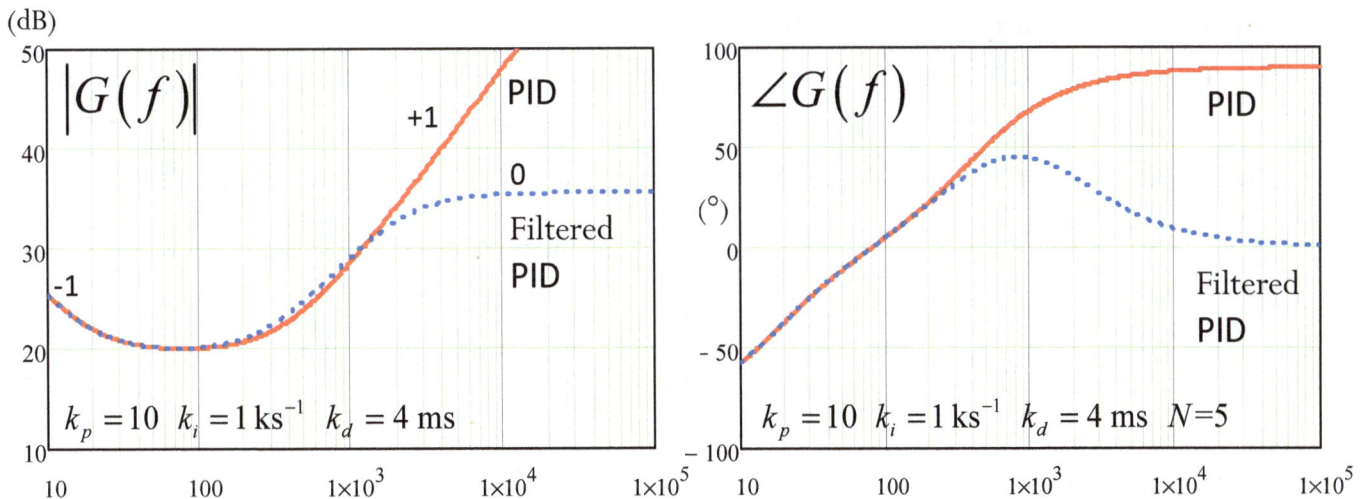

With the filtered version, the high-frequency pole breaks the $+1$-slope and forces magnitude flatness, improving on gain margin and spurious noise pickup. N in the below expression sets the pole position with respect to the zero:

$$G_{PIDF}(s) = k_p \left(1 + \frac{1}{\tau_i s} + \frac{s\tau_d}{1 + \frac{s\tau_d}{N}}\right) = G_0 \frac{\left(1 + \frac{\omega_{z_1}}{s}\right)\left(1 + \frac{s}{\omega_{z_2}}\right)}{1 + \frac{s}{\omega_p}} \qquad \omega_p = \frac{N}{\tau_d} = \frac{N \cdot k_p}{k_d}$$

added pole

From PID to Type 3

WHEN I WAS a student, my teacher was referring to the PID coefficients k_p, k_d and k_i to stabilize a loop. I honestly had difficulty to grasp the concept, having no clue how these terms affect the frequency response. How to tweak them efficiently was a mystery. I needed a link to compare the PID response with that of a known transfer function. As a matter of fact, if you compare the equation of a *filtered*-PID and a type 3 filter, the missing part is the second high-frequency pole in the PID. Let's add it now!

Filtered PID:

$$G_{FPID}(s) = k_p \left(1 + \frac{1}{sk_i} + \frac{s\tau_d}{1+\dfrac{s}{\omega_{p_1}}} \right) \frac{1}{1+\dfrac{s}{\omega_{p_2}}}$$

$$\omega_{p_1} = \frac{N}{\tau_d}$$

Filtering pole ⬆   ⬆ Extra pole

Similar ac responses ⬅➡

Type 3:

$$G_{T3}(s) = G_0 \frac{\left(1+\dfrac{\omega_{z_1}}{s}\right)\left(1+\dfrac{s}{\omega_{z_2}}\right)}{\left(1+\dfrac{s}{\omega_{p_1}}\right)\left(1+\dfrac{s}{\omega_{p_2}}\right)}$$

↑ gain

Assume we have a control-to-output transfer function and we must stabilize it with a filtered-PID. Luckily, we can still add an extra pole via a simple *RC* filter. In this case, rather than painfully tweaking the PID parameters, think with poles and zeroes and reflect them to k_p, k_i, k_d and N values instead:

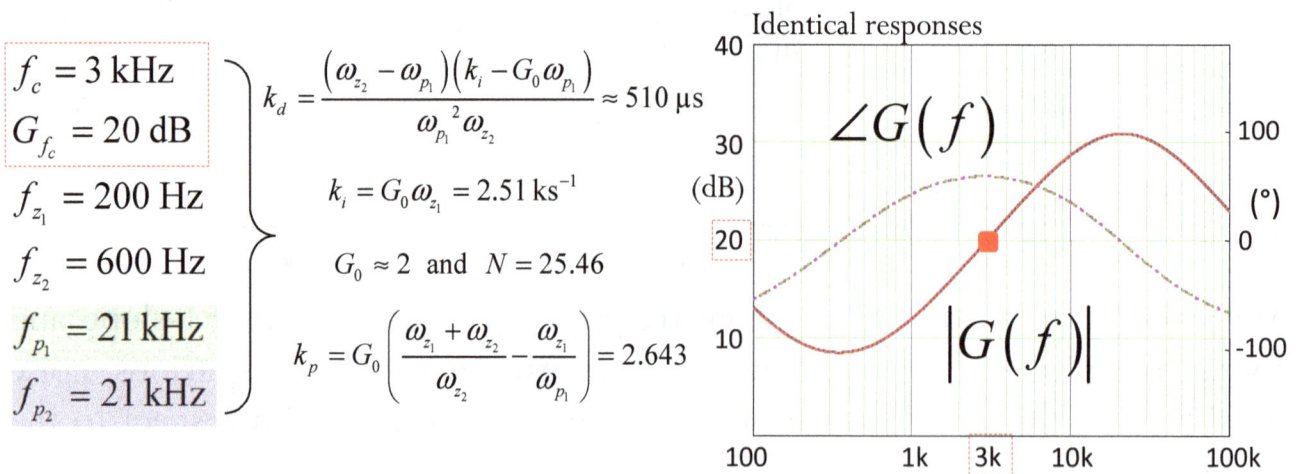

$$f_c = 3 \text{ kHz}$$
$$G_{f_c} = 20 \text{ dB}$$
$$f_{z_1} = 200 \text{ Hz}$$
$$f_{z_2} = 600 \text{ Hz}$$
$$f_{p_1} = 21 \text{ kHz}$$
$$f_{p_2} = 21 \text{ kHz}$$

$$k_d = \frac{(\omega_{z_2}-\omega_{p_1})(k_i - G_0\omega_{p_1})}{\omega_{p_1}^2\omega_{z_2}} \approx 510 \text{ }\mu s$$

$$k_i = G_0\omega_{z_1} = 2.51 \text{ ks}^{-1}$$

$$G_0 \approx 2 \text{ and } N = 25.46$$

$$k_p = G_0\left(\frac{\omega_{z_1}+\omega_{z_2}}{\omega_{z_2}} - \frac{\omega_{z_1}}{\omega_{p_1}}\right) = 2.643$$

Identical responses

$\angle G(f)$

$|G(f)|$

40 / 30 / 20 / 10 (dB)

100 / 0 / -100 (°)

100 1k 3k 10k 100k

With this method, it becomes easier to think of a compensation strategy involving poles and zeroes then translating it into PID coefficients. It is more rigorous than tweaking the parameters with arbitrarily-selected initial values. I wanted a 20-dB gain at a 3-kHz crossover and the computed parameters lead me to this value straight away.

Op-Amp and PID

THIS EXAMPLE is often shown in textbooks. It is the *filtered*-PID we already referred to, with one pole at the origin, two zeroes and an additional pole. The transfer function can be obtained using the fast analytical circuits techniques or via the classic way.

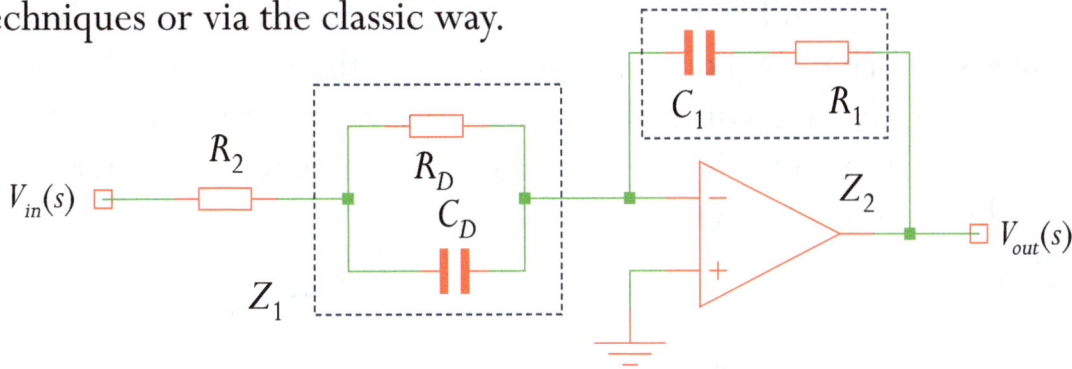

Using either approach, you should find the following expression:

$$G(s) = \frac{R_1}{R_2 + R_D} \frac{\left(1 + \dfrac{1}{sR_1C_1}\right)(1 + sR_DC_D)}{1 + sC_D\left(R_D \parallel R_2\right)} = G_0 \frac{\left(1 + \dfrac{\omega_{z_1}}{s}\right)\left(1 + \dfrac{s}{\omega_{z_2}}\right)}{1 + s/\omega_p}$$

By equating this expression to that of the filtered-PID, you should obtain:

$$\tau_d = \frac{\left(\omega_p - \omega_{z_1}\right)\left(\omega_p - \omega_{z_2}\right)}{\omega_p\left(\omega_p\omega_{z_1} + \omega_p\omega_{z_2} - \omega_{z_1}\omega_{z_2}\right)} \quad N = \frac{\omega_p}{\omega_{z_1} + \omega_{z_2} - \omega_{z_1}\omega_{z_2}/\omega_p} - 1 \quad \tau_i = \frac{\omega_{z_1} + \omega_{z_2}}{\omega_{z_1}\omega_{z_2}} - \frac{1}{\omega_p}$$

$$k_p = \frac{\omega_{po}}{\omega_{z_1}} - \frac{\omega_{po}}{\omega_p} + \frac{\omega_{po}}{\omega_{z_2}} \qquad k_d = k_p\tau_d \quad k_i = \frac{k_p}{\tau_i} \quad \omega_{po} = \frac{1}{C_1\left(R_2 + R_D\right)}$$

Assume we want a 20-dB gain at 3 kHz with the following pole and zeroes:

$$f_c = 3 \text{ kHz} \quad G_{f_c} = 20 \text{ dB} \quad f_{z_1} = 200 \text{ Hz}$$

$$f_{z_2} = 600 \text{ Hz} \quad f_p = 21 \text{ kHz}$$

$$G_{PID}(s) = 2.617 \times \left(1 + \frac{1}{s \cdot 1.053 \text{ ms}} + \frac{s \cdot 192.8 \text{ μs}}{1 + \dfrac{s \cdot 192.8 \text{ μs}}{25.44}}\right)$$

Plotting the two expressions shows no difference: we have our PID parameters from the pole-zeroes placement.

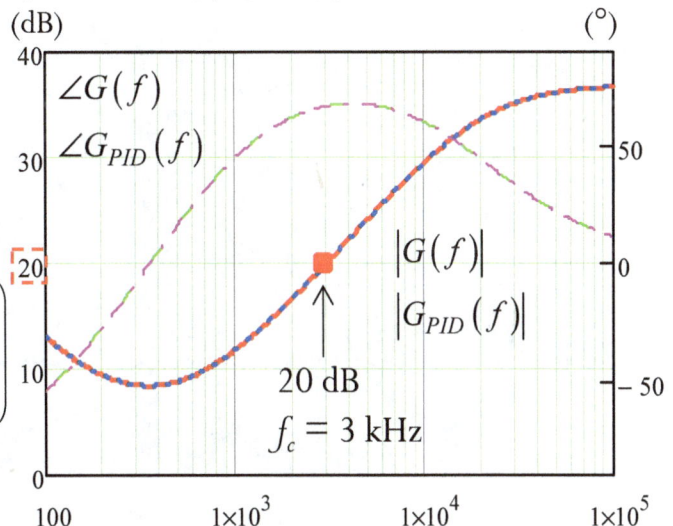

The Optocoupler

IN ISOLATED POWER SUPPLIES, like with ac-dc converters, there is a need for galvanic isolation separating the primary- and secondary-side grounds. The high-frequency transformer provides a first separation but the loop regulation information – which is often located in the isolated secondary side to observe V_{out} or I_{out} – needs to bring the control signal back to the primary side where the switching circuit resides. For that purpose, designers often resort to an optocoupler. This component is characterized by a current transfer ratio (CTR) and a pole that must be extracted on the bench:

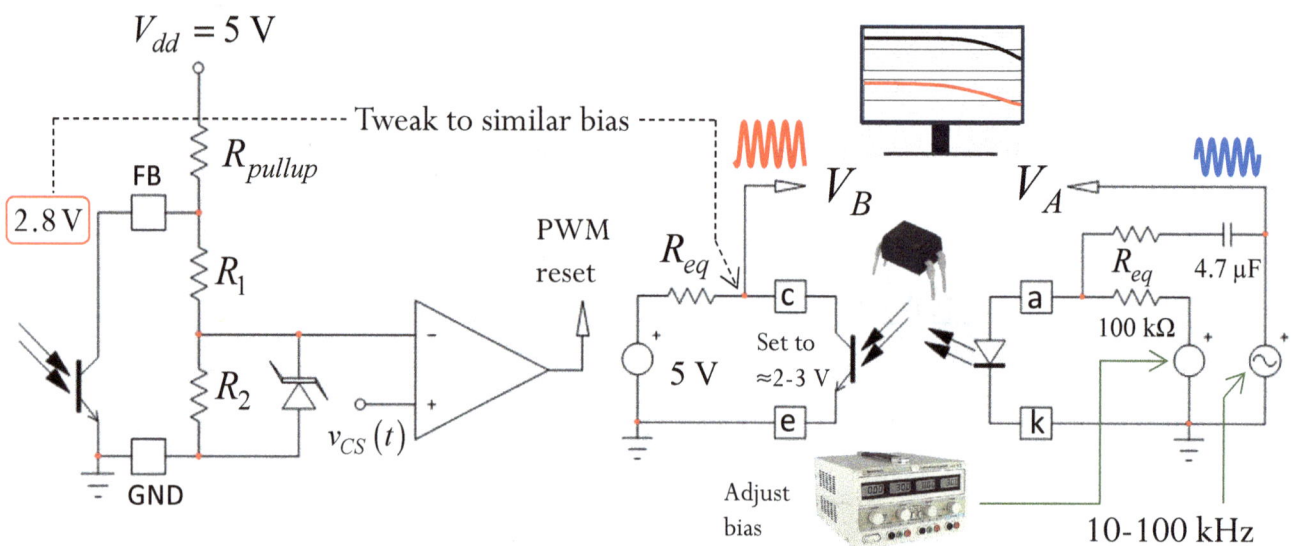

The idea behind the characterization is to reproduce what the collector will "see" when connected to the feedback pin of the controller. In the example, R_{eq} is the Thévenin resistance at the FB pin which drives the collector.

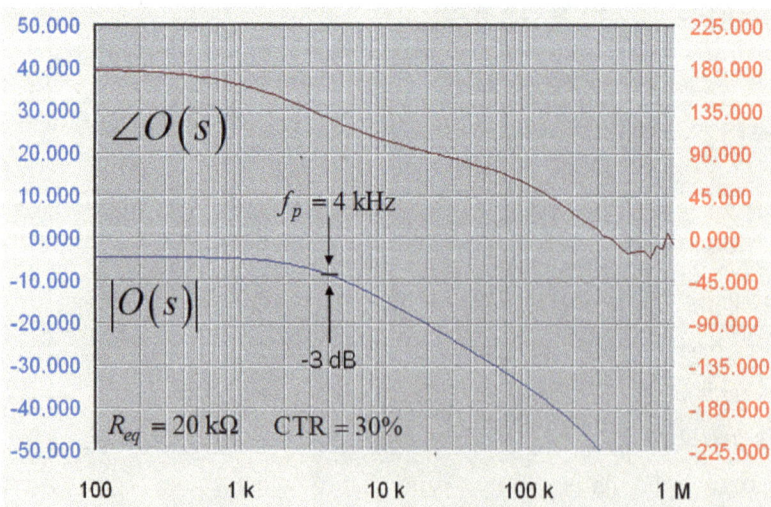

Once the bias point is set, ac-sweep the LED current and you obtain the transfer function of the optocoupler. A 4-kHz pole and a 20-kΩ pull-up R_{eq} resistance imply a parasitic capacitance C_{opto} placed between collector and emitter equal to:

$$C_{opto} = \frac{1}{6.28 \times 4k \times 20k} \approx 2 \text{ nF}$$

Pushing the Optocoupler Pole

IF YOU WANT TO PUSH crossover in isolated converters, you will quickly realize how the intrinsic optocoupler pole affects phase margin. You can reduce the pull-up or -down resistance but you may quickly hit a limit where the overall consumption, particularly in light-load conditions, is degraded. A solution to extend the optocoupler response is to resort to a cascode configuration. In this approach, the optocoupler is driven by a low-impedance source which keeps its collector voltage constant and minimizes the Miller effect. The current transmitted by the optocoupler translates into a voltage at the collector of the driving transistor. Two configurations are described below:

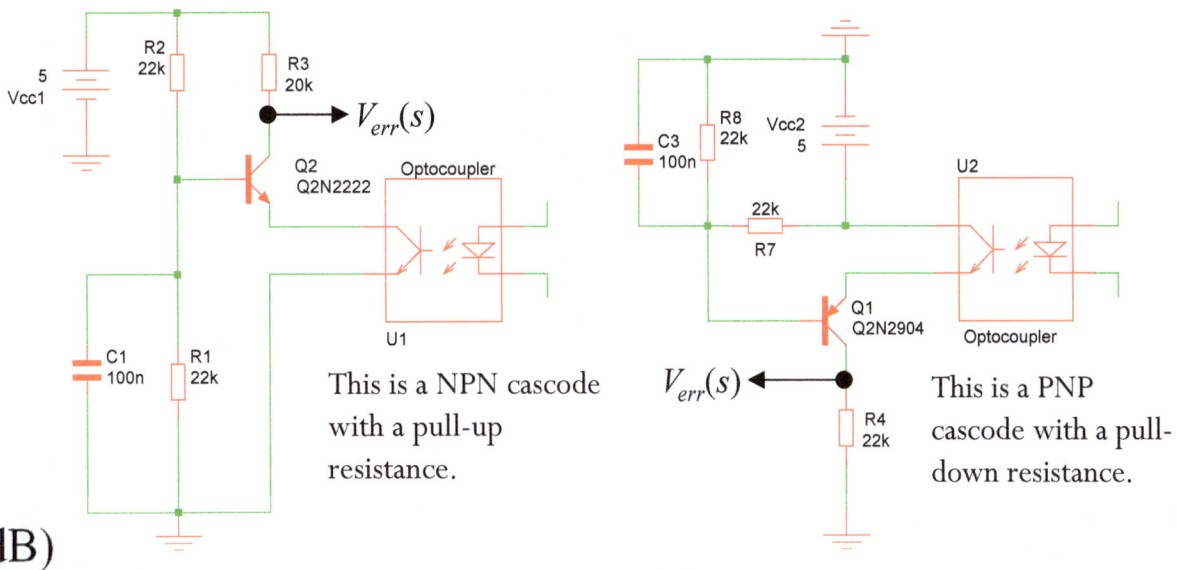

This is a NPN cascode with a pull-up resistance.

This is a PNP cascode with a pull-down resistance.

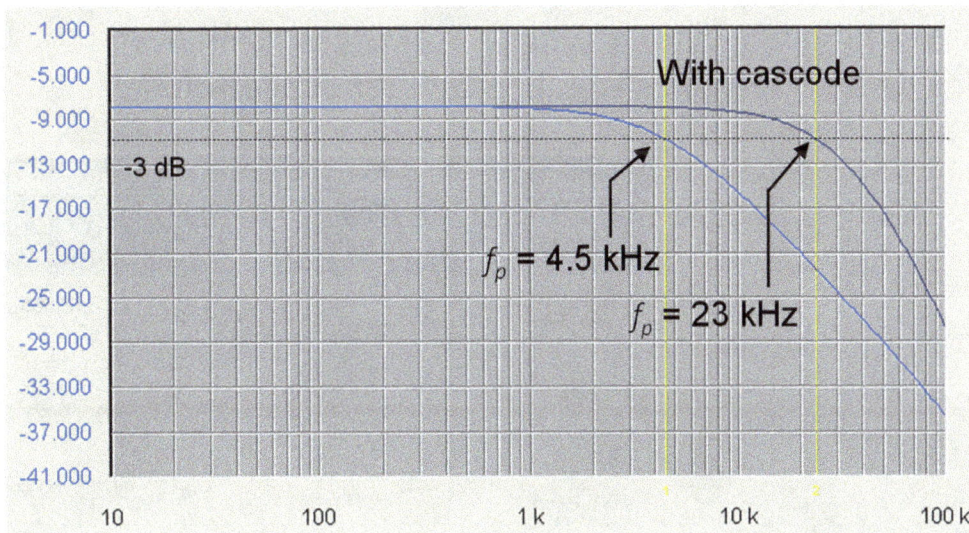

You can see the difference between an optocoupler driven by a classical pull-up resistance and when it is driven in a cascode configuration.

With the latter, the pole is pushed 5 times above its original position.

Optocoupler and TL431

ONCE YOU HAVE determined the low-frequency pole of the optocoupler, you can compute the parasitic capacitance the part brings between its collector-emitter connections. It is C_{opto} in the below simple model:

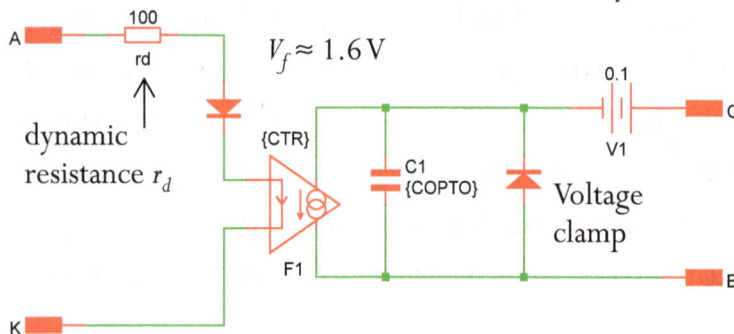

The macro will calculate the parasitic capacitance and pass it to the model:

```
* Optocoupler specifications *
*
.GLOBALVAR Fopto=15k
.GLOBALVAR Copto=1/(2*pi*Fopto*Rpullup)
.GLOBALVAR CTR=1
*
```

This extra capacitance is coming in parallel with the one needed for a type 1, 2 or 3 compensator built with a TL431: the final calculation must account for its contribution. You will also note the presence of a resistance in series with the diode symbol which is 100 Ω in the schematic. This element models the dynamic resistance r_d of the LED and depends on the operating point (forward current). Its value is important as it comes in series with the resistance that is externally added and affects the gain. Gain mismatch between calculations and measurements often finds it root in r_d, which hasn't been characterized at the expected operating point.

The TL431 is a popular 3-leg active component which includes an op-amp and a 2.5-V reference voltage. The part is self-supplied from its cathode current and you must inject a minimum bias current to make it operate correctly. If 2.5 V is not low enough, the TLV431 brings a 1.25-V reference voltage.

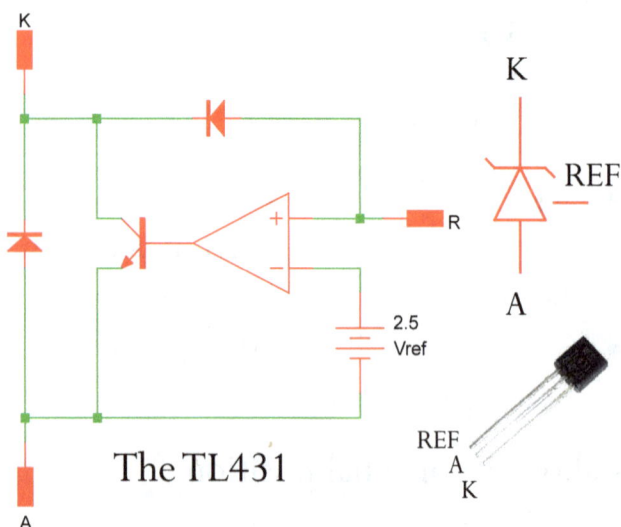

The TL431

Part	Max BV (V)	Min bias (µA)	V_{ref} (V)	Package
TL431	37	1000	2.5	TO92, SO9, µ8, DIP8
TLV431	18	100	1.25	TO92, TSOP5, SOT-23
NCP431	37	60	2.5	TO92, TSOP5

Type 1 with the TL431

When associated with an optocoupler, the TL431 lets you build type 1, 2 and 3 compensators but with some limits as we will see. The type 1 is the simplest one and is shown below:

The LED series resistance R_2 or R_{LED} sets the gain but also the bias point on the TL431. By construction, the minimum K-A operating voltage of the TL431 must obviously be above its 2.5-V reference voltage. Therefore, there is an upper limit for R_2 that you must first determine:

$$R_{LED} < \frac{\overset{\text{Regulated } V_{out}}{V_{out}} - \overset{\approx 1\,V}{V_f} - \overset{\text{e.g. 2.5 V}}{V_{TL431min}}}{\underset{\text{IC}}{V_{dd}} - \underset{\text{Opto sat}}{V_{CE,sat}} + \underset{\text{Bias}}{I_b} \cdot \underset{\text{Opto}}{CTR_{min}} R_{pullup}} R_{pullup} CTR_{min}$$

IC internal   Opto sat voltage   Bias current   Opto CTR   Internal pull-up

- This formula is also valid for other compensator types: if you choose a higher value, regulation can be lost.

- I_b is the extra $\approx$1-mA provided by R_5

The transfer function shows a zero and a pole which neutralize to keep ω_{po} alone:

$$G(s) = -\frac{1 + \dfrac{s}{\omega_z}}{\dfrac{s}{\omega_{po}}\left(1 + \dfrac{s}{\omega_p}\right)}$$

$$\omega_z = \frac{1}{R_1 C_1}$$

$$\omega_p = \frac{1}{R_{pullup} C_2}$$

$$\omega_{po} = \frac{R_{pullup} CTR}{R_{LED} R_1 C_1}$$

Simulating the Type 1 – the TL431

THE DESIGN METHODOLOGY for this type 1 starts with the placement of the 0-dB crossover pole, f_{po}. Assuming we read a gain of 20 dB at $f_c = 100$ Hz from the power stage Bode plot, we position the pole at:

$$f_{po} = G \cdot f_c = 10^{-\frac{20}{20}} \times 100 = 10 \text{ Hz} \quad \Longrightarrow \quad \begin{array}{l} \text{Attenuation at 100 Hz} \\[4pt] \text{will be 20 dB} \end{array}$$

The maximum value of the series LED resistance is computed based on the optocoupler characteristics and 1-mA extra bias I_b:

$$R_{LED} < \frac{12-1-2.5}{5-0.3+1m\times0.3\times20k} \, 20k\times0.3 \approx 4.8 \text{ k}\Omega \quad \overset{\substack{50\% \\ \text{margin}}}{\Longrightarrow} \quad R_{LED} = 2.4 \text{ k}\Omega$$

The adopted PWM controller features an internal pull-up resistance of 20 kΩ. For a 12-V output and a 250-µA bias current for the resistive divider, $R_1 = 38$ kΩ. We can now evaluate capacitors C_1 and C_2:

$$C_1 = \frac{R_{pullup}\text{CTR}}{2\pi R_{LED} R_1 f_{po}} = \frac{20k\times0.3}{6.28\times2.4k\times38k\times10} \approx 1 \text{ µF}$$

$$C_2 = \frac{\text{CTR}}{2\pi R_{LED} f_{po}} = \frac{0.3}{6.28\times2.4k\times10} \approx 2 \text{ µF}$$

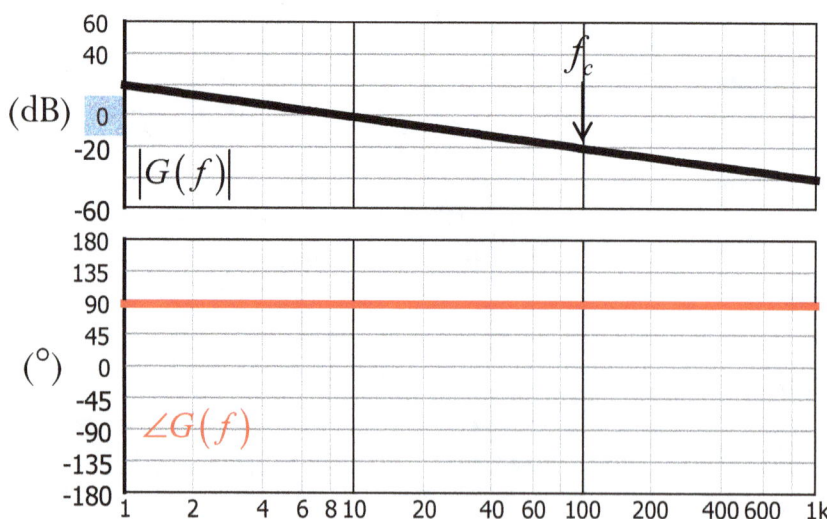

- The automated macro computes the values assigned to the components:

```
* Rupper = 38000
* Rlower = 10000
* Rmax = 4766.35514018692
* RLED = 2383.17757009346
* C2 = 2.00347987186268e-06
* C1 = 1.05446309045404e-06
* Ccol = 2.00215358067025e-06
* fpo = 10
```

The simulation confirms the 20-dB attenuation at 100 Hz as expected.

TL431 Fast and Slow Lanes

THE TL431 LENDS itself well to building a type 2 compensator. However, it is important to understand the regulation mechanism which is based on the LED current modulation. Considering the below circuit, you see that the LED current depends on V_{out} and on the cathode voltage v_k :

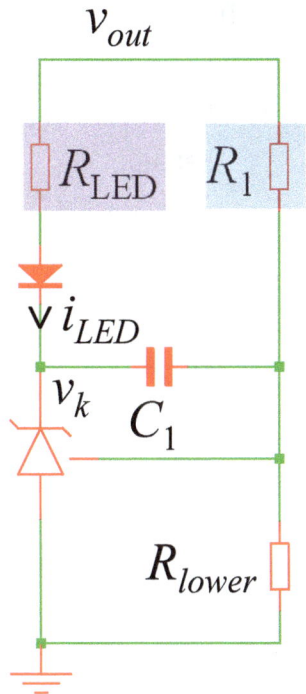

$$I_{LED}(s) = \frac{V_{out}(s) - V_K(s)}{R_{LED}} = \frac{V_{out}(s)}{R_{LED}} + \overbrace{\frac{V_{out}(s)\frac{1}{sR_1C_1}}{R_{LED}}}^{\text{Slow lane}}$$

As frequency increases, C_1 becomes a short circuit and the ac voltage at the cathode node becomes 0 V: the *slow lane* becomes ac silent. However, the dc regulation is kept of course and v_k is at a fixed dc level, imposed by the TL431. The LED current then becomes:

$$I_{LED}(s) = \frac{V_{out}(s) - \overbrace{V_K(s)}^{\approx 0}}{R_{LED}} = \underbrace{\frac{V_{out}(s)}{R_{LED}}}_{\text{Fast lane}}$$

$\Rightarrow$ R_{LED}, limited in value for bias reasons, sets a <u>minimum</u> gain by offering a second path from V_{out} to I_{LED}.

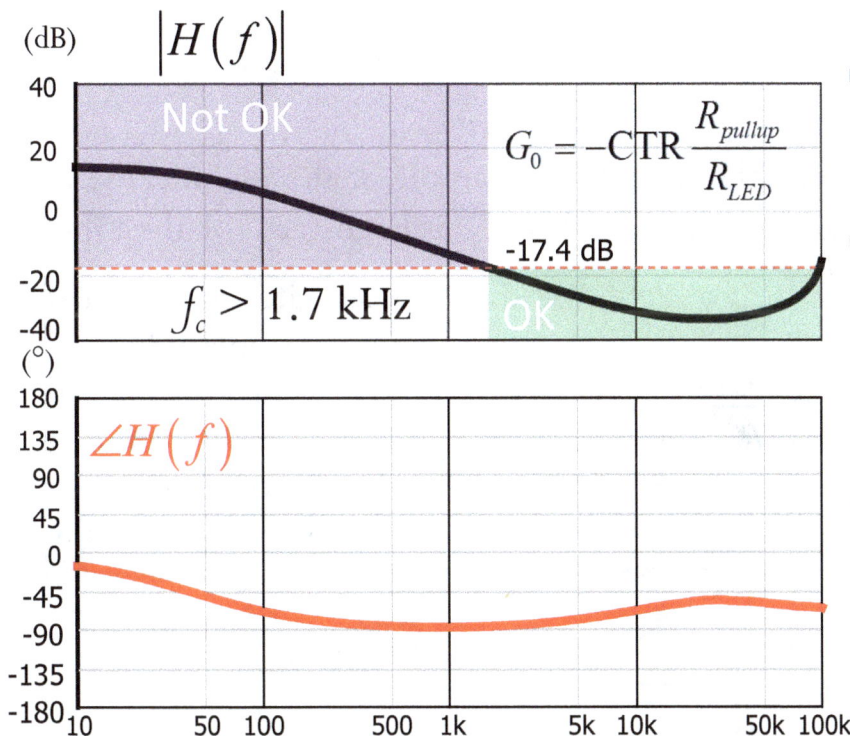

$$G_0 = -\text{CTR}\frac{R_{pullup}}{R_{LED}}$$

-17.4 dB

$f_c > 1.7\text{ kHz}$

- Assume R_{LED} cannot exceed 865 Ω for dc-bias reasons (V_{out} is 5 V)

- With a 20-kΩ pull-up resistor and a CTR of 32%, the minimum gain G_0 of the circuit is 17.4 dB.

- You must choose a crossover where, at least, you need a gain of 17.4 dB: f_c cannot be less than 1.7 kHz: this is the *fast-lane* action.

Type 2 with the TL431

THE FAST LANE effect must be well understood when designing a compensator with a TL431. If you try to increase the LED resistance beyond the calculated bias limit, or if you failed to include some design margin, you may lose regulation when the CTR or other parameters move during production. Beside this typical problem, the TL431 lends itself well for the design of a type 2 compensator:

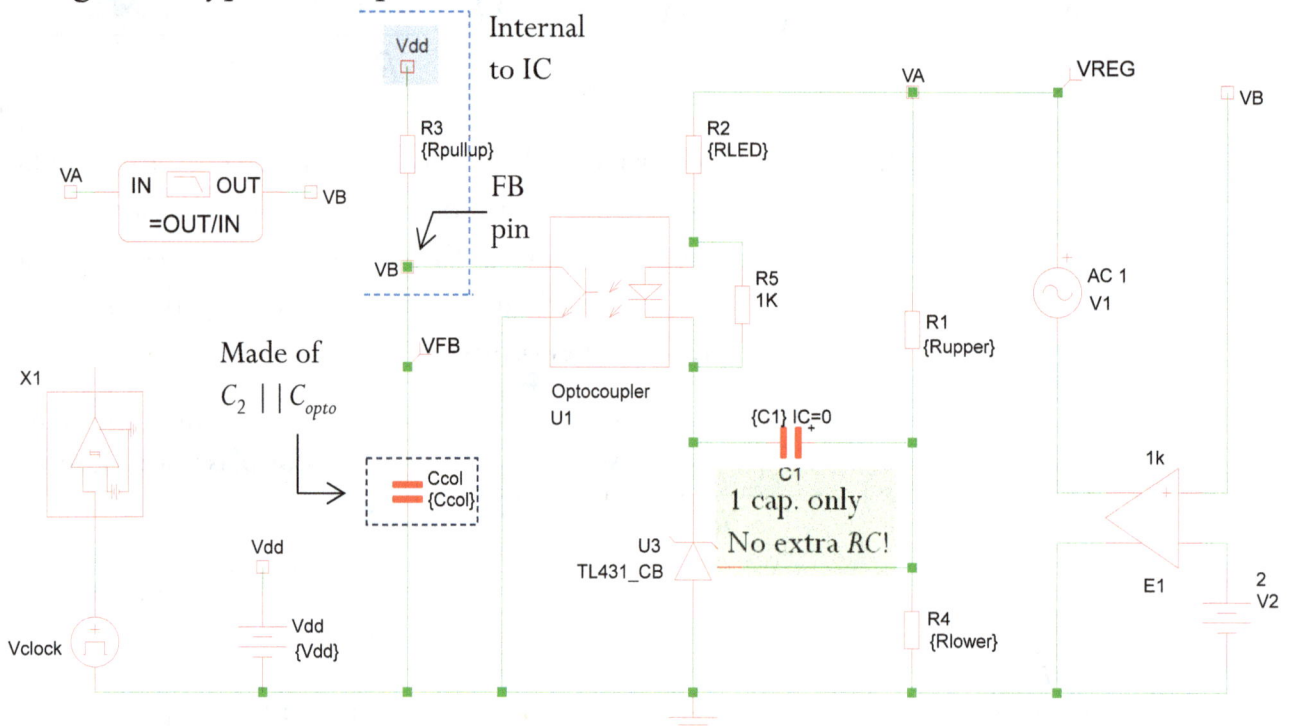

The transfer function highlights the presence of a pole at the origin, one zero and one pole. Please note that a *single* capacitor, C_1, is necessary for the type 2.

$$G(s) = -G_0 \frac{1 + \dfrac{\omega_z}{s}}{1 + \dfrac{s}{\omega_p}}$$

Mid-band gain — Inverted zero

$$G_0 = \frac{R_{pullup}}{R_{LED}} CTR$$

$$\omega_z = \frac{1}{R_1 C_1}$$

$$\omega_p = \frac{1}{R_{pullup} C_2}$$

■ C_2 is the capacitor you calculate to position the pole. However, the optocoupler parasitic capacitor C_{opto} will come in parallel. You have to account for its presence in the final selection:

$$C_{col} = C_2 - C_{opto}$$

Final value — Calculated for the pole — Parasitic from the optocoupler

Designing a Type 2 with the TL431

FOR THIS DESIGN example, we assume a power stage H exhibiting a gain of -30 dB and a phase lag of 85° at a selected crossover of 2 kHz. First, we determine the necessary phase boost for a phase margin target of 70°:

$$boost = \varphi_m - \angle H(f_c) - 90° = 70 - (-85°) - 90° = 65°$$

What is the needed gain to compensate for the 30-dB attenuation?

$$|G(f_c)| = 10^{\frac{G_{f_c}}{20}} = 10^{\frac{-30}{20}} \approx 31.6 \longleftarrow \text{ Necessary mid-band gain at 2 kHz}$$

The transfer function of a type 2 compensator based on the TL431 is here:

Mid-band gain
$$\downarrow$$

$$G(s) = -G_0 \frac{1 + \dfrac{\omega_z}{s}}{1 + \dfrac{s}{\omega_p}} \qquad G_0 = \frac{R_{pullup}\,CTR}{R_{LED}}$$

$$\omega_z = \frac{1}{R_1 C_1} \qquad \omega_p = \frac{1}{R_{pullup} C_2}$$

- For this example, we assume R_{pullup} is 10 kΩ, V_{out} = 12 V, R_1 = 38 kΩ, the CTR is 85% and the parasitic capacitor C_{opto} is 1 nF.

$$\Rightarrow \quad R_{LED} = \frac{R_{pullup}\,CTR}{G_0} = \frac{10k \times 0.85}{31.6} \approx 269\ \Omega$$

Before proceeding, make sure this LED resistance is compatible with the bias requirements otherwise regulation won't be ensured:

$$R_{LED} < \frac{12 - 1 - 2.5}{5 - 0.3 + 1m \times 0.85 \times 10k} 10k \times 0.85 \approx 5.5\ k\Omega \quad \checkmark \text{ Plenty of margin!}$$

You can now place the pole-zero pair manually or via the k factor:

$$f_p := \left[\tan(boost) + \sqrt{(\tan(boost))^2 + 1} \right] \cdot f_c = 9.02\,kHz \qquad k_1 := \tan\left(\frac{boost}{2} + \frac{\pi}{4} \right) = 4.51$$

$$f_z := \frac{f_c^2}{f_p} = 443.39\,Hz \qquad\qquad f_p := f_c \cdot k_1 = 9.02\,kHz$$

Value at the FB pin

From the optocoupler

$$f_z := \frac{f_c}{k_1} = 443.39\,Hz$$

$$C_1 := \frac{1}{2 \cdot \pi \cdot R_1 \cdot f_z} = 143.58\,nF$$

$$\downarrow \qquad\qquad \downarrow$$

$$C_2 := \frac{1}{2 \cdot \pi \cdot f_p \cdot R_{pullup}} = 1.76\,nF \qquad \Rightarrow \quad C_{col} := C_2 - C_{opto} = 764.19\,pF$$

Simulating the Type 2 with a TL431

NOW THAT WE HAVE the capacitors and resistances values for the compensator, we can run a simulation and check the ac response:

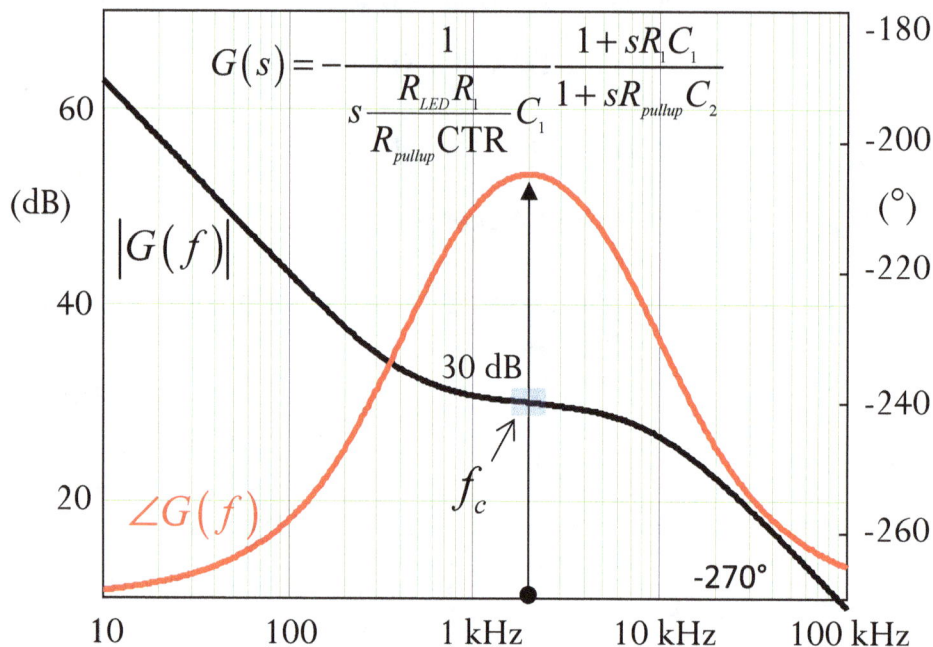

$$G(s) = -\cfrac{1}{s\dfrac{R_{LED}R_1}{R_{pullup}\,\mathrm{CTR}}C_1}\cdot\dfrac{1+sR_1C_1}{1+sR_{pullup}C_2}$$

The simulation in SIMPLIS® confirms the response with a 30-dB gain at 2 kHz and the phase boost of 65° as expected.

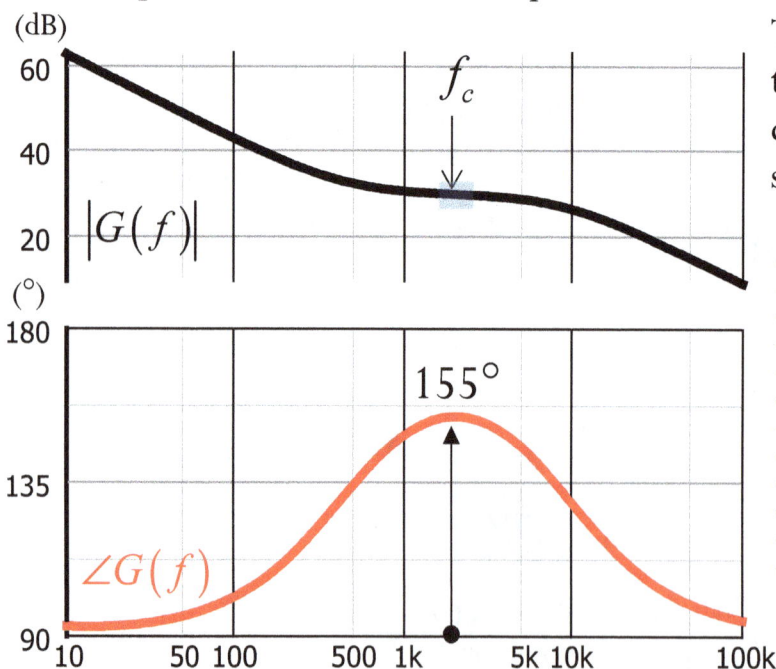

The automated macro leads to the same values we have calculated with the Mathcad® sheet:

```
*  Rupper = 38000
*  Rlower = 10000
*  RLED = 268.793601114312
*  RMAX = 5473.48484848485
*  C1 = 9.44607309479557e-09
*  C2 = 1.7641900708357e-09
*  Copto = 1.06103295394597e-09
*  Ccol = 7.03157116889733e-10
*  Boost = 65
*  Fz = 443.38932528588
*  Fp = 9021.41700732411
*
```

The performance of the TL431 depends on its bias current. You must ensure a proper level is injected in the part, otherwise ac response will suffer.

A Type 2 without the Fast Lane

IN SOME APPLICATIONS, you may not be able to connect the fast lane for instance if the regulated output voltage is too low or, on the contrary, too high and exceeds the TL431 breakdown voltage. In these cases, you may need to open the fast lane and bias it independently from the regulated V_{out}. Using a Zener diode is an option – and I did document it in the type 3 example – but connecting the LED series resistance to a fixed, well-regulated 5- or 12-V rail is also a possibility.

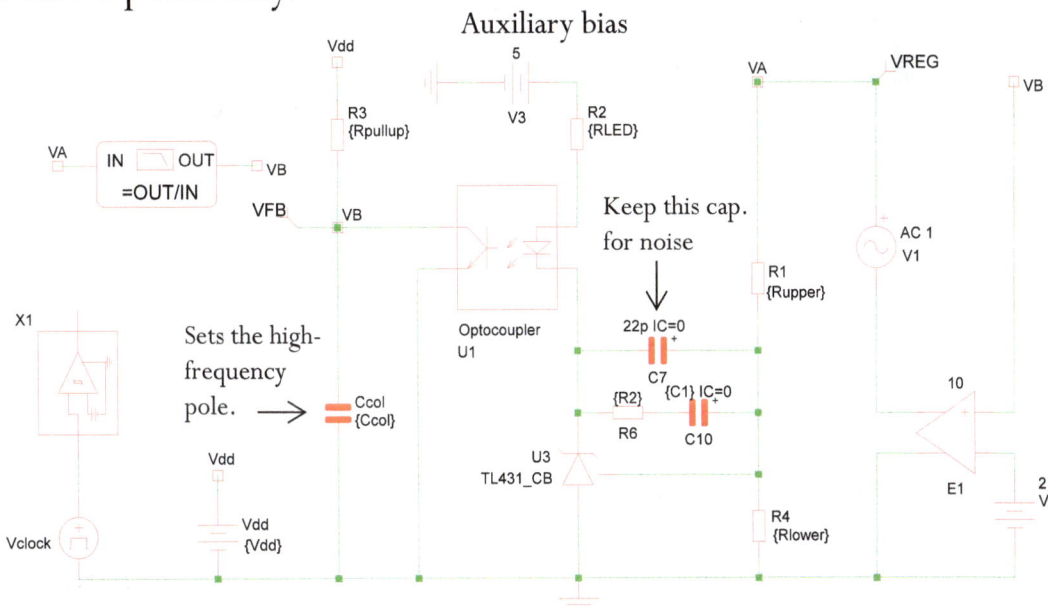

The LED current is decoupled from the regulated variable in this example. Calculations now account for the cascaded gains made of the TL431 and the optocoupler.

```
.VAR Vaux=5
.VAR VL=2.5
.VAR VCEsat=0.3
.VAR Vdd=5
.VAR Ib=1m
.VAR Vf=1
.VAR A=Vaux-Vf-VL
.VAR B=Vdd-VCEsat+Ib*CTR*Rpullup
.VAR Rmax=(A/B)*Rpullup*CTR
.VAR RLED=0.8*Rmax
*
* Do not edit the below lines *
*
.VAR G1={Rpullup*CTR/RLED}
.VAR G2={10^(-Gfc/20)}
.VAR G={G2/G1}
.VAR d={(fz^2+fc^2)*(fp^2+fc^2)}
.VAR c={(fz^2+fc^2)}
.VAR R2={(sqrt(d)/c)*G*fc*Rupper/fp}
.VAR C1={1/(2*pi*fz*R2)}
.VAR C2={1/(2*pi*fp*Rpullup)}
.VAR Ccol={C2-Copto}
```

The crossover frequency is set to 1 kHz and the phase boost amounts to 65°. The gain deficiency was set to 20 dB in this example. The LED resistance is arbitrarily selected here to cope with bias requirements.

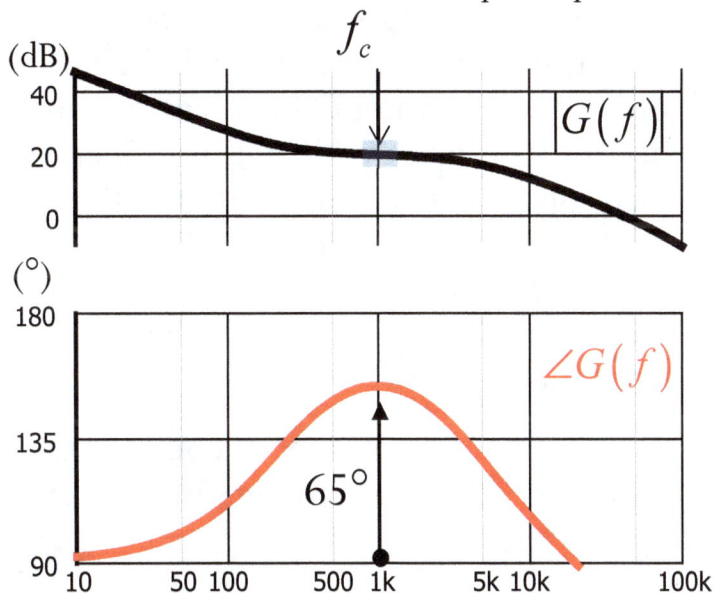

A Type 3 with the TL431

THE FAST LANE presence complicates the design of a type 3 compensator. This is because the extra RC filter – R_3C_3 in the op-amp design – should take place across the LED resistor and *not* across the upper side resistance R_1:

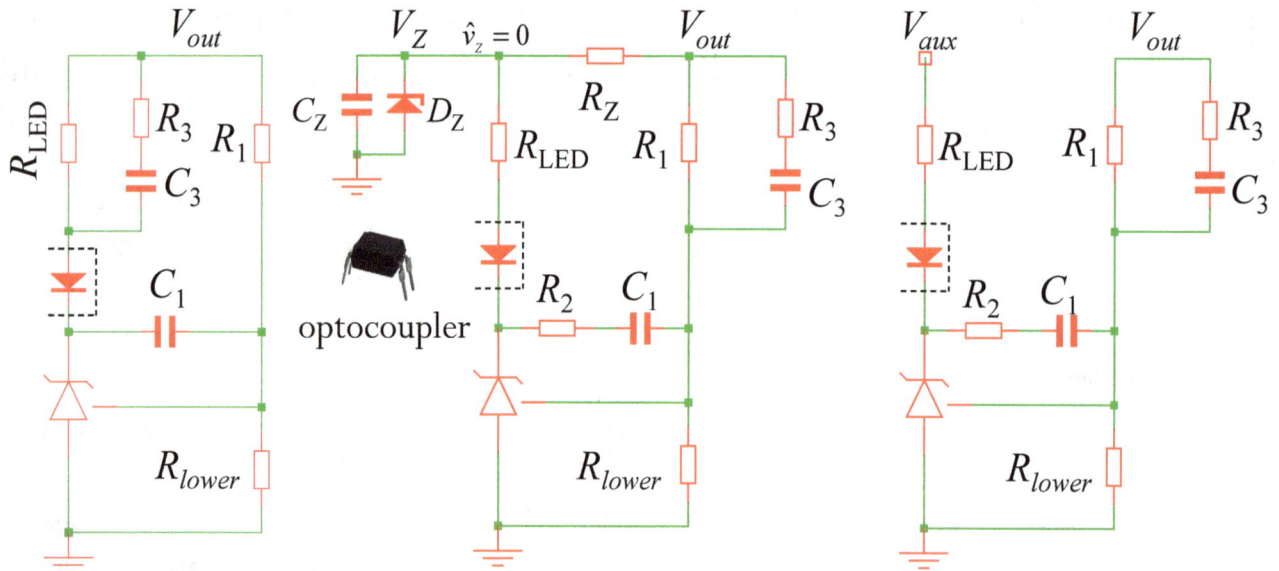

Fast lane is on - bad idea!   Fast lane is off via a Zener diode   An auxiliary voltage is good.

As a result, R_{LED} plays a role in the zero of the transfer function but also in the gain definition. If you add an upper limit of R_{LED} for bias considerations, the reliable design of this type of compensator can quickly turn into a nightmare.

One option is to disable the fast lane via a Zener-based regulator or, even better, by using a well-regulated auxiliary rail. The key is to cut the direct ac modulation of the LED current coming from V_{out}. When it is done, the circuit becomes an open-collector amplifier and the design gains in flexibility. If you plan on using a Zener diode, the best results are obtained when V_Z is substantially lower than V_{out}. For instance, a 12-V *regulated* voltage and a 5-V Zener (V_Z) ensure a good rejection of the regulated 12-V rail, even more if you comfortably bias the Zener diode (I_{Zbias}) to minimize its dynamic resistance r_d.

The design is slightly more complicated but not insurmountable. Let's use the following design parameters for our example:

$V_{TL431,min} = 2.5\,\text{V}$, $I_b = 1\,\text{mA}$, $V_Z = 5\,\text{V}$, $I_{Zbias} = 2\,\text{mA}$, $V_f = 1\,\text{V}$

$V_{CE,sat} = 0.3\,\text{V}$, $V_{dd} = 5\,\text{V}$, $V_{out} = 12\,\text{V}$, $R_{pullup} = 20\,\text{k}\Omega$, $\text{CTR}_{min} = 30\%$

$R_1 = 38\,\text{k}\Omega$ and $C_{opto} = 1\,\text{nF}$.

Designing a Type 3 with TL431

REGARDLESS of the configuration, you still need to check for the proper TL431 bias current with the minimum optocoupler CTR. The expression changes as V_{out} is replaced by V_Z:

Include margin

$$R_{LED} < \frac{5-1-2.5}{5-0.3+1m\times0.3\times20k}20k\times0.3 \approx 840\,\Omega \implies R_{LED} = 680\,\Omega$$

Now determine the dropping resistance to bias the Zener diode but also the TL431:

$$R_z = \frac{\left(V_{out}-V_z\right)R_{pullup}\text{CTR}_{min}}{\left(V_{dd}-V_{CE,sat}\right)+\left(I_b+I_{Zbias}\right)\text{CTR}_{min}R_{pullup}}$$

$$= \frac{(12-5)\times20k\times0.3}{(5-0.3)+(1m+2m)\times0.3\times20k} \approx 1.8\,\text{k}\Omega$$

✔ A 100-nF capacitor or more will decouple the Zener diode.

D_Z C_Z

Assume we read the Bode plot of the power stage and at a 5-kHz crossover frequency, the gain is -20 dB with a phase shift of -125°. We need one zero at 800 Hz, the second at 2 kHz and the high-frequency pole at half the 100-kHz switching frequency. We want a phase margin of 70°, what is the phase boost?

$$boost = \varphi_m - \angle H\left(f_c\right) - 90° = 70 - \left(-125°\right) - 90° = 105°$$

We want the gain at f_c = 5 kHz to compensate the 20-dB attenuation, therefore:

$$\left|G\left(f_c\right)\right| = 10^{\frac{G_{f_c}}{20}} = 10^{\frac{-20}{20}} = 10$$

The magnitude and phase of a TL431 in a type 3 compensator without fast lane is as follows:

$$G(s) = -G_0\frac{\left(1+\dfrac{\omega_{z_1}}{s}\right)\left(1+\dfrac{s}{\omega_{z_2}}\right)}{\left(1+\dfrac{s}{\omega_{P_1}}\right)\left(1+\dfrac{s}{\omega_{P_2}}\right)}$$

$$G_0 = \text{CTR}\frac{R_{pullup}}{R_{LED}}\frac{R_2}{R_1} \qquad \omega_{z_1} = \frac{1}{R_2C_1} \qquad \omega_{z_2} = \frac{1}{C_3\left(R_1+R_3\right)}$$

$$\omega_{P_1} = \frac{1}{R_{pullup}C_2} \qquad \omega_{P_2} = \frac{1}{R_3C_3}$$

C_2 is the required capacitor for placing the pole, it is adjusted based on C_{opto}.

Determining Components Values

WITH THE PHASE BOOST in hand, we can place the pole at the following position:

$$f_{z1} := 800\,Hz \qquad\qquad f_{z2} := 2\,kHz \qquad\qquad f_{p2} := 50\,kHz$$

$$f_{p1} := \frac{f_c}{\tan\left(\operatorname{atan}\left(\dfrac{f_c}{f_{z1}}\right) + \operatorname{atan}\left(\dfrac{f_c}{f_{z2}}\right) - \operatorname{atan}\left(\dfrac{f_c}{f_{p2}}\right) - boost\right)} = 6.31 \cdot kHz$$

The LED resistance being fixed, we have to determine the value of R_2:

Gain
of 10 ↘

$$R_2 := \frac{G_1 \cdot R_1 \cdot R_{LED}}{R_{pullup} \cdot CTR} \cdot \frac{\sqrt{1 + \left(\dfrac{f_c}{f_{p1}}\right)^2} \cdot \sqrt{1 + \left(\dfrac{f_c}{f_{p2}}\right)^2}}{\sqrt{1 + \left(\dfrac{f_{z1}}{f_c}\right)^2} \cdot \sqrt{1 + \left(\dfrac{f_c}{f_{z2}}\right)^2}} = 20.25\,k\Omega$$

The rest of the components are easily obtained based on the poles and zeroes:

$$R_3 := \frac{R_1 \cdot f_{z2}}{f_{p2} - f_{z2}} = 1.58\,k\Omega \qquad\qquad C_1 := \frac{1}{2 \cdot \pi \cdot f_{z1} \cdot R_2} = 9.82 \cdot nF$$

$$C_2 := \frac{1}{2 \cdot \pi \cdot f_{p1} \cdot R_{pullup}} = 1.26\,nF \qquad C_3 := \frac{f_{p2} - f_{z2}}{2 \cdot \pi \cdot R_1 \cdot f_{p2} \cdot f_{z2}} = 2.01 \cdot nF$$

➡ $C_{col} := C_2 - C_{opto} = 261.34\,pF$

Capacitor C_{col} must be closely located to the controller and connected between its FB and GND pins:

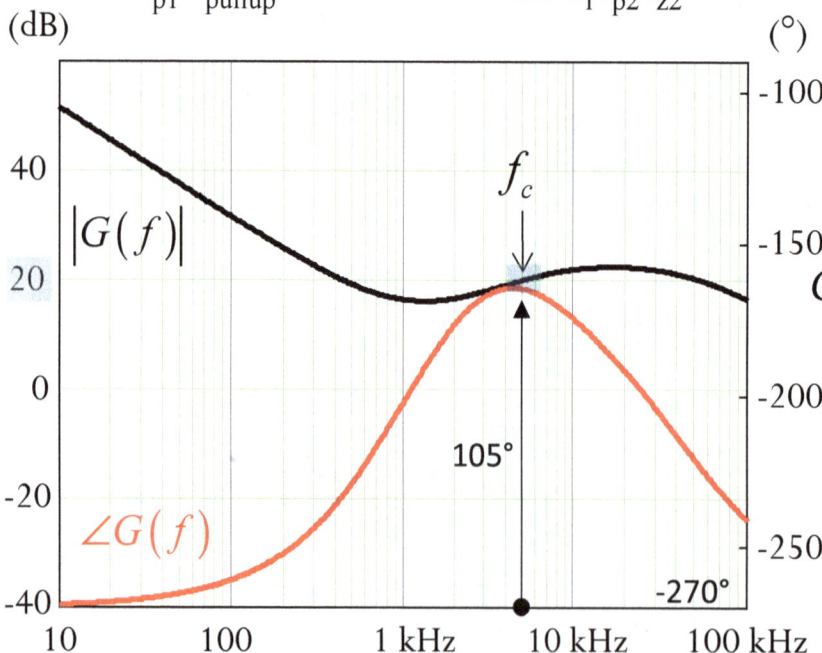

The emitter of the optocoupler goes straight to the controller GND pin: do not connect it to a noisy ground!

The theoretical response exactly meets our target of 20 dB gain at 5 kHz.

Simulating without the Fast Lane

THE TYPE 3 WITHOUT fast lane can also be simulated in SIMPLIS® by adding the Zener diode or providing a separate auxiliary rail. The performance of the compensator depends on the ac separation between the auxiliary bias (V_Z or the aux. winding) and the regulated V_{out}. Care must be taken to offer the best possible isolation between the two, or the magnitude and phase response will be distorted:

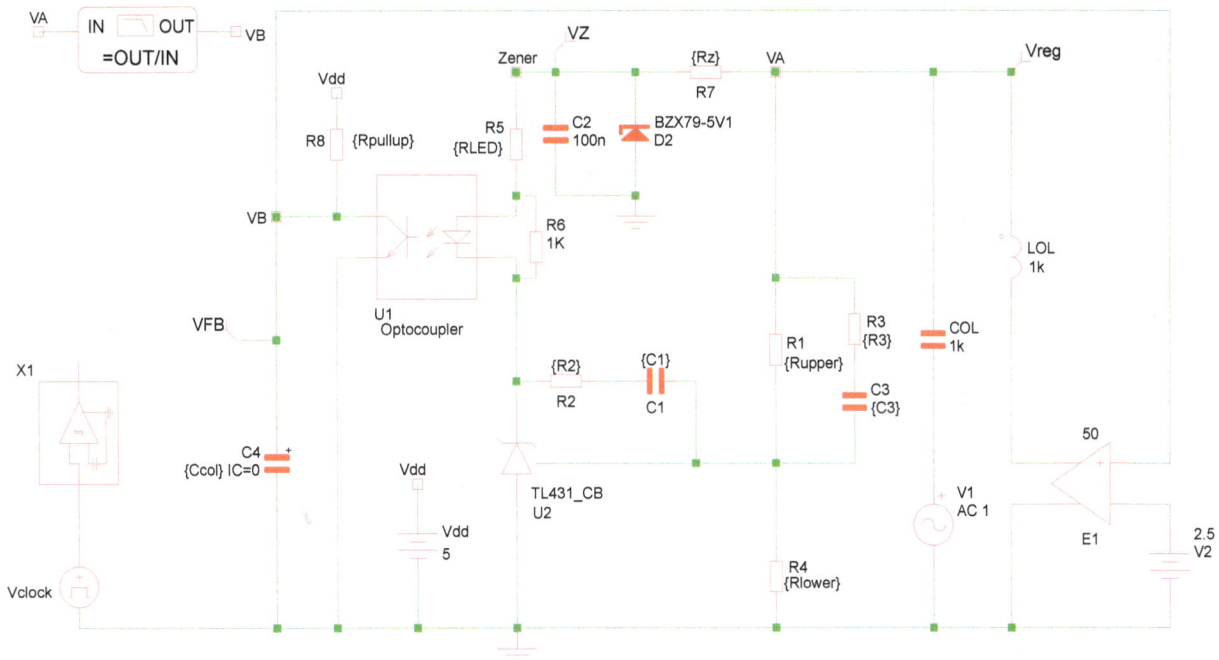

The ac response brings a 19-dB gain with the wanted boost at 5 kHz.

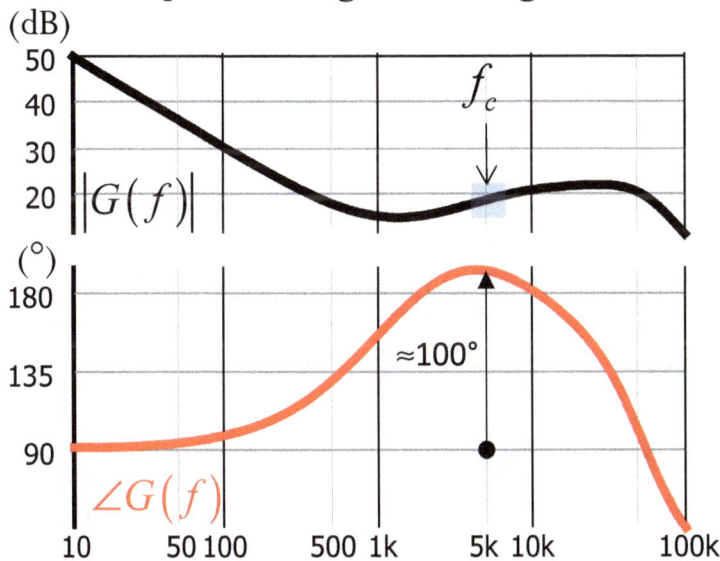

```
* Rupper = 38000
* Rlower = 10000
* R2 = 20252.7508976602
* R3 = 1583.33333333333
* C1 = 9.82304477402393e-09
* C2 = 1.26134343952181e-09
* C3 = 2.0103782285292e-09
* Rmax = 841.121495327103
* Rz = 1850.22026431718
* Ccol = 2.66625045197464e-10
* Boost = 105
* Fz1 = 800
* Fz2 = 2000
* Fp1 = 6308.94560930339
* Fp2 = 50000
```

The overall shape is linked to the imperfect ac isolation of the fast lane but also because of the TL431 own response and limited open-loop gain.

An OTA for Compensation

THE OPERATIONAL TRANSCONDUCTANCE amplifier or OTA is often found in integrated switching controllers. One reason behind this choice lies in the small die area occupied on the silicon compared to a traditional op-amp. The part lends itself well to building type 1 and 2 but I do not recommend the type 3. The absence of virtual ground forces the inclusion of the low-side resistance, R_{lower}, previously ignored with classical op-amp design and it imposes a limit in the way the second pole-zero pair is spread. In certain cases, this severely limits the maximum achievable phase boost.

Types 1 and 2 built with an OTA appear below. As underlined, there is no virtual ground anymore – no local feedback as with an op-amp – and the division ratio k with the transconductance value g_m now enter the equation.

- The transconductance g_m is variable and not production-tested which can bring variations in the ac response.

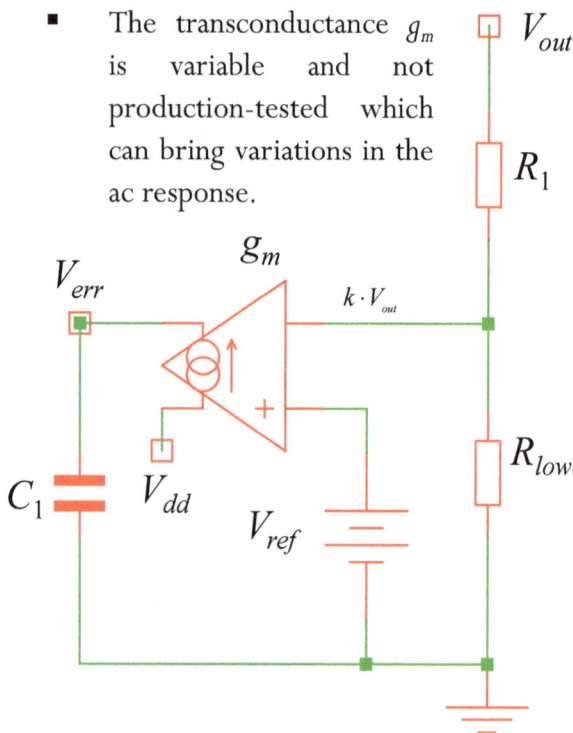

- There is no virtual ground here and the division ratio involving R_1 and R_{lower} is now part of the transfer function.

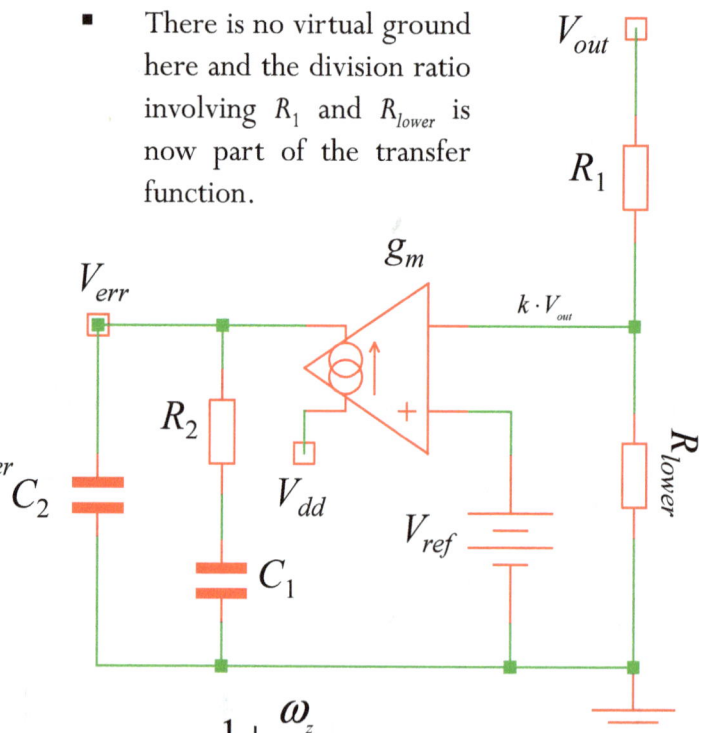

$$G(s) = -\frac{1}{\dfrac{s}{\omega_{po}}}$$

$$\omega_{po} = \boxed{\frac{R_{lower}}{R_{lower} + R_1}} \cdot \frac{g_m}{C_1}$$

Division ratio k

$$G(s) = -G_0 \frac{1 + \dfrac{\omega_z}{s}}{1 + \dfrac{s}{\omega_p}} \qquad \omega_z = \frac{1}{R_2 C_1}$$

$$G_0 = \frac{R_{lower} g_m}{R_1 + R_{lower}} \cdot \frac{R_2 C_1}{C_1 + C_2} \qquad \omega_p = \frac{1}{R_2 \dfrac{C_1 C_2}{C_1 + C_2}}$$

Type 1 Design with an OTA

IN THIS EXAMPLE, we will stabilize a power factor correction (PFC) circuit with a gain of 25 dB at 10 Hz. The output voltage is 400 V which, together with a 2.5-V reference voltage, leads to a resistive divider made of a 3.975-MΩ upper resistance and a low-side value of 25 kΩ. The selected OTA features a transconductance of 100 µS. Considering the needed attenuation of -25 dB at 10 Hz, we will place the 0-dB crossover pole at:

$$f_{po} = f_c \cdot G = 10 \times 10^{-\frac{25}{20}} = 562 \text{ mHz}$$

We can now determine the value of capacitor C_1:

$$C_1 = \frac{R_{lower}}{R_{lower} + R_1} \cdot \frac{g_m}{2\pi f_{po}} = \frac{25k}{25k + 3.975Meg} \cdot \frac{100u}{6.28 \times 562m} \approx 177 \text{ nF}$$

> A real OTA will have an output resistance r_0, affecting the maximum open-loop gain. Its role is neglected here.

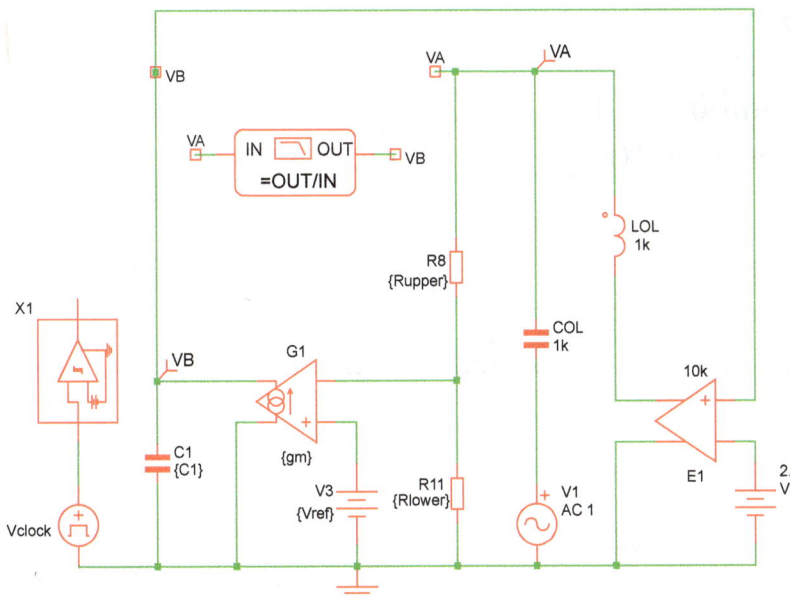

The macro automates the capacitor calculation based on the needed gain and crossover.

```
*
.VAR Gfc=25 * magnitude at crossover *
*
* Enter Design Goals Information Here *
*
.VAR fc=10 * targetted crossover *
*
* Enter the Values for Vout and Bridge Bias Current *
*
.VAR Vout=400
.VAR Ibias=100u
.VAR Vref=2.5
.VAR Rlower=Vref/Ibias
.VAR Rupper=(Vout-Vref)/Ibias
*
* Choose OTA characteristics *
*
.VAR gm=100u * transconductance in Siemens *
*
*
* Do not edit the below lines *
.VAR G=10^(-Gfc/20)
.VAR fp0=fc*G
.VAR C1=gm*Rlower/(2*pi*(Rlower+Rupper)*fp0)|
*
```

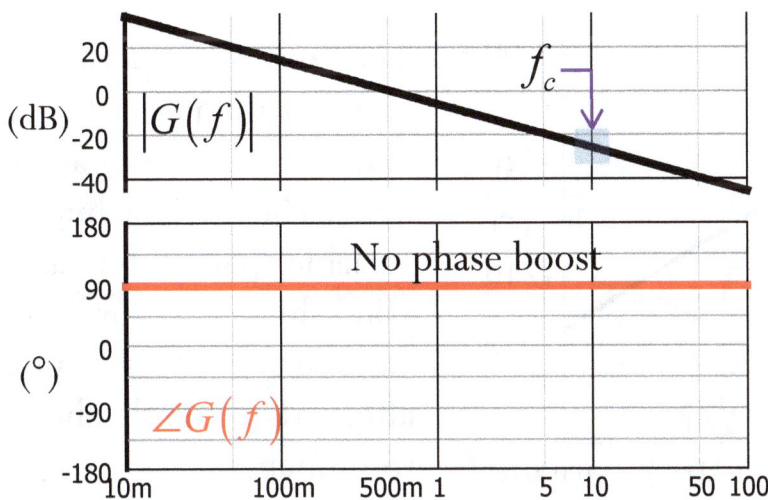

```
*  Rupper = 3975000
*  Rlower = 25000
*  C1 = 1.76888723941396e-07
*  G = 0.0562341325190349
*  fp0 = 0.562341325190349
```

Type 2 Design with an OTA

BY ADDING AN EXTRA *RC* network to the integrator, a type 2 circuit is created. The design of the filter is very similar to its counterpart with an op-amp. For this example, we assume a power stage having a gain of -15 dB at a selected 1-kHz frequency. The phase lag is 65° for this 12-V converter. The process starts with the needed phase boost and the pole-zero pair placement:

$$boost = 70 - (-65°) - 90° = 45°$$

$$k = \tan\left(\frac{boost}{2} + 45°\right) = 2.41$$

$$f_z = \frac{f_c}{k} = \frac{1k}{2.41} = 414 \text{ Hz}$$

$$f_p = k \cdot f_c = 2.41 \times 1k = 2.41 \text{ kHz}$$

A 15-dB attenuation requires an amplification by a factor of:

$$\left|G(f_c)\right| = 10^{-\frac{G_{f_c}}{20}} = 10^{-\frac{-15}{20}} \approx 5.6$$

Resistance R_2 is obtained by combining the pole-zero pair and the needed gain at f_c. The OTA transconductance g_m is 100 µS for this example.

10 kΩ 38 kΩ

5.6

$$R_2 := \frac{f_p \cdot G_0}{f_p - f_z} \cdot \frac{R_{lower} + R_1}{R_{lower} \cdot g_m} \cdot \frac{\sqrt{1 + \left(\frac{f_c}{f_p}\right)^2}}{\sqrt{1 + \left(\frac{f_z}{f_c}\right)^2}} = 325.83k\Omega$$

$$C_2 := \frac{R_{lower} \cdot g_m}{2 \cdot \pi \cdot f_p \cdot G_0 \cdot (R_{lower} + R_1)} \cdot \frac{\sqrt{1 + \left(\frac{f_z}{f_c}\right)^2}}{\sqrt{1 + \left(\frac{f_c}{f_p}\right)^2}} = 244.23pF$$

$$C_1 := \frac{1}{2 \cdot \pi \cdot f_z \cdot R_2} = 1.18nF$$

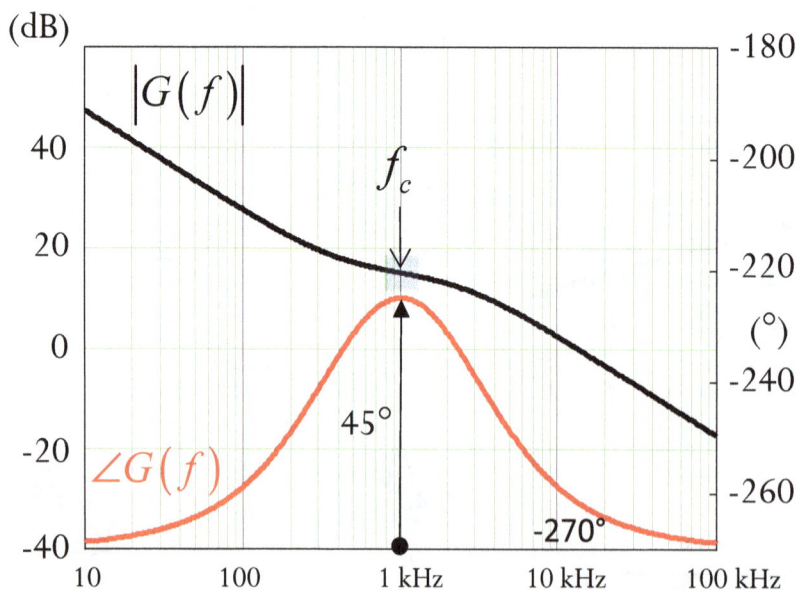

■ The response perfectly matches the needed gain and phase boost at 1 kHz. The g_m is a highly variable parameter and you should ensure its variability does not affect the overall shape too much.

Simulating the OTA in a Type 2

THE TYPE 2 with an OTA uses a simple voltage-controlled current source affected by the right transconductance value. The bias point is classically controlled by the E_1 amplifier. Please note that all these structures can equally be simulated with a SPICE engine like SIMetrix® or LTspice®.

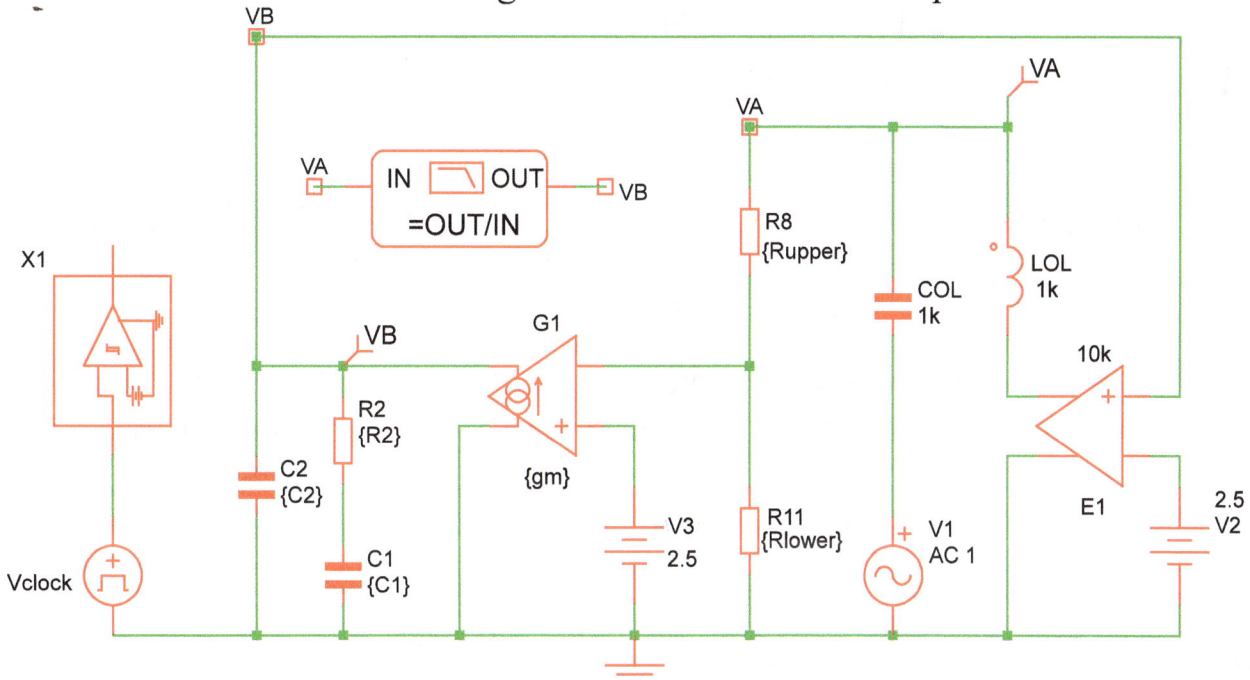

The ac response is instantly obtained and confirms the 15-dB mid-band gain, well centered at 1 kHz.

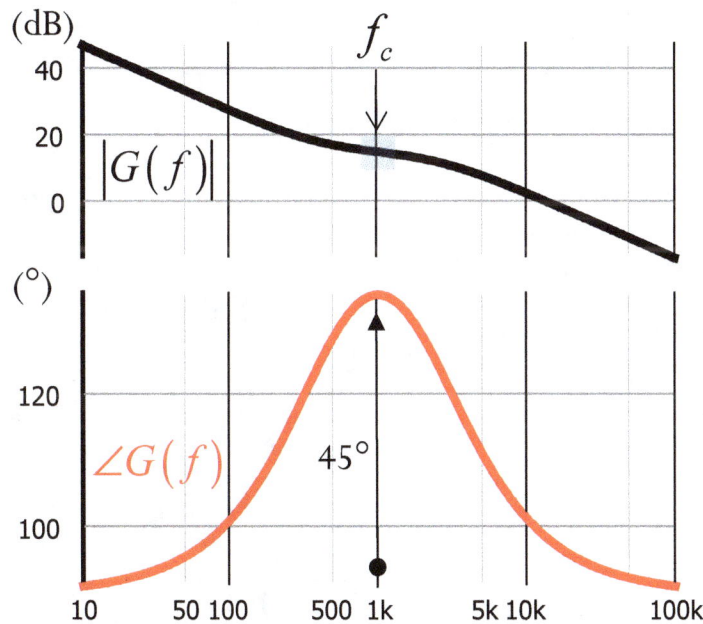

The macro automates components calculations:

```
* Choose OTA characteristics *
*
.VAR gm=100u * transconductance in Siemens *
*
.VAR boost=PM-PS-90
.VAR G=10^(-Gfc/20)
.VAR k=tan((boost/2+45)*pi/180)
.VAR fp=fc*k
.VAR fz=fc/k
.VAR a=sqrt((fc^2/fp^2)+1)
.VAR b=sqrt((fz^2/fc^2)+1)
*
.VAR R2=(a/b)*(fp*G)*(Rlower+Rupper)/((fp-fz)*Rlower*gm)
.VAR C1=1/(2*pi*R2*fz)
.VAR C2=(Rlower*gm/(2*pi*fp*G*(Rlower+Rupper)))(b/a)
*
* Simpler approach if C2 << C1 *
*
* .VAR R2=G*(Rlower+Rupper)/(Rlower*gm)
* .VAR C1=1/(2*pi*R2*fz)
* .VAR C2=1/(2*pi*R2*fp)
*
```

```
*
* Rupper = 38000
* Rlower = 10000
* R2 = 325826.892949776
* C2 = 2.44232361624656e-10
* C1 = 1.17925815960931e-09
* k = 2.41421356237309
* Boost = 45
* Fz = 414.213562373095
* Fp = 2414.2135623731
*
```

Digital Compensators

THIS SMALL SECTION on digital control will quickly show you how to determine the coefficients for the main compensator types 2 and 3. With these coefficients in hand, you will then have to code the difference equation to perform the expected filtering function. We start with a biquad filter whose coefficients are obtained by mapping the type 2 transfer function expressed in the Laplace domain to the z domain:

$$G(s) = G_0 \frac{1 + \dfrac{\omega_z}{s}}{1 + \dfrac{s}{\omega_p}} \Bigg] \underset{s = \frac{2}{T_s} \frac{1 - z^{-1}}{1 + z^{-1}}}{} \Rightarrow G(z) = \frac{G_0 T_s \omega_p (1 + z)(2z + T_s \omega_z + T_s \omega_z z - 2)}{4z^2 - 8z - 2T_s \omega_p + 2T_s \omega_p z^2 + 4}$$

Then identify the coefficients of a biquad filter with the expression of G you will have rearranged by dividing the numerator and the denominator by z^2.

$$G(z) = \frac{a_0 + a_1 z^{-1} + a_2 z^{-2}}{1 + b_1 z^{-1} + b_2 z^{-2}} \qquad y[n] = a_0 x[n] + a_1 x[n-1] + a_2 x[n-2] - b_1 y[n-1] - b_2 y[n-2]$$

$$a_0 = \frac{G_0 T_s \omega_p (T_s \omega_z + 2)}{2(2 + T_s \omega_p)} \qquad a_1 = \frac{G_0 T_s^2 \omega_p \omega_z}{2 + T_s \omega_p}$$

$$a_2 = \frac{G_0 T_s \omega_p (T_s \omega_z - 2)}{2T_s \omega_p + 4} \qquad b_1 = -\frac{2}{1 + 0.5 T_s \omega_p}$$

$$b_2 = \frac{2}{1 + 0.5 T_s \omega_p} - 1$$

↑ Sampling period

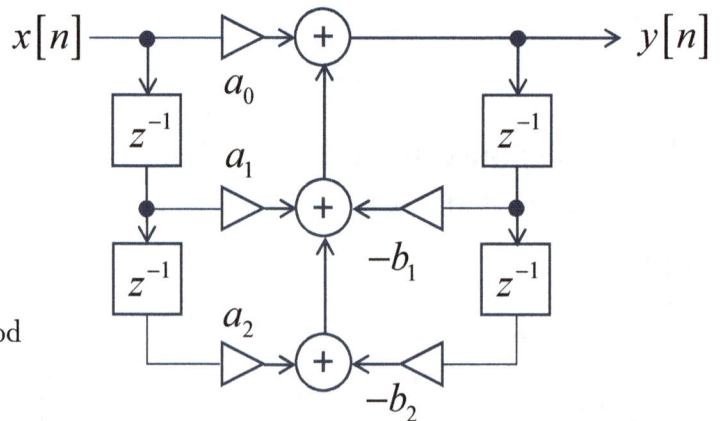

Assume we have the following specifications for the type 2:

$f_c = 1 \text{ kHz}$

$G_0 = 20 \text{ dB}$

$Boost = 50°$

$k = \tan\left(\dfrac{boost}{2} + \dfrac{\pi}{4}\right) = 2.74$

$f_z = 364 \text{ Hz}$

$f_p = 2.74 \text{ kHz}$

$F_s = 1 \text{ MHz}$

$a_0 = 0.0857$

$a_1 = 1.957 \cdot 10^{-4}$

$a_2 = -0.0855$

$b_1 = -1.9829$

$b_2 = 0.9829$

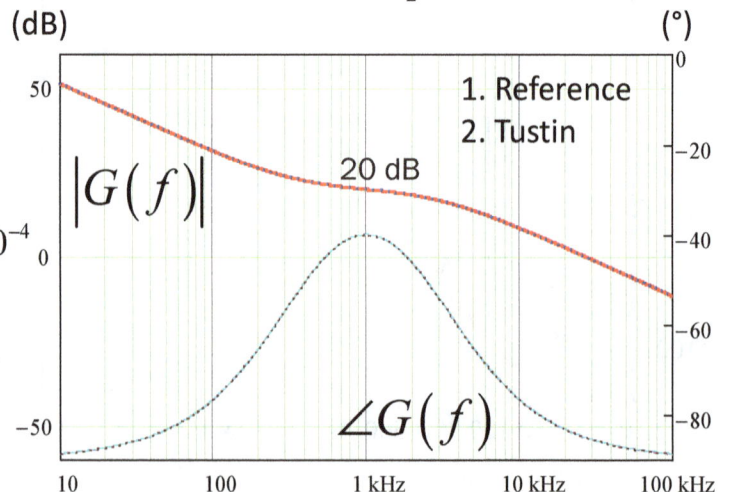

Ac response of this type 2 filter

Simulating the Digital Type 2

BEFORE GOING STRAIGHT to coding, it is a good practice to verify that the obtained coefficients are correct and will lead to the expected response. Mathcad® can do it but we can also resort to a simulation model which we can later use for simulating the entire converter. SPICE lends itself well for this exercise and whether it is SIMetrix® or LTspice®, both have access to a delay line.

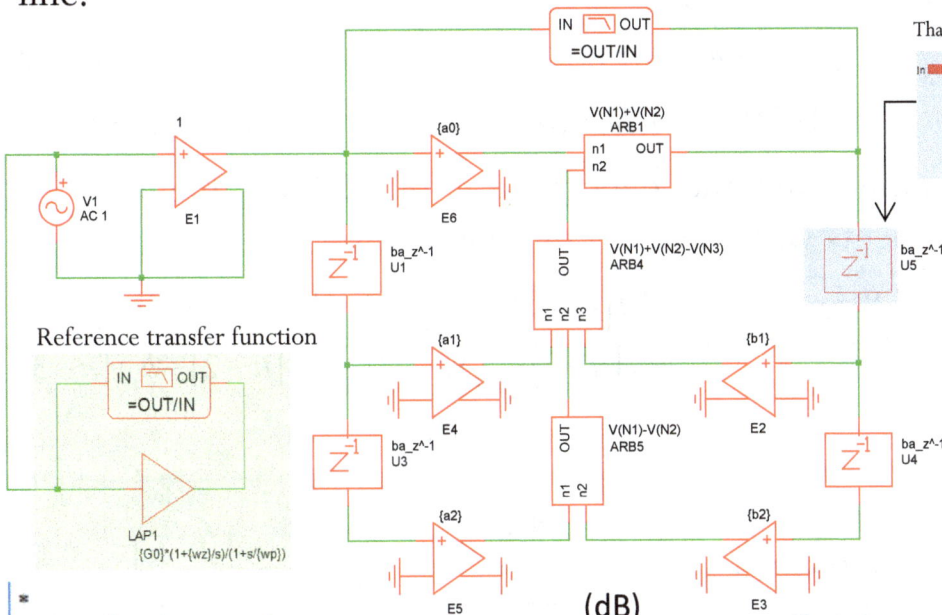

Thank you Basil!

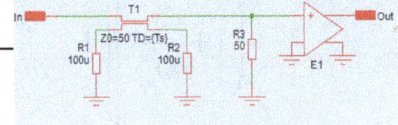

In this circuit, the delay line models a z^{-1} function and you assemble the blocks following the difference equation I gave for $G(z)$.

Reference transfer function

```
*
.PARAM Gfc=-20 ; magnitude at crossover
.PARAM PS=-80 ; phase lag at crossover
*
* Enter Design Goals Information Here *
*
.PARAM fc=1k ; targetted crossover *
.PARAM PM=60 ; choose phase margin at crossover *
*
* Do not edit the below lines *
.PARAM boost={PM-PS-90}
.PARAM G={10^(-Gfc/20)}
.PARAM k={tan((boost/2+45)*pi/180)}
.PARAM fp={fc*k}
.PARAM fz={fc/k}
*
*
.PARAM Fsw=1Meg
.PARAM Tsw={1/Fsw
.PARAM Ts=Tsw
*
.PARAM wz={2*pi*fz
.PARAM wp={2*pi*fp
.PARAM G0=G
*
.PARAM a0={G0*Tsw*wp*(Tsw*wz+2)/(2*Tsw*wp+4)}
.PARAM a1={G0*Tsw^2*wp*wz/(Tsw*wp+2)}
.PARAM a2={G0*Tsw*wp*(Tsw*wz-2)/(2*Tsw*wp+4)}
.PARAM b1={-8/(4+2*Tsw*wp)}
.PARAM b2={(4/(Tsw*wp+2))-1}
*
```

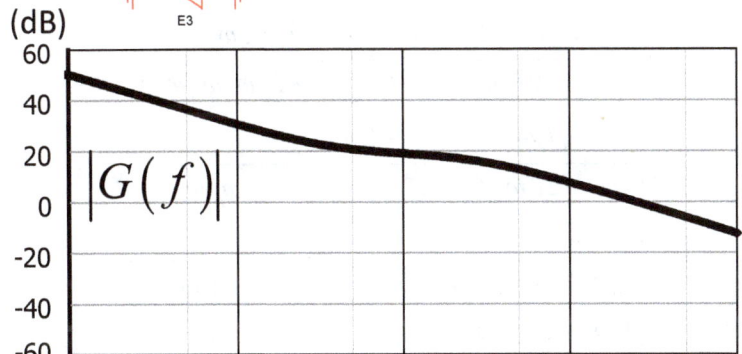

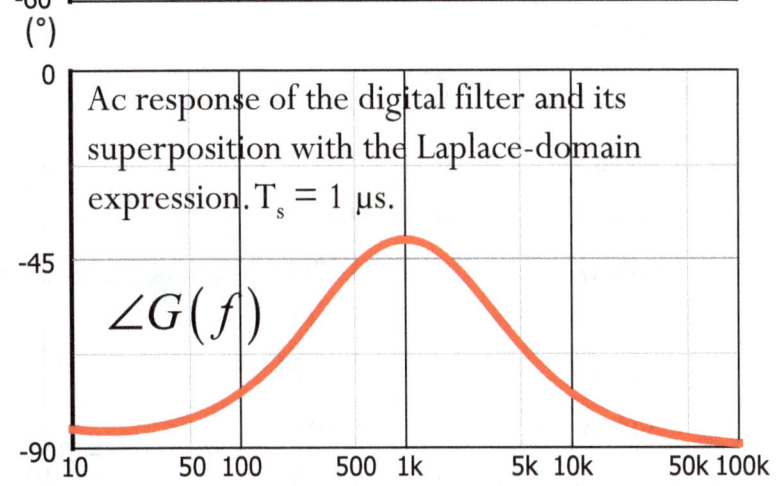

Edit *design.out* file:

```
a0  =  85.673755657m
a1  =  195.70288738u
a2  =  -85.47805277m
b1  =  -1.9828848192
b2  =  982.88481916m
LAP1.{wz} = 2.2868924282k
LAP1.{wp} = 17.262909754k
LAP1.{G0} = 10
```

All coefficients are automatically calculated.

Ac response of the digital filter and its superposition with the Laplace-domain expression. $T_s = 1\ \mu s$.

A Digital Type 3 Compensator

SIMILARLY TO THE TYPE 2, we can assemble a type 3 compensator starting from the Laplace-domain expression and translating it into the discrete-time domain. It is a bit more complicated with 3rd-order coefficients but nothing insurmountable:

$$G(s) = G_0 \frac{\left(1+\dfrac{\omega_{z_1}}{s}\right)\left(1+\dfrac{s}{\omega_{z_2}}\right)}{\left(1+\dfrac{s}{\omega_{p_1}}\right)\left(1+\dfrac{s}{\omega_{p_2}}\right)} \qquad s = \frac{2}{T_s}\frac{1-z^{-1}}{1+z^{-1}}$$

$$N(z) = G_0 T_s \omega_{p_1} \omega_{p_2}(z+1)\left(2z+T_s\omega_{z_1}+T_s\omega_{z_1}z-2\right)\left(2z+T_s\omega_{z_2}+T_s\omega_{z_2}z-2\right)$$

$$D(z) = 2\omega_{z_2}(z-1)\left(2z+T_s\omega_{p_1}+T_s\omega_{p_1}z-2\right)\left(2z+T_s\omega_{p_2}+T_s\omega_{p_2}z-2\right)$$

Rearrange the above numerator and denominator by dividing them by z^3. Then identify the coefficients with the 3rd-order digital filter expression:

$$G(z) = \frac{a_0 + a_1 z^{-1} + a_2 z^{-2} + a_3 z^{-3}}{1 + b_1 z^{-1} + b_2 z^{-2} + b_3 z^{-3}}$$

$$y[n] = a_0 x[n] + a_1 x[n-1] + a_2 x[n-2]$$
$$+a_3[n-3] - b_1 y[n-1] - b_2 y[n-2] - b_3 y[n-3]$$

$$a_0 = \frac{G_0 T_s \omega_{p_1}\omega_{p_2}\left(T_s\omega_{z_1}+2\right)\left(T_s\omega_{z_2}+2\right)}{2\left(4\omega_{z_2}+2T_s\omega_{p_1}\omega_{z_2}+2T_s\omega_{p_2}\omega_{z_2}+T_s^2\omega_{p_1}\omega_{p_2}\omega_{z_2}\right)}$$

$$a_1 = \frac{G_0 T_s \omega_{p_1}\omega_{p_2}\left(2T_s\omega_{z_1}+2T_s\omega_{z_2}+3T_s^2\omega_{z_1}\omega_{z_2}-4\right)}{2\left(4\omega_{z_2}+2T_s\omega_{p_1}\omega_{z_2}+2T_s\omega_{p_2}\omega_{z_2}+T_s^2\omega_{p_1}\omega_{p_2}\omega_{z_2}\right)}$$

$$a_2 = -\frac{G_0 T_s \omega_{p_1}\omega_{p_2}\left(2T_s\omega_{z_1}+2T_s\omega_{z_2}-3T_s^2\omega_{z_1}\omega_{z_2}+4\right)}{2\left(4\omega_{z_2}+2T_s\omega_{p_1}\omega_{z_2}+2T_s\omega_{p_2}\omega_{z_2}+T_s^2\omega_{p_1}\omega_{p_2}\omega_{z_2}\right)}$$

$$a_3 = \frac{G_0 T_s \omega_{p_1}\omega_{p_2}\left(T_s\omega_{z_1}-2\right)\left(T_s\omega_{z_2}-2\right)}{2\left(4\omega_{z_2}+2T_s\omega_{p_1}\omega_{z_2}+2T_s\omega_{p_2}\omega_{z_2}+T_s^2\omega_{p_1}\omega_{p_2}\omega_{z_2}\right)}$$

$$b_1 = \frac{T_s\omega_{p_2}-2}{T_s\omega_{p_2}+2}-\frac{4}{T_s\omega_{p_1}+2} \qquad b_2 = \frac{16}{\left(T_s\omega_{p_1}+2\right)\left(T_s\omega_{p_2}+2\right)}$$

$$b_3 = -\frac{\left(T_s\omega_{p_1}-2\right)\left(T_s\omega_{p_2}-2\right)}{\left(T_s\omega_{p_1}+2\right)\left(T_s\omega_{p_2}+2\right)}$$

Sampling period

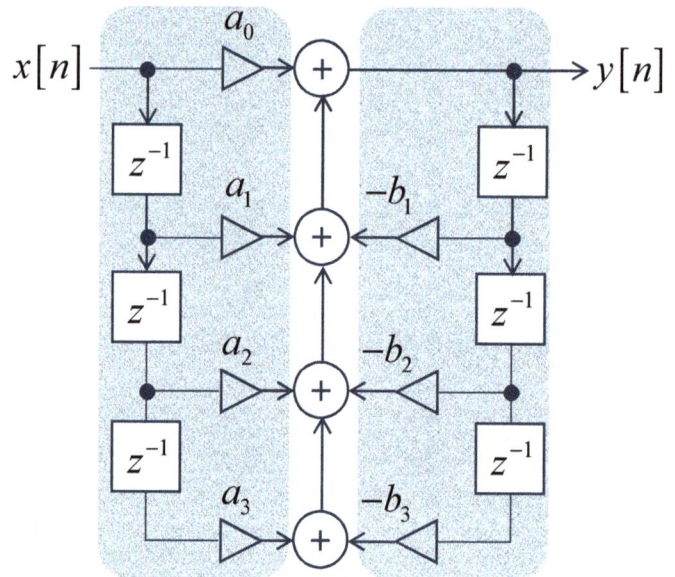

Assume the following specifications:

$f_c = 10$ kHz	$a_0 = 1.0451$
$G_{mb} = 0$ dB	$a_1 = -0.9553$
$f_{z_1} = 1.8$ kHz	$a_2 = -1.0433$
$f_{z_2} = 1$ kHz	$a_3 = 0.9571$
$f_{p_1} = 56$ kHz	$b_1 = -1.4234$
$f_{p_2} = 30$ kHz	$b_2 = 0.4464$
$F_s = 250$ kHz	$b_3 = -0.023$

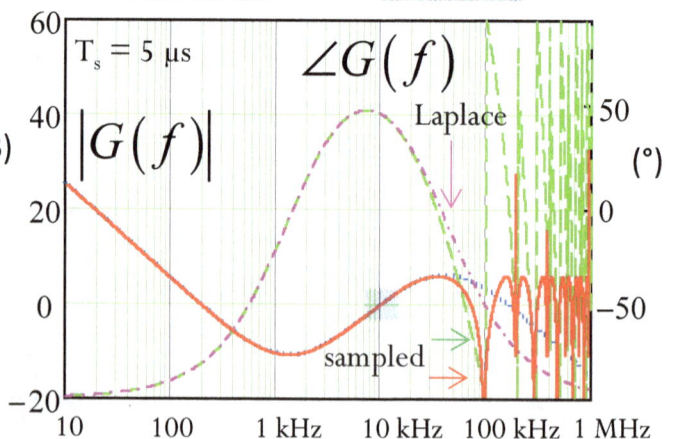

Simulating the Digital Type 3

THE SIMULATION OF A TYPE 3 does not differ from that of the type 2 besides a larger number of delay lines:

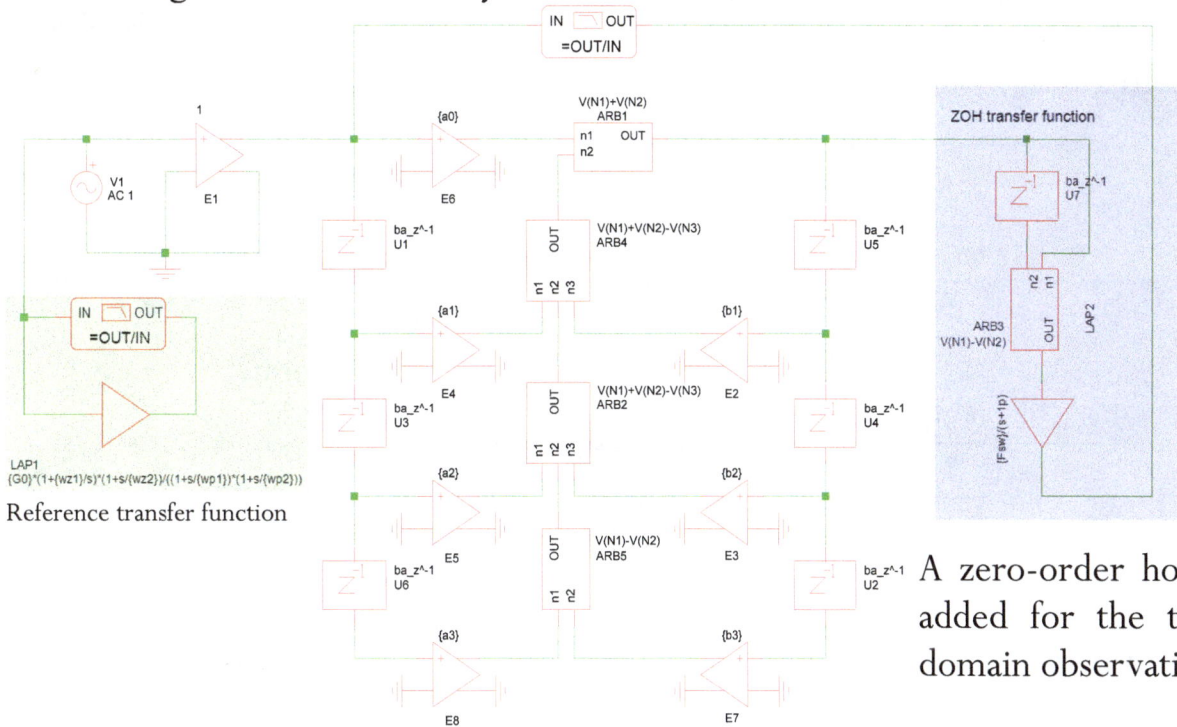

Reference transfer function

A zero-order hold is added for the time-domain observation.

The automated macro is a bit more dense to determine all the coefficients:

```
*
.PARAM a0={(G0*Tsw*wp1*wp2*(Tsw*wz1+2)*(Tsw*wz2+2))/(2*(4*wz2+2*Tsw*wp1*wz2+2*Tsw*wp2*wz2+Tsw^2*wp1*wp2*wz2))}
.PARAM a1={(G0*Tsw*wp1*wp2*(2*Tsw*wz1+2*Tsw*wz2+3*Tsw^2*wz1*wz2-4))/(2*(4*wz2+2*Tsw*wp1*wz2+2*Tsw*wp2*wz2+Tsw^2*wp1*wp2*wz2))}
.PARAM a2={(-G0*Tsw*wp1*wp2*(2*Tsw*wz1+2*Tsw*wz2-3*Tsw^2*wz1*wz2+4))/(2*(4*wz2+2*Tsw*wp1*wz2+2*Tsw*wp2*wz2+Tsw^2*wp1*wp2*wz2))}
.PARAM a3={(G0*Tsw*wp1*wp2*(Tsw*wz1-2)*(Tsw*wz2-2))/(2*(4*wz2+2*Tsw*wp1*wz2+2*Tsw*wp2*wz2+Tsw^2*wp1*wp2*wz2))}
.PARAM b1={((Tsw*wp2-2)/(Tsw*wp2+2))-4/(2+Tsw*wp1)}
.PARAM b2={(16/((Tsw*wp1+2)*(Tsw*wp2+2)))-1}
.PARAM b3={-(Tsw*wp1-2)*(Tsw*wp2-2)/((Tsw*wp1+2)*(Tsw*wp2+2))}
*
```

```
fz1 = 1.8k
fz2 = 1k
fp1 = 56k
fp2 = 30k
Fsw = 200k
Tsw = 5u
Ts  = 5u
G0  = 104.8601054m
wz1 = 11.309733553k
wp1 = 351.8583772k
wz2 = 6.2831853072k
wp2 = 188.49555922k
a0  = 1.04510049032
a1  = -955.3014493m
a2  = -1.043322815
a3  = 957.07912467m
b1  = -1.4234287073
b2  = 446.44105798m
b3  = -23.01235066m
```

The file *design.out* contains the data.

Ac response of the digital filter and its superposition with the Laplace-domain plot

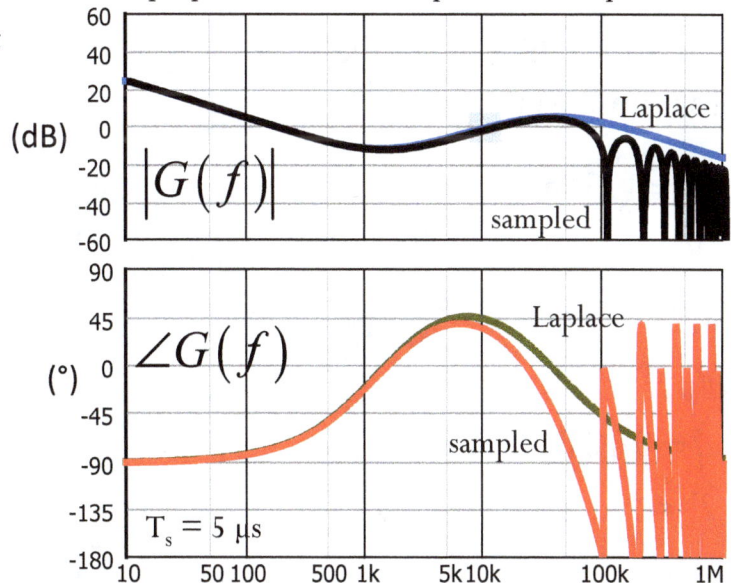

Building a Discrete-Time PID

THE FILTERED PID in discrete-time version can be assembled by separately converting the basic functions like integration, differentiation and gain. In this approach, we've started from the Laplace-domain and applied the backward-Euler mapping:

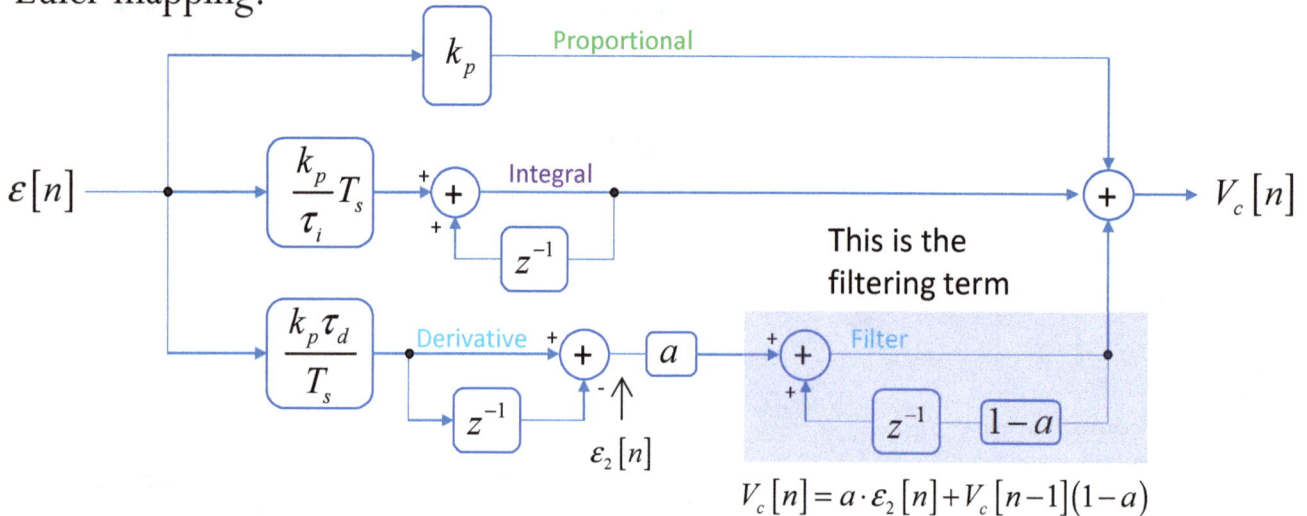

$$V_c[n] = a \cdot \varepsilon_2[n] + V_c[n-1](1-a)$$

The sampled equation is obtained by combining the three paths:

$$\frac{V_c(z)}{\varepsilon(z)} = k_p + \frac{k_p T_s}{\tau_i}\frac{1}{1-z^{-1}} + \frac{k_p \tau_d}{T_s}\left(1-z^{-1}\right)\frac{a}{1-z^{-1}(1-a)} \qquad a = \frac{T_s}{\dfrac{\tau_d}{N}+T_s}$$

This expression translates into a difference equation that needs to be coded:

$$V_c[n] = k_p \varepsilon[n] + \frac{k_p T_s}{\tau_i}\varepsilon[n] + V_c[n-1] + \left(a \cdot \frac{k_p \tau_d}{T_s}\left(\varepsilon[n]-\varepsilon[n-1]\right)+V_c[n-1](1-a)\right)$$

I used the below coefficients for these Mathcad® plots:

$$\tau_d = 193\ \text{ms}$$
$$N = 26.439$$
$$\tau_i = 1.053\ \text{ms}$$
$$k_p = 2.643$$
$$a = 0.121$$
$$T_s = 1\ \mu\text{s}$$

20-dB gain at 3 kHz

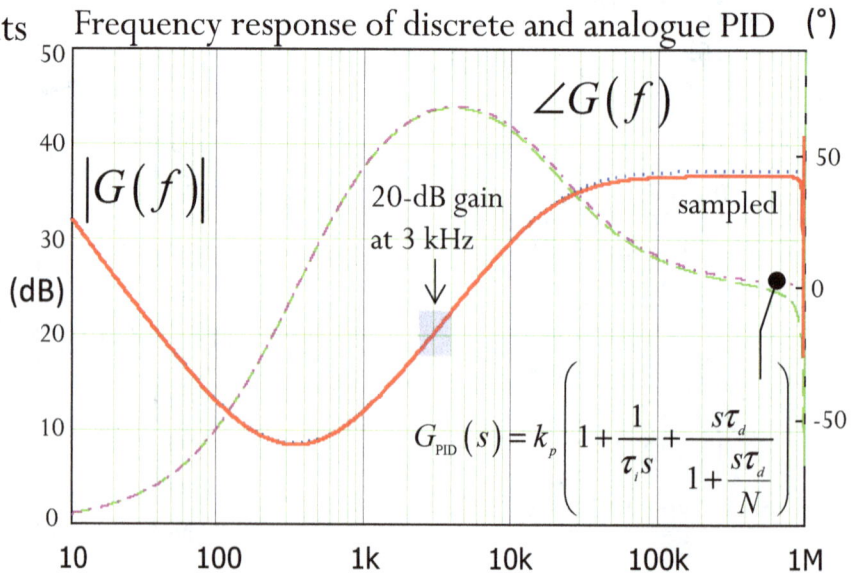

Frequency response of discrete and analogue PID (°)

$$G_{\text{PID}}(s) = k_p\left(1+\frac{1}{\tau_i s}+\frac{s\tau_d}{1+\dfrac{s\tau_d}{N}}\right)$$

Simulating a Digital Filtered-PID

SIMILARLY to what we did with the classical type 2 and 3 compensators, we can simulate a complete filtered PID with SIMetrix®. Please note that I purposely assembled gains and delay lines in my digital subcircuits for the sake of illustrating the structure that can be ported to any SPICE engine. SIMPLIS® provides ready-made digital filters, including a PID block ready for immediate inclusion in a switching converter.

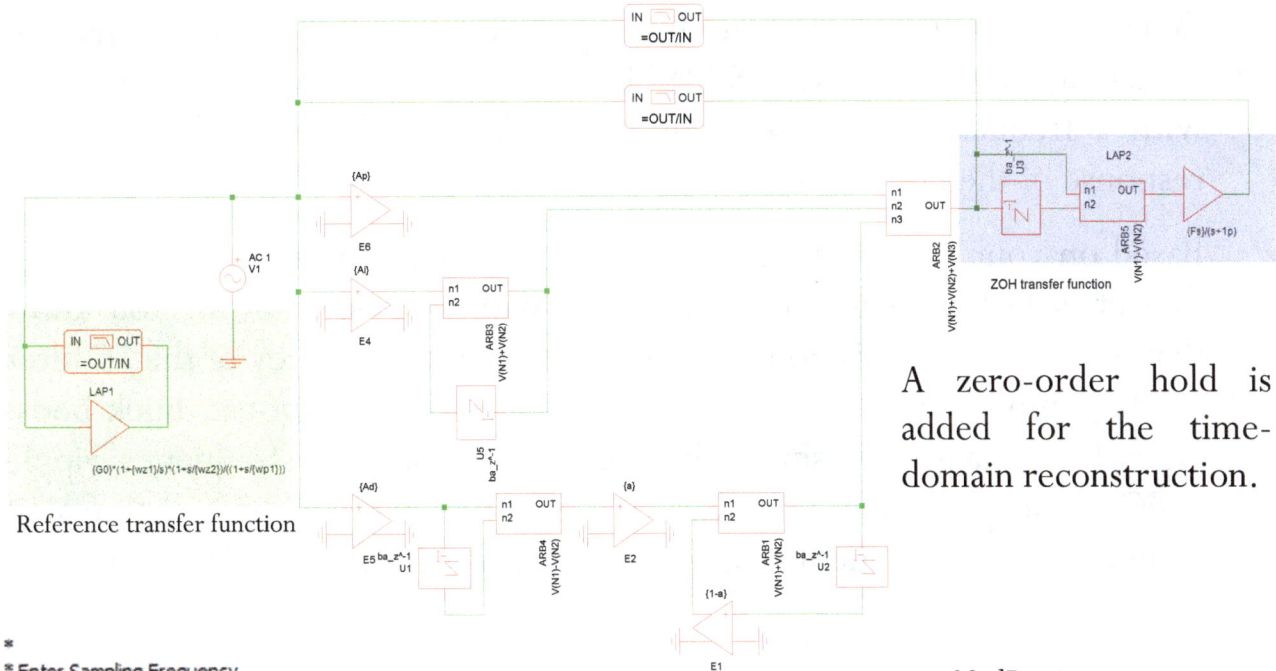

A zero-order hold is added for the time-domain reconstruction.

Reference transfer function

```
*
* Enter Sampling Frequency
*
.PARAM Fs=1Meg
.PARAM Ts={1/Fs}
*
* Enter PID parameters
*
.PARAM kp=2.643
.PARAM ki=2.509k
.PARAM kd=509.6u
.PARAM fp=21.8k ; pole for the filtered PID
*
.PARAM Td={kd/kp}
.PARAM Ti={kp/ki}
*
.PARAM N={2*pi*fp*Td}
.PARAM Ad={kp*Td/Ts}
.PARAM a={Ts/(Ts+Td/N)}
.PARAM Ap={Kp}
.PARAM Ai={kp*Ts/Ti}
*
* Equivalent poles-zeroes computation *
*
.PARAM wz1={(1/Td)-(sqrt(((Ti-4*Td)/Ti))+1)/(2*Td)}
.PARAM wz2={(sqrt(((Ti-4*Td)/Ti))+1)/(2*Td)}
.PARAM wpo={ki}
.PARAM wp1={N/Td}
.PARAM G0={wpo/wz1}
*
.PARAM fz1={wz1/(2*pi)}
.PARAM fz2={wz2/(2*pi)}
.PARAM fpo={wpo/(2*pi)}
.PARAM fp1={wp1/(2*pi)}
```

You enter the selected PID parameters and the macro computes the digital coefficients for you.

20-dB gain at 3 kHz

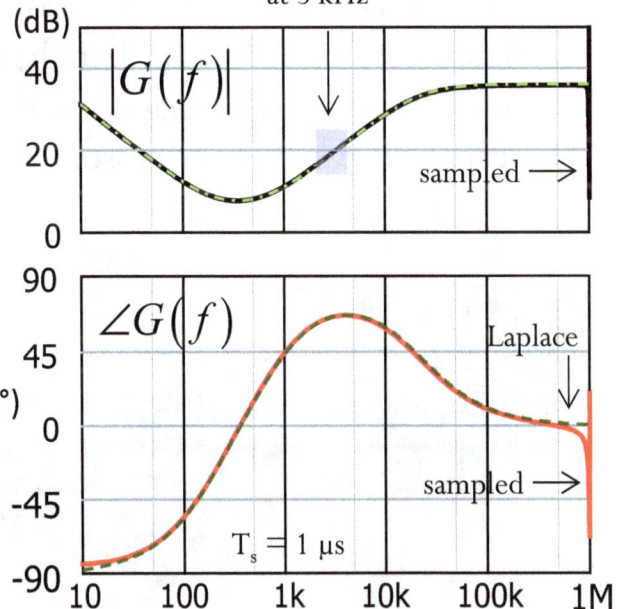

The macro reconstructs the poles and zero location too.

20-dB gain at 3 kHz

Stabilizing Switching Converters

NOW THAT WE HAVE reviewed many of the compensators found in the design of switching converters, we can start applying our knowledge to practical cases. The methodology I adopted in this book will follow the below steps:

1. Obtain the *control-to-output* transfer function of the converter you need to compensate. This is the *mandatory* starting point and you cannot evade it. Whether you determine it by equations, simulation (averaged or switching model like with SIMPLIS®) or measurement on a prototype with a frequency-response analyzer (FRA), you have to know the ac response of the power stage.

2. Based on your specifications, the type of converter and the response you expect, select a crossover frequency f_c with a phase margin. You know that in some cases, f_c is bounded by the resonant frequency of the *LC* filter (voltage-mode control), the presence of a RHP zero (boost, buck-boost types for instance) or simply because of a strong low-frequency ripple (PFC). The phase margin selection depends on the expected transient response and components variability but I usually shoot between 60 and 70°.

3. The simulator is your friend but you have to feed it with well-characterized components. For stability analysis, extract parasitics from datasheets or from the bench, in inductors, transformers but also, and most importantly, capacitors. If you do a good job, bench and modeled ac responses can be very close.

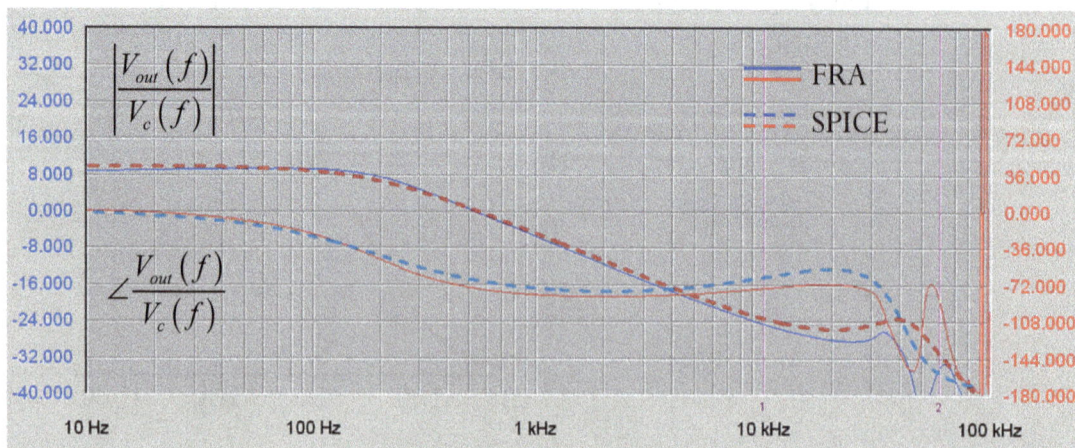

Control-to-output transfer function of a CM buck for which parasitics have been extracted.

What Compensator to Pick?

4. THE POWER STAGE ac response is obtained and you know your goals in terms of crossover and phase margin. Based on the control scheme and the converter, you select a compensator type:

- For voltage-mode-controlled CCM switching cells like buck, boost or buck-boost, the response is of second-order and the power stage phase lag approaches $180°$. For building phase margin, a type 3 compensator is necessary. You will usually place two zeroes at the resonant frequency f_0 – which is either fixed (buck) or variable (boost and buck-boost) – to cancel the double poles. However, it is often interesting to place one zero at f_0 but the second at a lower point to stay away from conditional stability in light load. Then a pole is placed at half the switching frequency F_{sw} while the second is adjusted to set the phase margin. The macro will do it for you in the SIMPLIS® simulation examples.

- For current-mode-controlled CCM switching cells, the response is of 1st-order in the low-frequency part but exhibits a 3rd-order response when you involve the two sub-harmonic poles located at $F_{sw}/2$. You will cross over way before the poles with a type 2 compensator but also ensure proper damping to avoid instability in worst-case conditions. You damp the poles by mixing the current sense information with a small amount of artificial ramp. This has the effect of lowering the inner current-loop gain magnitude and ensures stability of the converter.

- In some applications where no phase boost is necessary, a simple integrator like a type 1 can do the job. You simply determine the capacitor value and that is all.

- If you select crossover frequencies in the kHz area or so, you usually do not need to care about the ac response of the active element used in the compensator. On the contrary, if you plan on pushing f_c to 10-20 kHz or beyond, then make sure to include the op-amp or OTA frequency response in your model otherwise you may not reach your goals in terms of gain and phase boost at the selected crossover. It is a classic error and often explains discrepancies between simulations and measurements.

Reliability

5.   THE POWER STAGE small-signal response is usually plotted in worst-case conditions, e.g. minimum input voltage and maximum output current. However, the power supply will experience many different operating points during its lifetime and you must ensure stability in all cases. Once your compensation strategy gives the expected results in terms of loop gain and transient response, run simulations at both input voltage extremes and vary the output current along its minimum and maximum values. The stability criteria must always be met during these experiments. After simulation, running endurance tests with the prototype immersed in a thermal chamber is a good way to further test the robustness of your design at temperature extremes. You know that all parasitics will move during these tests and it is important to verify stability. If you can run the FRA and confirm sound loop gains in these conditions, then this is a good approach to confirm the design is reliable. However, this is sometimes inconvenient. As a possible alternative, you superimpose transient steps over the one captured at room temperature and check any flagrant discrepancy. If you find a suspicious response, it may reveal a potential weakness you must investigate. Yes, these are long and tedious tests but you can't avoid them if you want to sleep well before and after start of mass production.

6.   Sweeping all pertinent parameters via Monte Carlo analysis is another good solution to cover many different situations with parasitics. It is easy to assign tolerances to the components identified as stability offenders and pseudo-randomly sweep them while plotting loop gains. Collecting crossover frequency, phase and gain margins variations will reveal whether the adopted strategy resists natural components variability due to production, age and temperature. That is the reason why these elements need to be well characterized in the first place. In a CM flyback converter, for instance, the elements affecting the loop gain are the primary inductance, the transformer turns ratio, the sense resistance, the load of course, the output capacitor, its ESR and, obviously, all the elements around the compensator like the optocoupler, the tolerance on resistance and capacitors, the internal components in the PWM controller etc.

Monte Carlo

7. TRYING TO SWEEP all these elements by hand is an impossible task not only because there are numerous values to change but also because you ignore the *sensitivity* of your design to a particular component or set of components. You could analytically run a *sensitivity analysis* and identify, in the system, the components which truly affect stability when they deviate from their nominal specifications too much. Designers for aerospace or military applications often run this type of exercise which consists of identifying *worst-case* scenarios and making sure the converter survives them. One advantage is that you do not need statistical data of the components – you push them to the minimum or the maximum of their limits – but it often leads to an over-conservative design which can conflict with cost targets in consumer and industrial applications.

8. Monte Carlo, on the other hand [4], does not need sensitivity analysis data but requires the distribution laws affecting tolerances of the components you plan on sweeping. With this technique, you will obtain the most realistic estimate of the true worst case. Needless to say that statistical analysis is a specialty in itself and we will limit ourselves to running simple cases. Nevertheless, having simulations showing that over thousands of samples, crossover frequency and margins remain in reasonable limits, is a good indication that you have something robust. This is how SIMPLIS® handles Monte Carlo analysis:

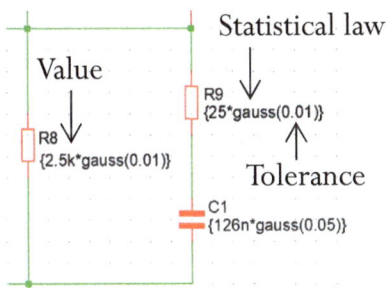

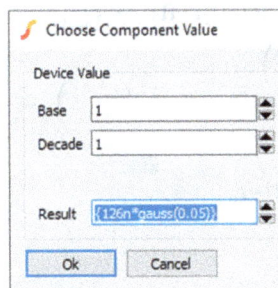

- You choose the distribution law and assign a tolerance to the component. A probe setting a *goal function* is then setup to collect crossover frequency and margins.

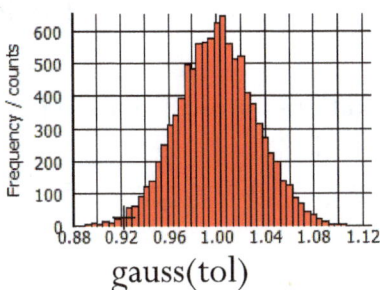

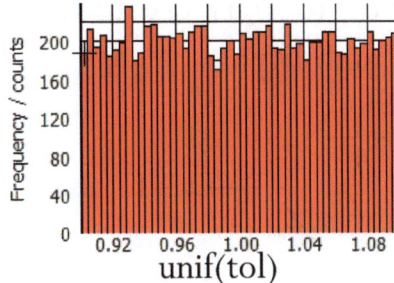

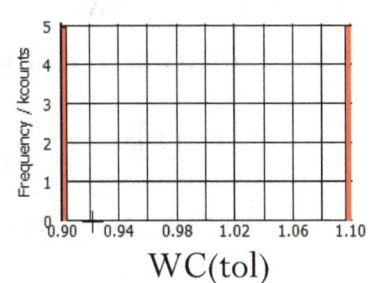

Front-End Filter Interaction

A SWITCHING CONVERTER is a noisy circuit in essence, considering the fast voltage and current disruptions brought by the power switches. As such, it produces a rich spectrum of conducted and radiated noise that needs to be attenuated. For this purpose, a filter is inserted in front of the converter and its role is to block the high-frequency current pulses which would otherwise pollute the source. The subject is vast and I am not going to explore it here. However, it is important that you understand the potential risk of blindly inserting an EMI filter.

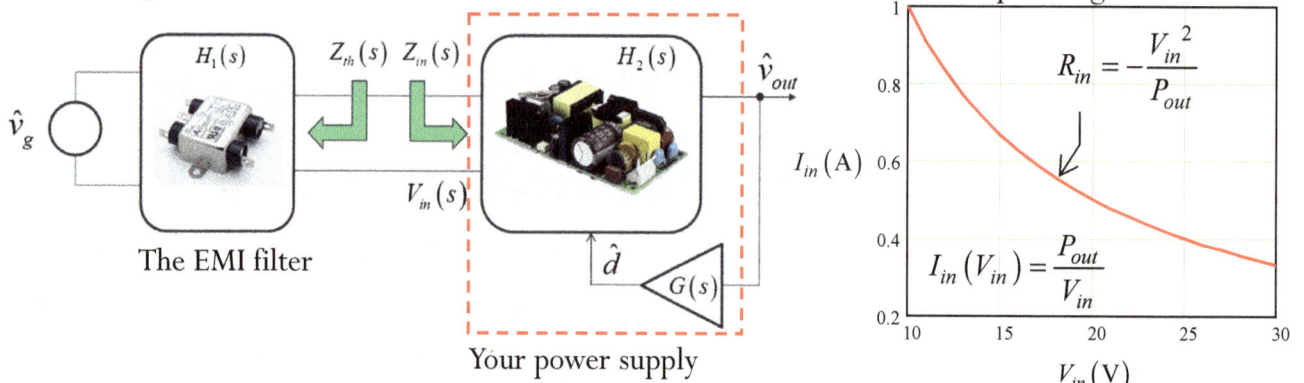

The EMI filter — Your power supply

A filter is usually built around energy-storing elements such as inductors and capacitors. When this filter is loaded by a switching converter, problems can arise. Indeed, the *incremental resistance* of a converter operated in closed-loop conditions, is negative as shown in the above graph. For a constant output power, the input current increases for a decreasing input voltage and vice versa. Loading an *LC* filter with a negative resistance, can bring conditions for instability as the below drawing illustrates [5]:

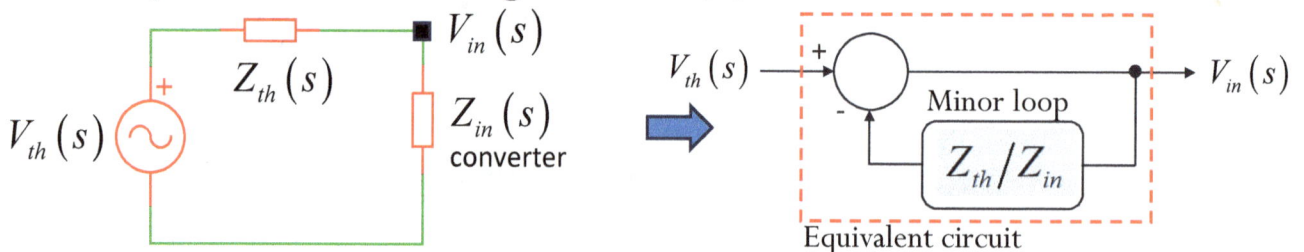

Equivalent circuit

You can see how a minor loop is created, making the Nyquist criterion a pertinent tool to analyze stability:

$$V_{in}(s) = V_{th}(s)\frac{1}{\underbrace{1+\dfrac{Z_{th}(s)}{Z_{in}(s)}}_{\substack{\text{Minor} \\ \text{loop}}}}\Bigg\} \quad \frac{Z_{th}(s)}{Z_{in}(s)} = -1 \implies$$

Conditions for oscillations

$$\left|\frac{Z_{th}(s)}{Z_{in}(s)}\right| = 1 \text{ and } \angle\frac{Z_{th}(s)}{Z_{in}(s)} = -180°$$

Impedance Characterization

ONCE THE FILTER has been designed, you will need to plot its output impedance and check how the magnitude peaks. You can do it with a classical SPICE-based engine or SIMPLIS®. It is important to include the parasitics which dampen the system:

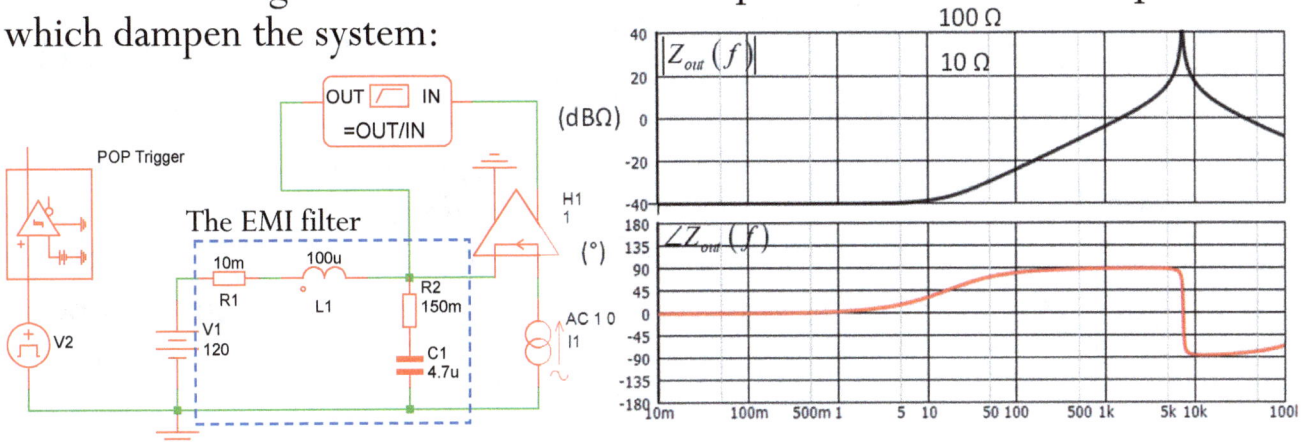

Then, you will plot the input impedance of the power converter and superimpose its magnitude over that of the above filter's output impedance. Any overlap between the two implies a minor loop gain of 1 and the phase must be carefully checked.

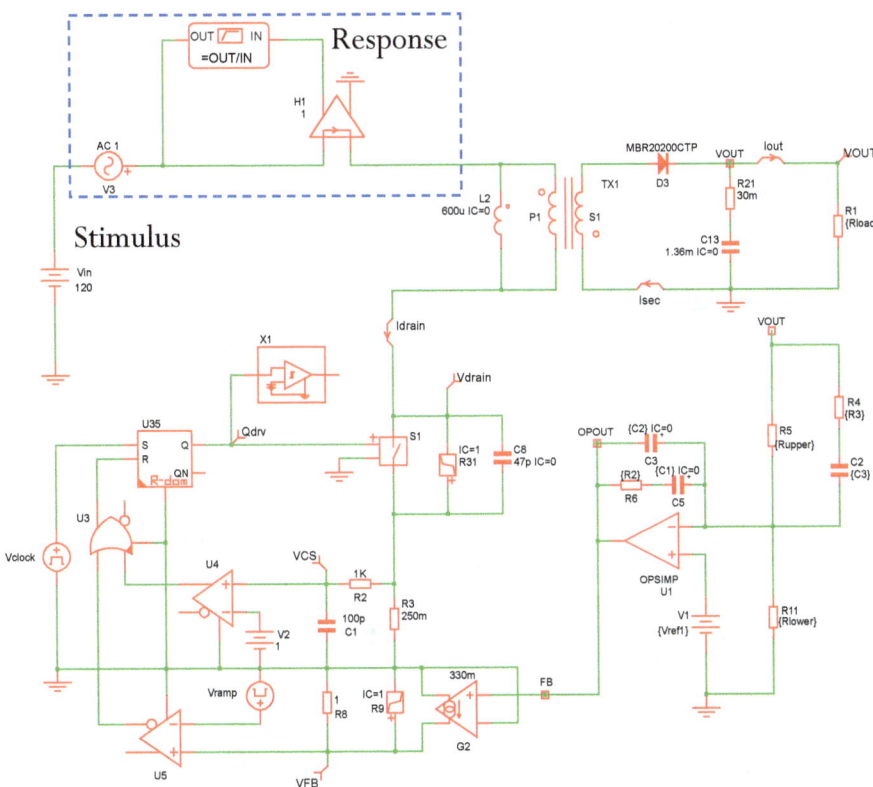

The flyback converter is operated at the minimum input voltage and maximum power. An ac voltage source is inserted in series with the dc source and sweeps the input. The converter has been stabilized and operates in closed-loop conditions. It is important to verify the operating point before considering the results.

This is a current-mode converter and instabilities can arise from the EMI filter if precautions are not taken.

Watch for Overlap

WHEN THE CLOSED-LOOP input impedance is obtained, you can superimpose it over your filter's output impedance:

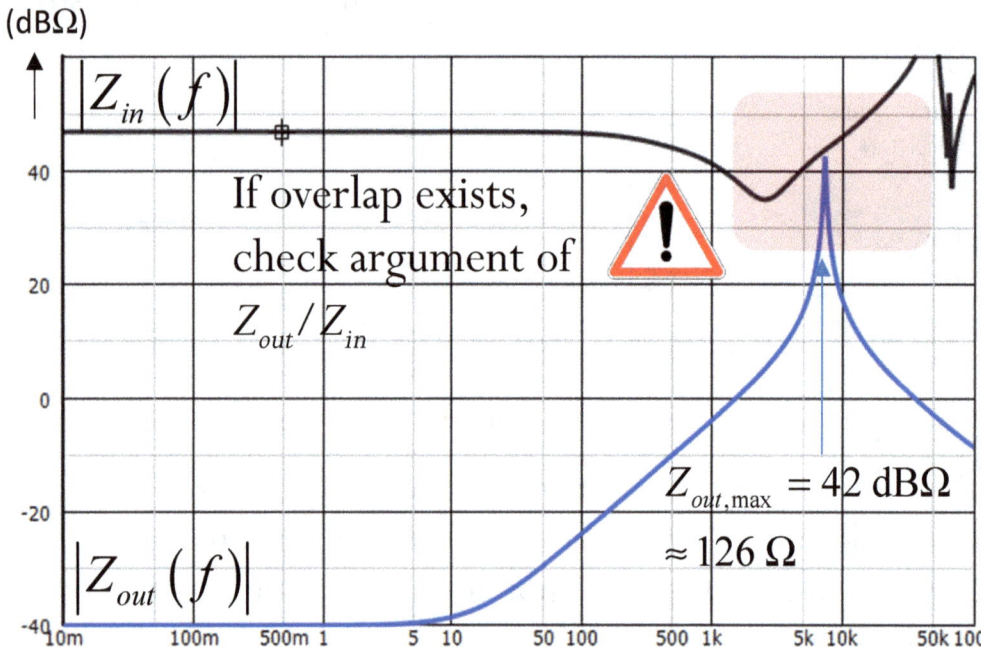

You can see on this graph how the two curves almost overlap. The output filter must be damped to reach a peak of 20 dBΩ, leading to a 15-20-dB margin.

If you install the *LC* filter, without damping, in front of the flyback circuit operated in current-mode control and run a load step, you may not observe an instability on the output voltage. Rather, if you probe the input port, after the filter, damped oscillations are present and must be treated.

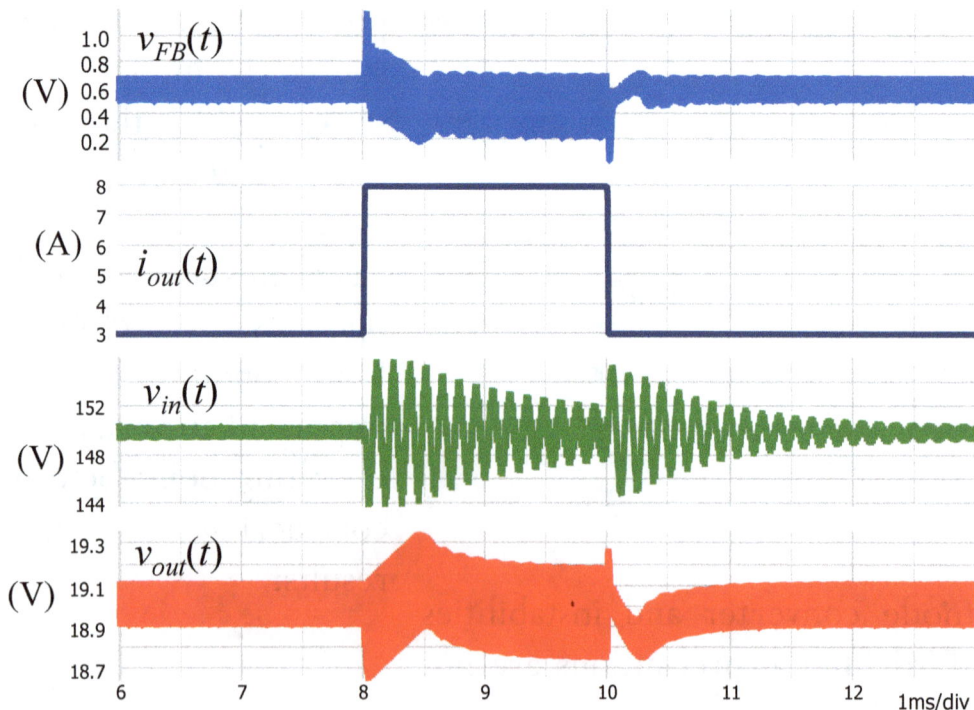

You can see the phenomenon clearly in this plot where lightly-damped oscillations appear on the filter output (V_{in} = 150 V). Reducing V_{in} brings a higher amplitude and possibly divergence.

Damping the Filter

DESCRIBING HOW TO damp the filter goes beyond the scope of this book but the method described in [5], leads you to the optimum selection of the RC damping elements. In this particular example, adding 6 Ω in series with a 5.5-µF capacitor will do the job of reducing the peak to exactly 10 Ω as expected:

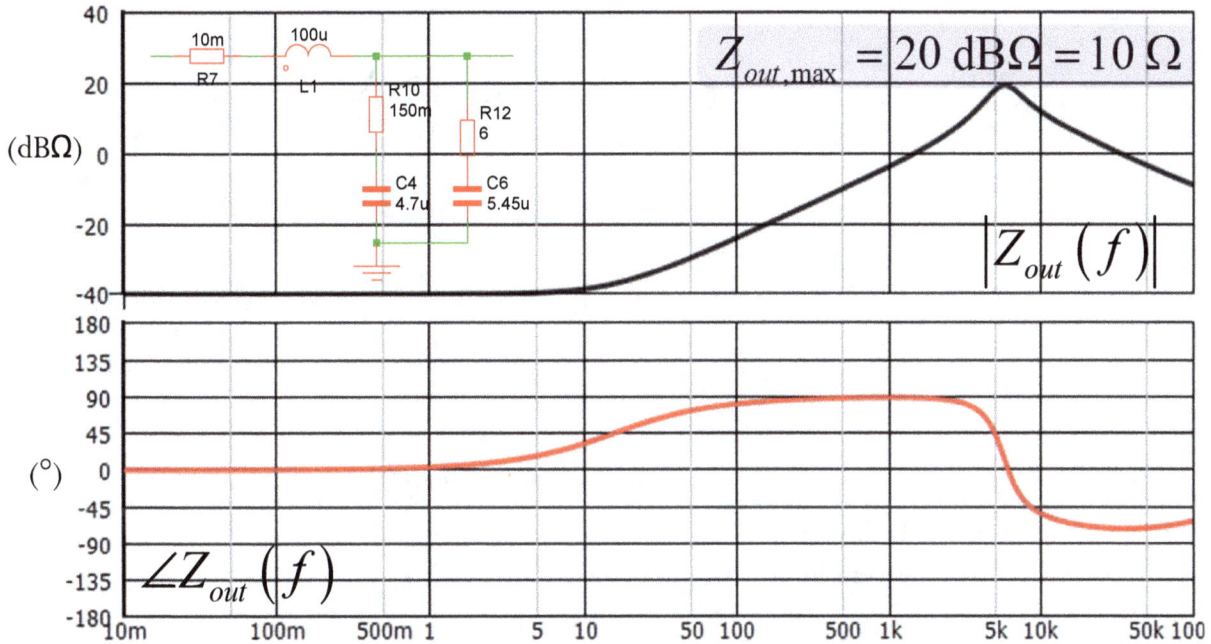

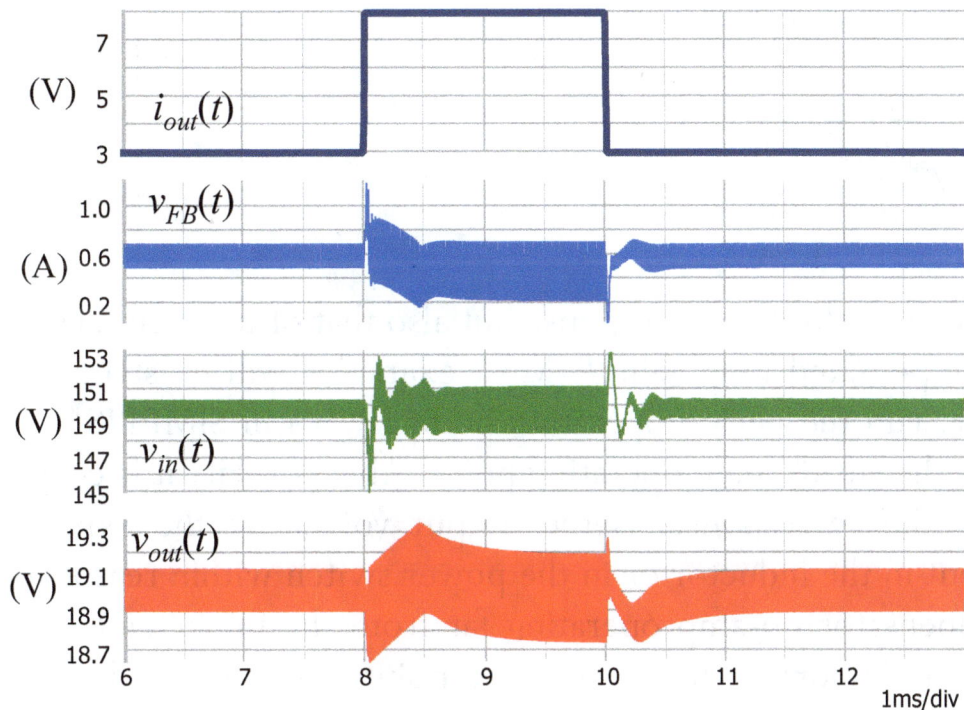

When the filter is properly damped, oscillations on the output filter quickly disappear, naturally making the converter operate in a more reliable way.

Buck Converter in VM

THIS FIRST example starts with the SIMPLIS® application circuit of a 100-kHz buck converter delivering 5 V/5 A from a 12-V input voltage. The first thing is to obtain the control-to-output transfer function, also called the power stage small-signal response.

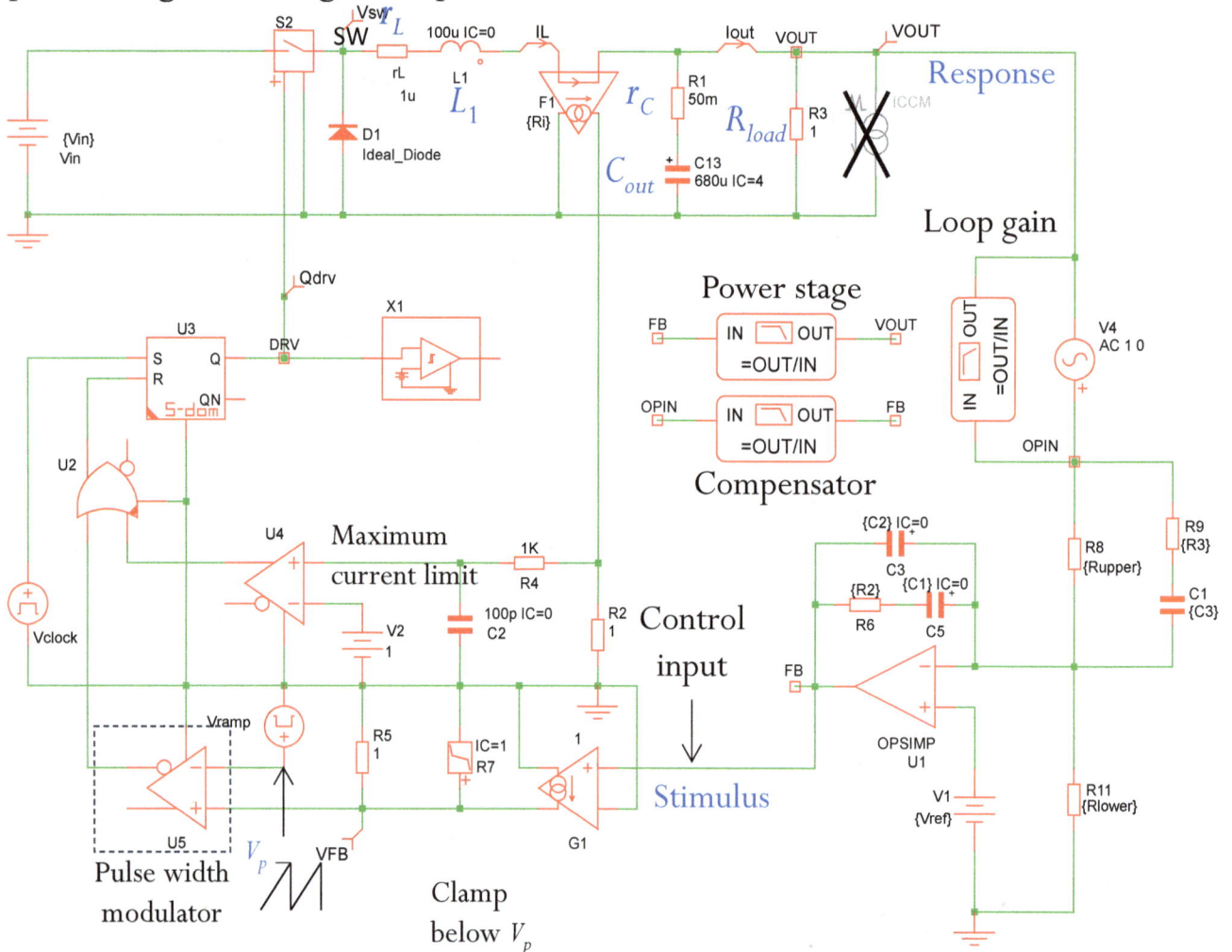

The template allows you plot this ac response but also that of the compensator alone, then the compensated loop gain T. Most examples will be structured this way. When you run the period operating point (POP) in SIMPLIS®, the program computes the steady-state operating point. Always verify it is correct before considering the ac response: is the output voltage at the expected value?, is the current in the inductor or in the power switch within reasonable limits?, is the compensator op-amp operating far from the upper and lower rails? and so on. It is important to check these points are respected before proceeding.

Power Stage and Compensation

WE START with the control-to-output transfer function of the CCM buck converter operated in voltage-mode control:

$$H(s) = \frac{V_{out}(s)}{V_{err}(s)} = H_0 \frac{1 + \dfrac{s}{\omega_z}}{1 + \dfrac{s}{\omega_0 Q} + \left(\dfrac{s}{\omega_0}\right)^2} \quad H_0 \approx \frac{V_{in}}{V_p} \quad \omega_z = \frac{1}{r_C C_{out}} \quad Q \approx R_{load}\sqrt{\frac{C_{out}}{L_1}} \quad \omega_0 \approx \frac{1}{\sqrt{L_1 C_{out}}}$$

Modulator peak voltage ↑

I used the approximation sign considering very low values for the parasitics r_L (inductor L_1 DCR) and r_C (output capacitor C_{out} ESR).

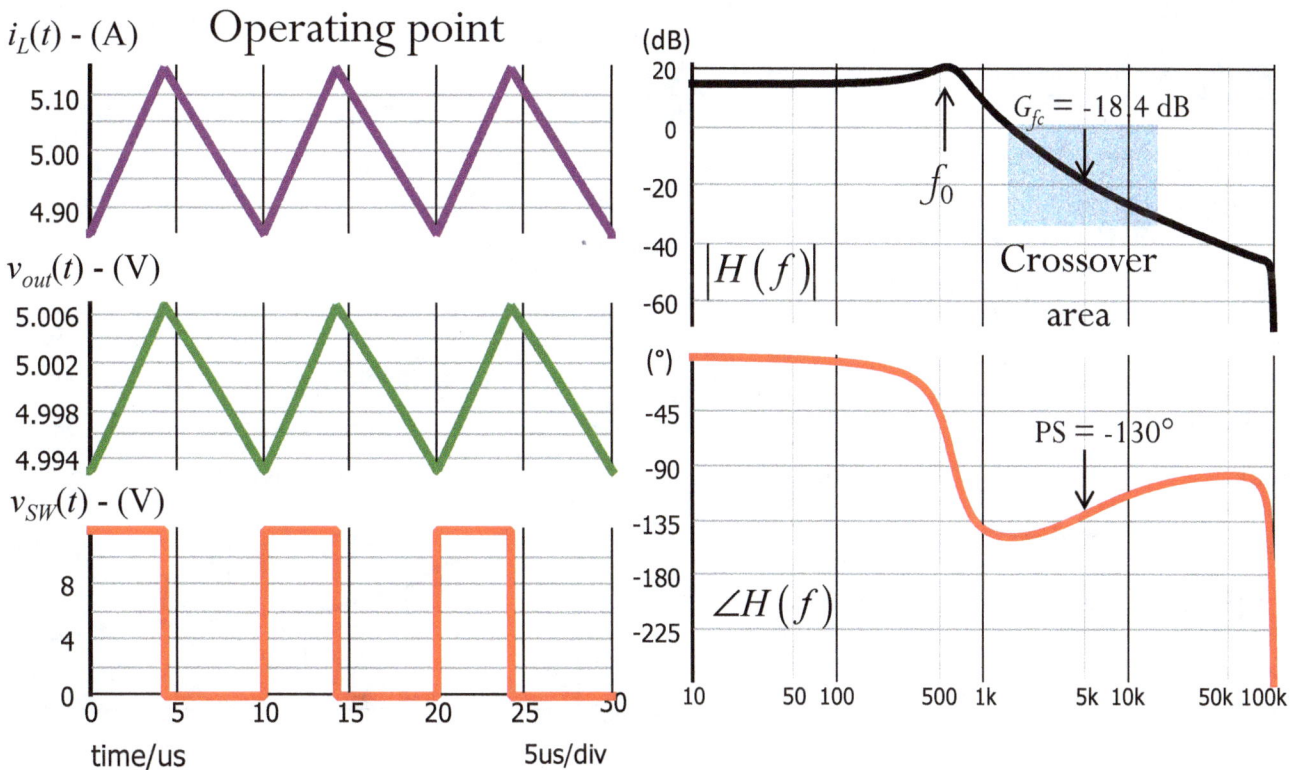

The operating point is correct (V_{out} is 5 V) and the Bode plot shows a peaking around 650 Hz. We need to pick a crossover frequency at least 3-5 times above f_0, implying a possible crossover at 5 kHz, for instance. We extract the magnitude and phase at 5 kHz: G_{fc} = -18.4 dB and PS = -130°. We are going to place one zero at 650 Hz and a second at a lower value, 200 Hz, to offer enough phase margin when the converter transitions into DCM. Then a pole is placed at $F_{sw}/2$ for good gain margin while the other one is adjusted to get the targeted phase margin of 70°: the macro will automate these calculations.

Compensated Loop Gain

THE COMPENSATOR for a CCM buck converter operated in voltage mode is a type 3. The macro starts by calculating the resistive divider for the 5-V output considering a 2.5-V reference voltage. Then, the components around the op-amp are determined and passed to the simulation engine. You can display the computed values in *Simulator>Edit Netlist* (*after preprocess*) by using the following syntax if you want to print the upper resistance value for instance: { '*' } Rupper = {Rupper}. The two loop gains *T* are shown below.

```
*
.VAR Vin=12
.VAR Vout=5
.VAR L=100u
.VAR Ri=160m
*
.VAR Gfc=-18.4 * magnitude at crossover *
.VAR PS=-130 * phase lag at crossover *
*
* Enter Design Goals Information Here *
*
.VAR fc=5k * targetted crossover *
.VAR PM=70 * choose phase margin at crossover *
*
* Enter the Values for Vout and Bridge Bias Current *
*
.VAR Ibias=1m      ⟵    Bias current
.VAR Vref=2.5           for the bridge
.VAR Rlower={Vref/Ibias}
.VAR Rupper={(Vout-Vref)/Ibias}
*
```

```
.VAR fz1=200    The two zeroes
.VAR fz2=650    and the pole         The 3rd pole is
.VAR fp2=50k                         adjusted to meet
*                                    the PM target
* Do not edit the below lines *
.VAR boost=PM-PS-90
.VAR G=10^(-Gfc/20)
.VAR fp1=fc/tan(atan(fc/fz1)+atan(fc/fz2)-atan(fc/fp2)-boost*pi/180)
*
* adjust second pole for targetted boost *
.VAR a=sqrt((fc^2/fp1^2)+1)
.VAR b=sqrt((fc^2/fp2^2)+1)
.VAR c=sqrt((fz1^2/fc^2)+1)
.VAR d=sqrt((fc^2/fz2^2)+1)
.VAR R2=((a*b/(c*d))/(fp1-fz1))*Rupper*G*fp1
.VAR C1=1/(2*pi*fz1*R2)
.VAR C2=C1/(C1*R2*2*pi*fp1-1)
.VAR C3=(fp2-fz2)/(2*pi*Rupper*fp2*fz2)
.VAR R3=Rupper*fz2/(fp2-fz2)
.VAR G0=((R2*C1)/(Rupper*(C1+C2)))*c*d/(a*b) * Gain at fc sanity check *
*
```

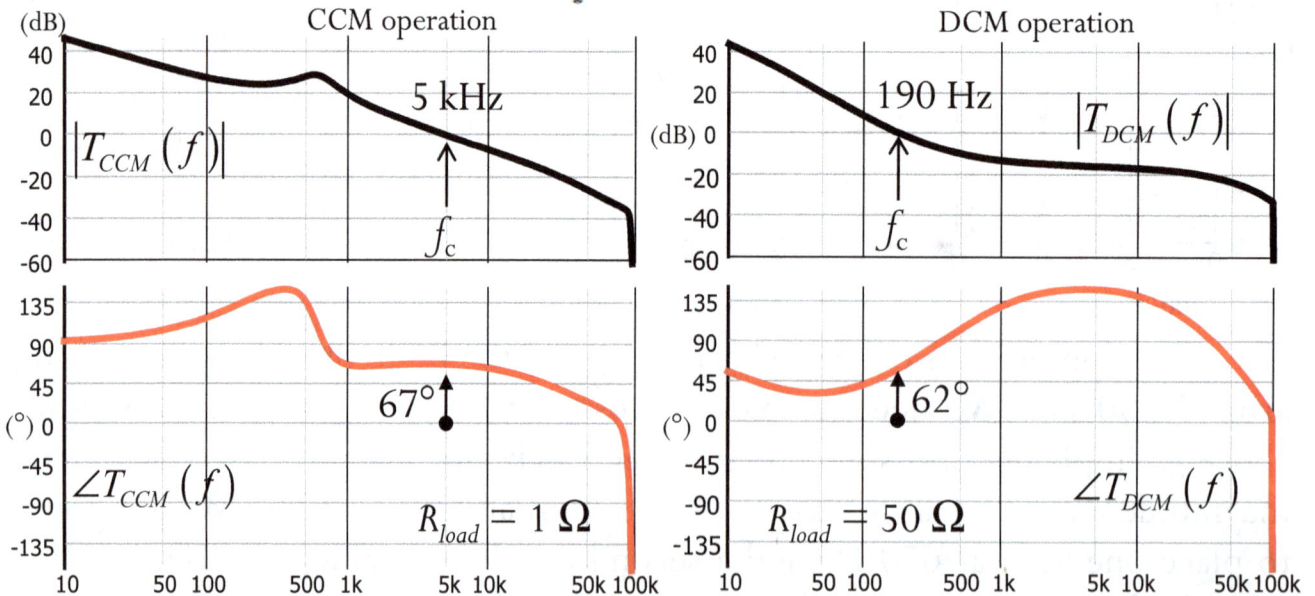

The compensated loop gain in CCM shows a good phase margin while crossover collapses in DCM but with an adequate phase margin though. Lowering the zero before resonance helps to keep a good phase margin in DCM.

Transient Response

NOW THAT THE CONVERTER is stabilized and meets the goals, you can run a load transient test, e.g. between 50 and 100% of the load. I usually adopt a current slope of 1 A/μs for the variation speed. Please note that an undershoot specified without the current slope is meaningless so always mention it in your results.

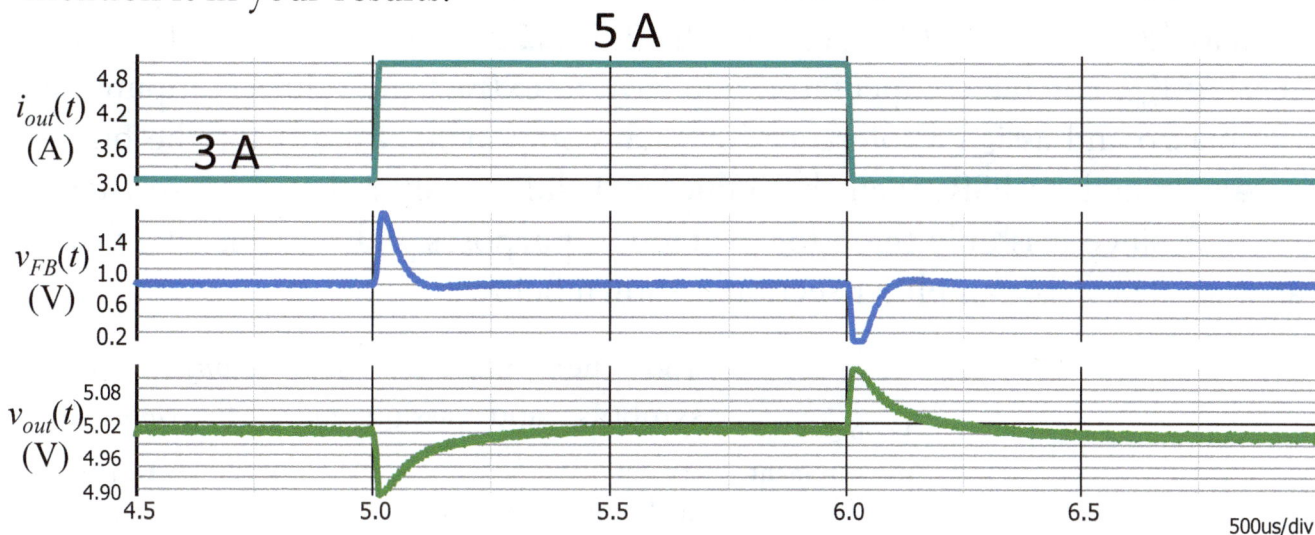

In DCM, the response time is slower but still the undershoot remains well controlled. You can now easily alter the zeroes position and see the effects on the transient response.

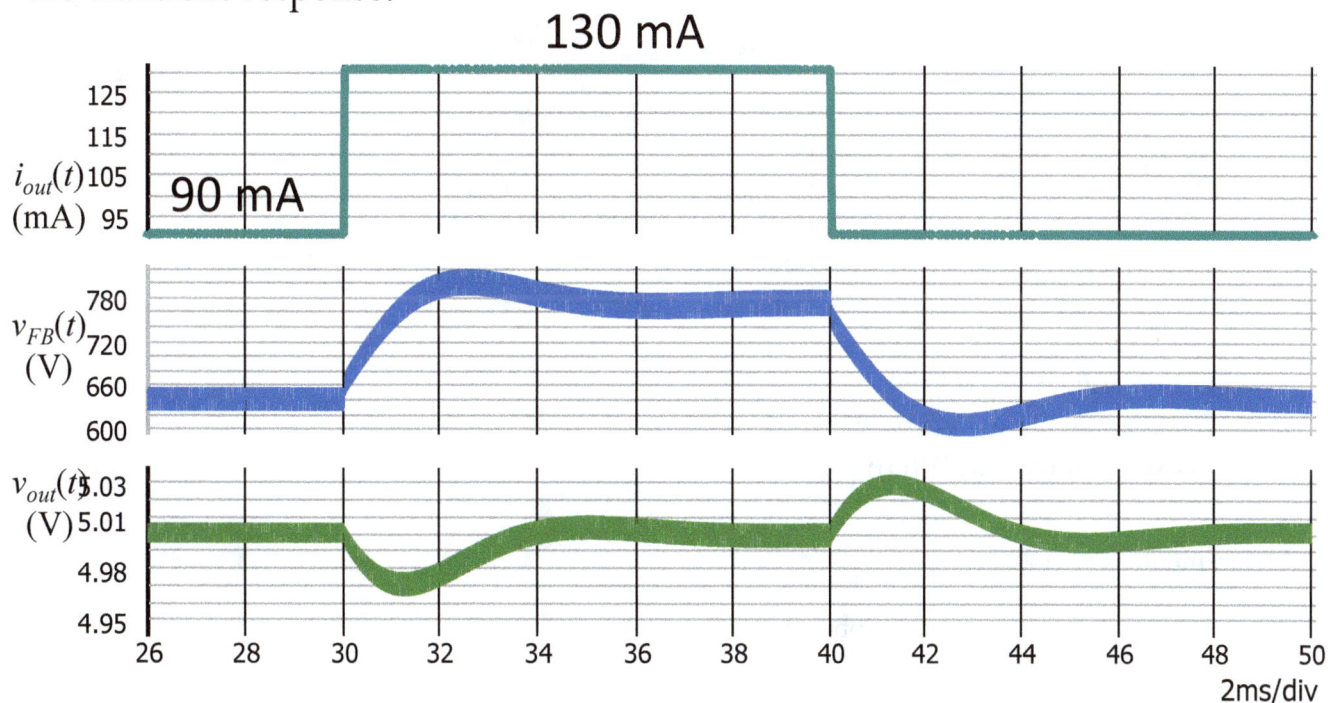

Assigning Components Tolerance

WE WILL SHOW in this example how to conduct a basic Monte Carlo analysis and check the sensitivity of the design to components variations. The first thing is to assign tolerance and a distribution type to the component. In this example, I have chosen a gaussian distribution with a specific tolerance for resistors and capacitors. In SIMPLIS®, you fill out the value field with a line such as {2.5k*gauss(0.01)}, e.g. for the 2.5-kΩ ±1% resistances of the bridge sensing V_{out}. For the compensator, you have to extract the values computed by the macro and assign a tolerance for each component. I did it for the output capacitor and its ESR, then the inductor. I did not alter the sense resistance value R_i since it is only there for protection purposes and remains silent in ac analysis. The error amplifier section is shown below:

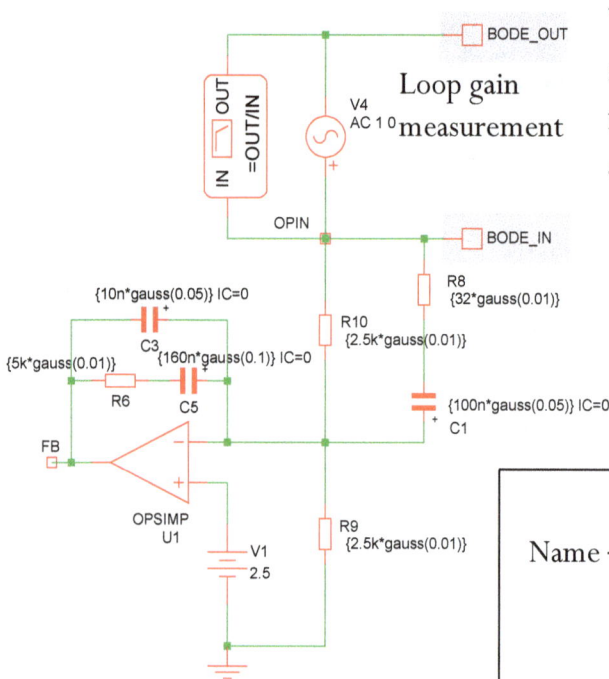

You then place 3 *fixed arbitrary probe* symbols and assign *goal functions* for measuring the crossover frequency, phase and gain margins:

PhaseMargin(V(out)/V(in))
GainMargin(V(out)/V(in))
XatNthYn(db(V(out)/V(in)) , 0 , 1)

Measurement nodes

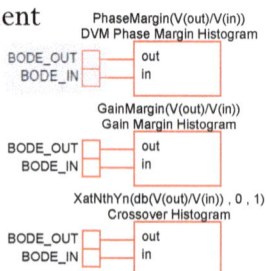

Arbitrary probes and goal functions

Visualizing Results

IN THIS EXAMPLE, I have selected 100 runs distributed in 6 bins. The program takes advantage of the 4 cores I have on my computer and results were delivered in less than 10 minutes. First, you can see how the transient waveforms move when the key components values are pseudo-randomly swept with a converter fed from a 12-V input and loaded by 1-Ω resistance:

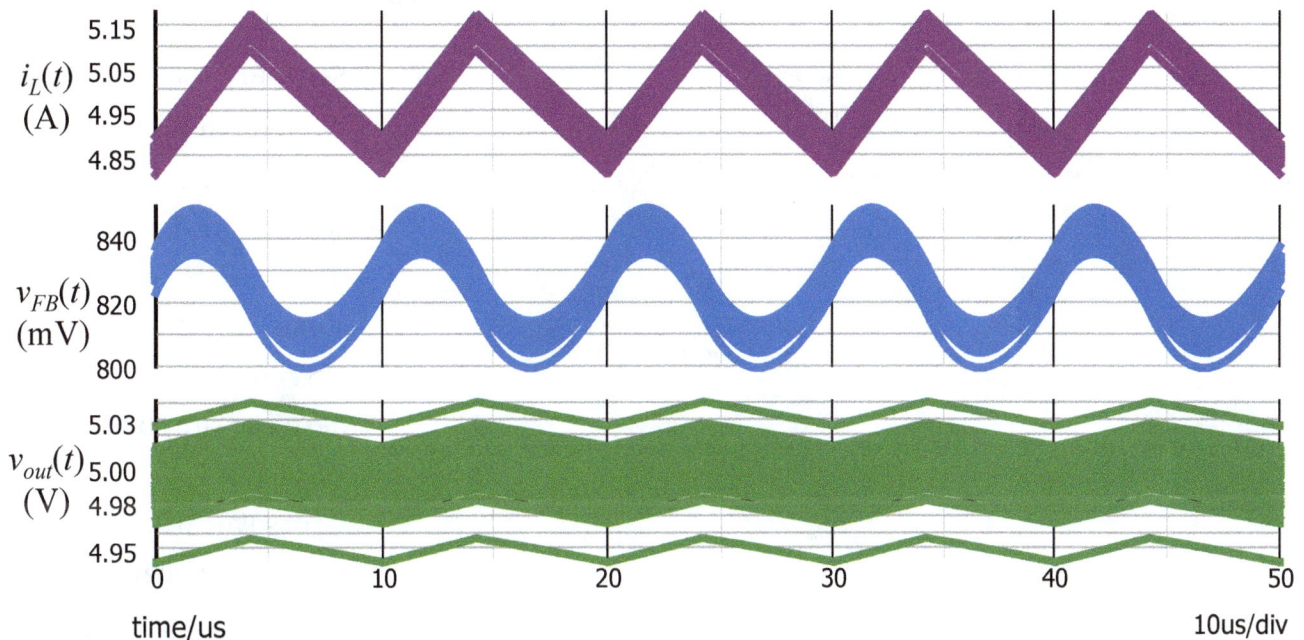

Our interest now lies in checking how crossover and margins are moving when all components values change. SIMPLIS® plots histograms which tell you how these parameters were affected during the runs:

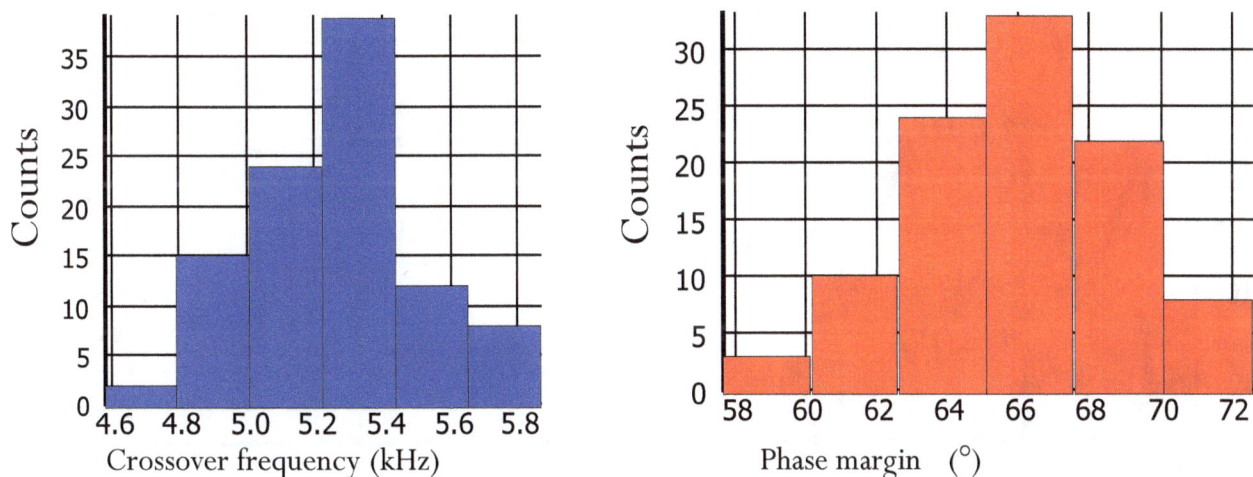

You can see fairly stable results with a crossover spread between 4.6 and 5.8 kHz and a phase margin above 58°. I did not include gain margin considering the layout but it is above 31 dB.

Synchronous Rectification

The freewheel diode forward drop V_f affects the efficiency of the buck converter during the off-time. It is possible to add a controlled power switch in parallel with the diode and activate it during the off-time. The efficiency is greatly improved while keeping the converter in CCM even in no-load conditions: the transfer function no longer changes with load variations.

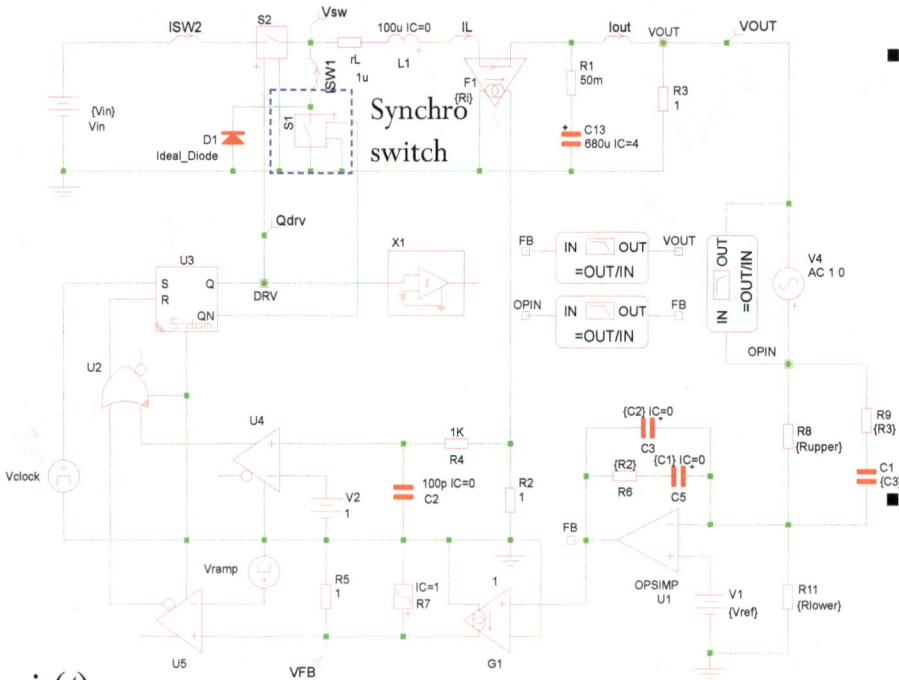

- Switches are controlled with complementary waveforms featuring a deadtime to avoid shoot-through currents. There is none here for simplicity of the circuit.
- Crossover frequency remains unchanged when transitioning to light-load operation.

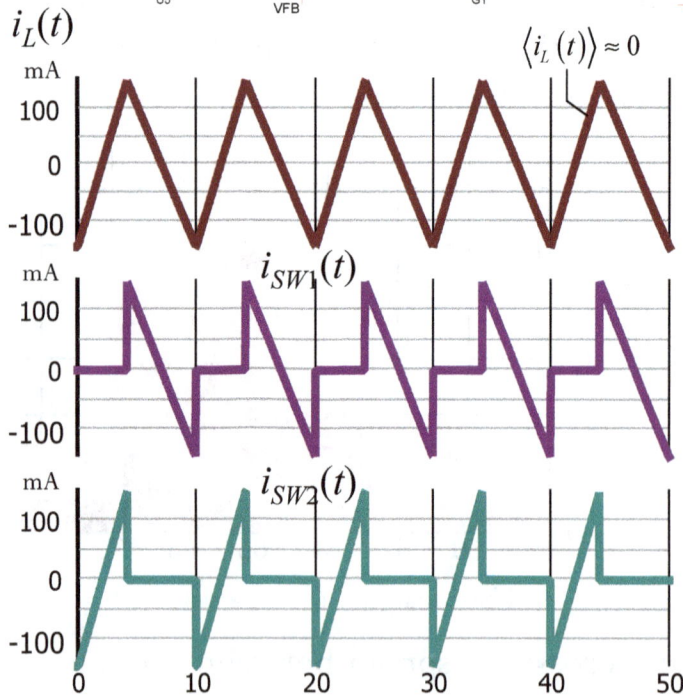

Light-load operation, $R_{load} = 10\ \text{k}\Omega$

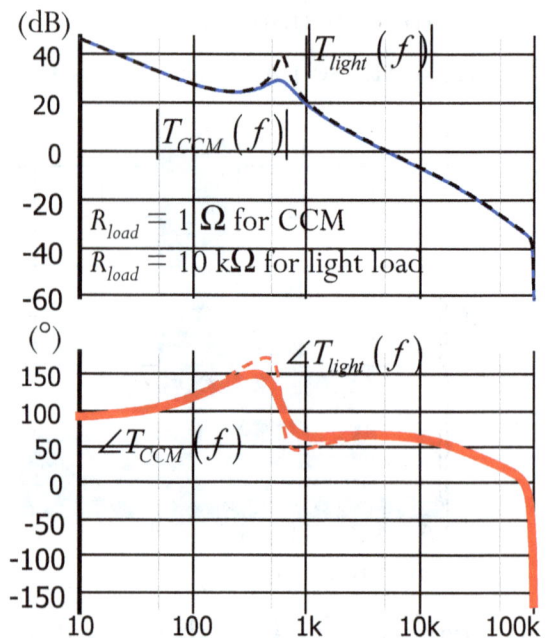

$R_{load} = 1\ \Omega$ for CCM
$R_{load} = 10\ \text{k}\Omega$ for light load

f_c in CCM = 5 kHz
f_c in light load = 5 kHz

Physically Opening the Loop

IN THE BUCK EXAMPLE and the upcoming circuits, we are keeping the loop closed in dc so that the simulation program determines the correct bias point according to input and output conditions. An ac source is then inserted in the return path and perturbs the system. A signal of sufficiently-high amplitude can be observed and collected on the output to deliver a Bode plot. The source amplitude is automatically adjusted by SIMPLIS® during the sweep so that the system always remains linear. In practice, you would need to modulate the injected signal to prevent saturation: a rather strong amplitude at low frequency (the loop gain is high and fights the perturbation) and then reduce it as you approach crossover (risks of saturation as the gain decreases and passes below unity). But what if the converter is not stable in the first place or you have no idea of what its response is? In this case, you run your converter in open-loop: a dc source drives the control pin to obtain the right V_{out} for a given input voltage and you superimpose the ac source to this bias:

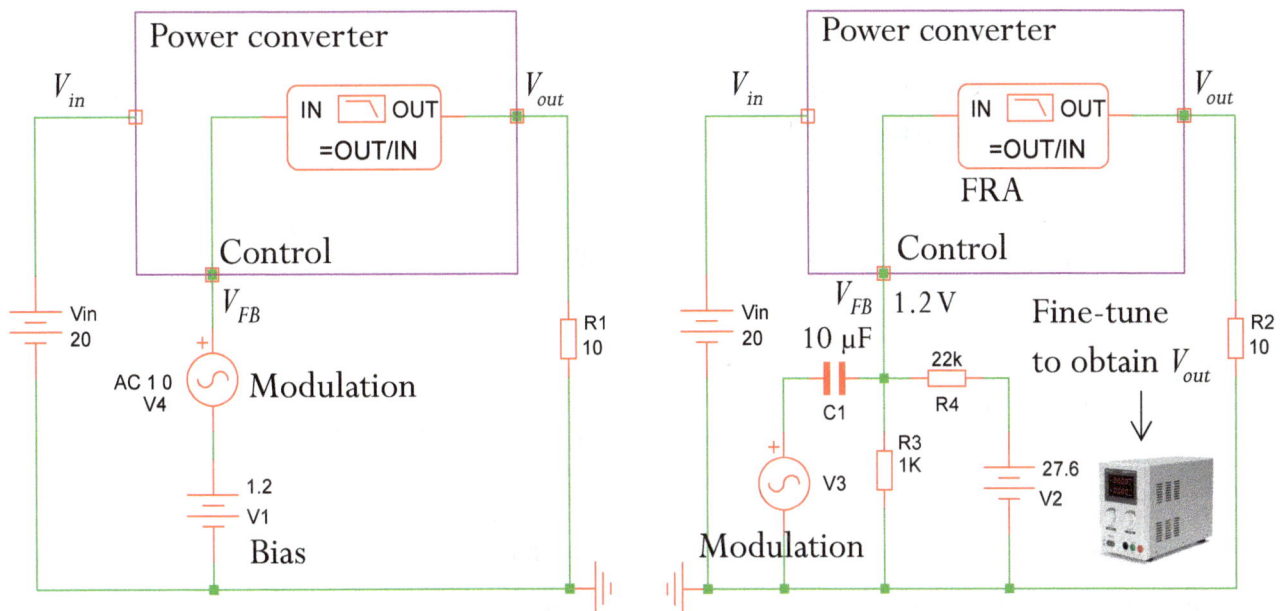

On the bench, I was often opening the loop with a dc bias *carefully* applied at the control input. This bias is provided via a resistive divider to offer fine tuning and reduce noise. The resistors have to be closely wired to the control input while twisted wires bring the bias. If opening the loop in simulation is inherently safe, be *very careful* when considering this experiment with a power converter and check galvanic isolation, start-up overshoot, protections, temperature drift and so on.

Buck Converter in CM

The buck converter now operates in current-mode control and involves an inner current loop sensing current via F_1 and its coefficient R_i. To damp the sub-harmonic poles located at $F_{sw}/2$, an artificial ramp V_{saw} is added to the current sense signal. You size its amplitude via the following expressions:

$$S_n = \frac{V_{in} - V_{out}}{L_1} R_i \qquad S_{ramp} = \frac{1\,\text{V}}{10\,\mu s} = 100\,\frac{\text{mV}}{\mu s} \qquad m_c = \frac{S_e}{S_n} + 1 \longrightarrow \begin{aligned} S_e &= (m_c - 1)S_n \\ k_r &= \frac{S_e}{S_{ramp}} \end{aligned}$$

On-time slope   Artificial slope   Compensation

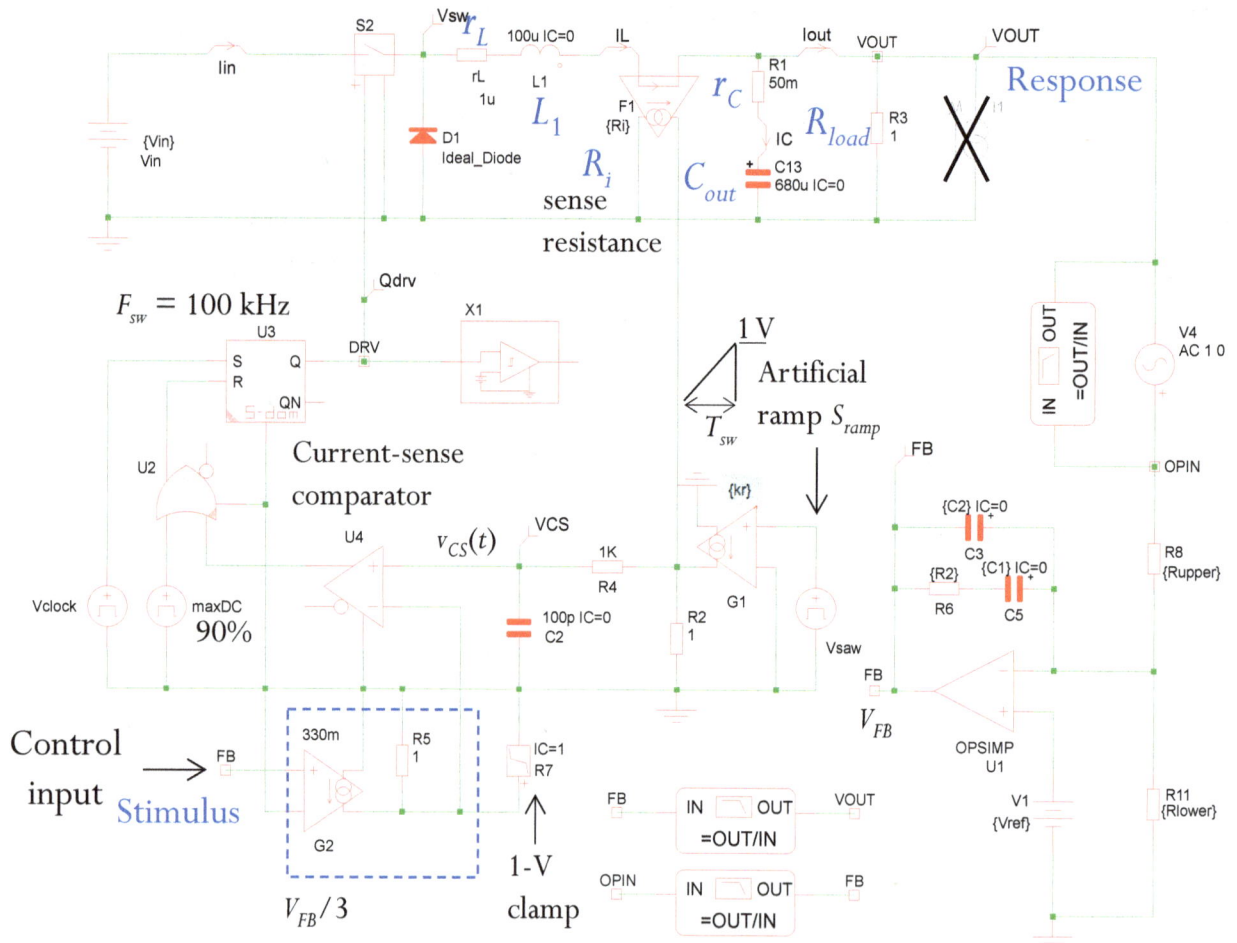

The feedback voltage delivered by the op-amp undergoes a division by 3 to augment its operating dynamics: if the current-sense comparator setpoint is 400 mV, the voltage at the op-amp output will be 1.2 V, offering a good noise immunity and operating the op-amp well within its linear range. The maximum setpoint is clamped to 1 V but can be altered based on the adopted controller. I added a source delayed by 9 μs with respect to the main clock which imposes a maximum duty ratio of 90% in case the current reset does not trip.

Power Stage and Compensation

WE START with the control-to-output transfer function of the CCM buck converter operated in current-mode control:

$$H(s) = \frac{V_{out}(s)}{V_{err}(s)} = H_0 \frac{1 + \frac{s}{\omega_z}}{1 + \frac{s}{\omega_p}} \frac{1}{1 + \frac{s}{\omega_n Q} + \left(\frac{s}{\omega_n}\right)^2} \qquad H_0 \approx \frac{R_{load}}{R_i} \frac{1}{1 + \frac{R_{load} T_{sw}}{L_1}\left[m_c(1-D) - 0.5\right]}$$

$$\omega_z = \frac{1}{r_C C_{out}} \qquad \omega_p \approx \frac{1}{R_{load} C_{out}} + \frac{T_{sw}}{L_1 C_{out}}\left[m_c(1-D) - 0.5\right] \qquad \omega_n = \frac{\pi}{T_{sw}} \qquad Q = \frac{1}{\pi\left[m_c(1-D) - 0.5\right]} \qquad m_c = 1 + \frac{S_e}{S_n}$$

We can now run the simulation circuit, check the operating point and ac plot.

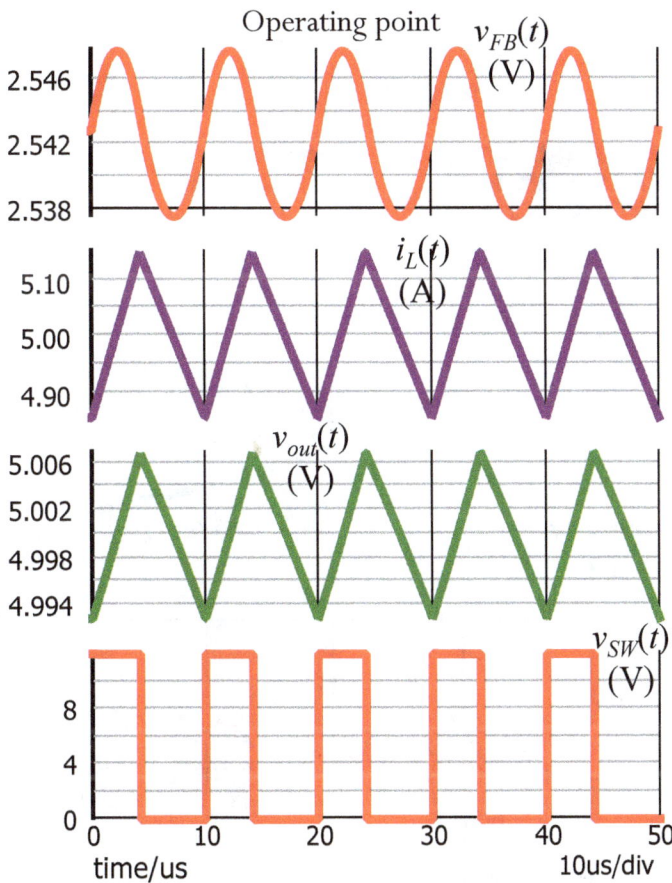

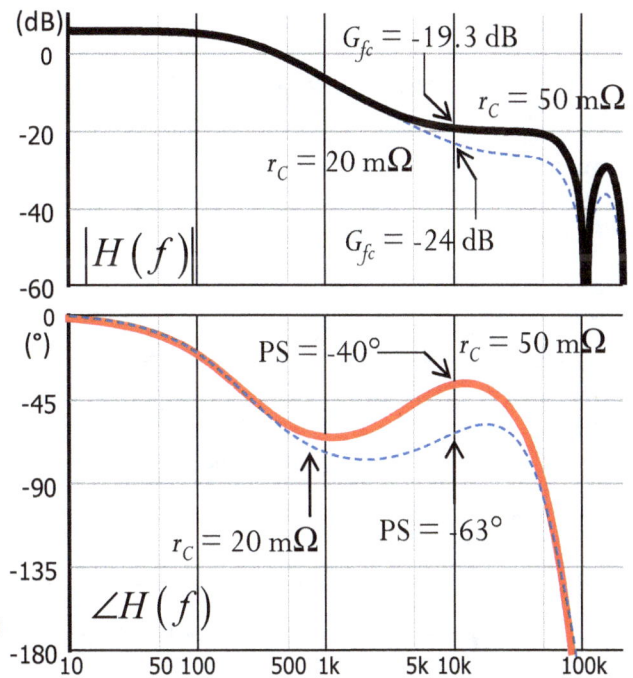

In the Bode plot, you see the phase heading down to -90° before going up again, owing to the ESR zero located slightly below 5 kHz

As it occurs in the crossover region – we will shoot for 10 kHz here – then the zero helps for the phase response. This zero will vary with temperature and age so make sure the compensator already factors this parameter in so that phase margin remains adequate at both ESR extremes. You see how the plot changes as the ESR varies from 50 to 20 mΩ, thus carefully check its impact on the ac response.

Compensated Loop Gain

WE WILL SELECT a 10-kHz crossover frequency here. The peaking linked to voltage-mode control is gone and we have a single low-frequency pole now. The sub-harmonic poles are well damped by the external ramp we have added.

```
*
.VAR Vin=12
.VAR Vout=5
.VAR L=100u
.VAR Ri=160m
.VAR Ts=10u * please update clock and ramp generators *
*
.VAR Gfc=-24 * magnitude at crossover *
.VAR PS=-70 * phase lag at crossover *
*
* Enter Design Goals Information Here *
*
.VAR fc=10k * targetted crossover *
.VAR PM=60 * choose phase margin at crossover *
*
.VAR Sn={((Vin-Vout)/L)*Ri}
.VAR Sramp={1/Ts}
.VAR mc=1.5 * set this value for ramp comp *
.VAR Se={(mc-1)*Sn}
.VAR kr={Se/Sramp}
*
```

```
* Enter the Values for Vout and Bridge Bias Current *
*
.VAR Ibias=1m
.VAR Vref=2.5
.VAR Rlower={Vref/Ibias}
.VAR Rupper={(Vout-Vref)/Ibias}
*
* Do not edit the below lines *
.VAR boost=PM-PS-90
.VAR G=10^(-Gfc/20)
.VAR fp=(tan(boost*pi/180)+sqrt((tan(boost*pi/180))^2+1))*fc
.VAR fz=fc^2/fp
.VAR a=sqrt((fc^2/fp^2)+1)
.VAR b=sqrt((fz^2/fc^2)+1)
.VAR R2=((a/b)*G*Rupper*fp)/(fp-fz)
.VAR C1=1/(2*pi*R2*fz)
.VAR C2=C1/(C1*R2*2*pi*fp-1)
*
```

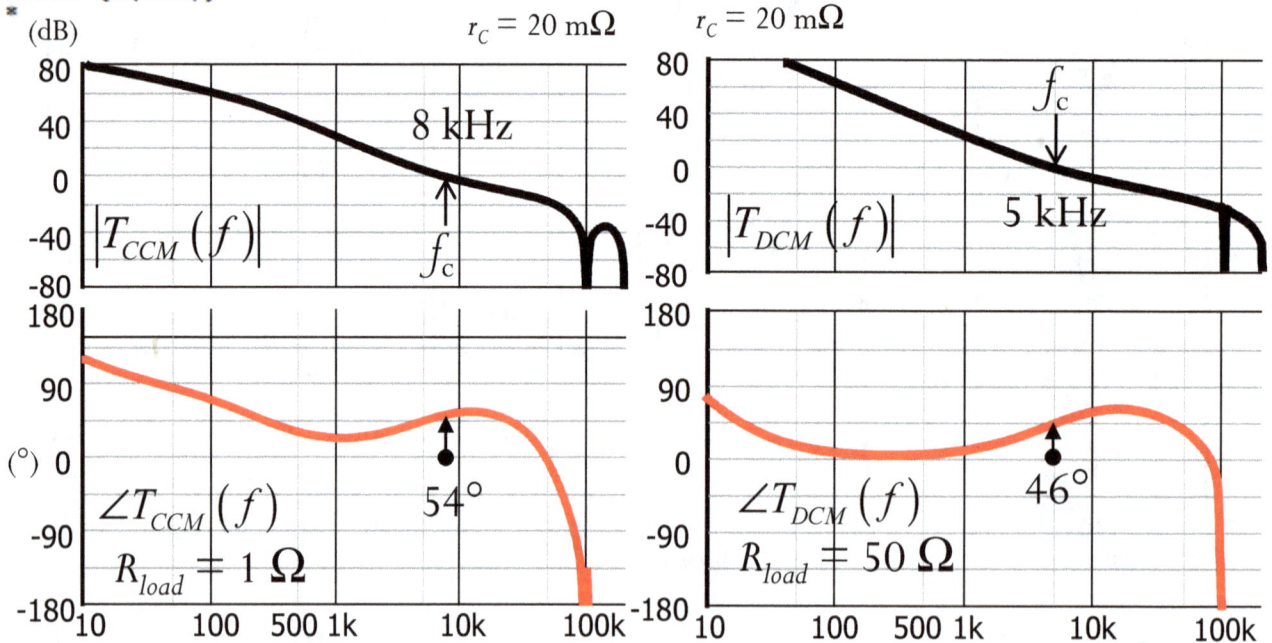

$r_C = 20\ \text{m}\Omega$

$r_C = 20\ \text{m}\Omega$

(dB) 8 kHz $|T_{CCM}(f)|$ f_c $\angle T_{CCM}(f)$ $R_{load} = 1\ \Omega$ 54°

f_c $|T_{DCM}(f)|$ 5 kHz $\angle T_{DCM}(f)$ $R_{load} = 50\ \Omega$ 46°

The DCM plot shows a good crossover unlike in the VM version. However, you may fear some conditional stability issue as the phase approaches 0° in the 100 Hz-1-kHz region. By lowering the zero to 2 kHz, it will help boost the phase in this area.

Transient Response

THE TRANSIENT RESPONSE to a load step is very stable and shows a fast recovery. The drop is only 30 mV with a 20-mΩ ESR for the output capacitor.

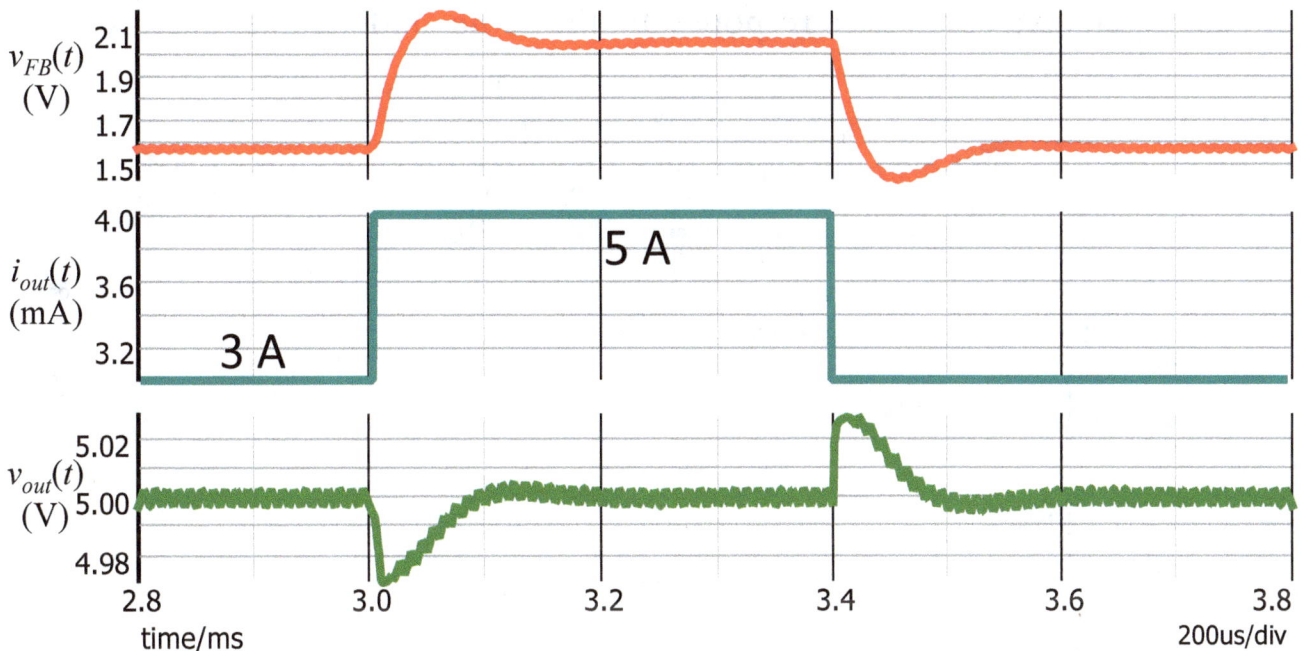

In DCM, the response time is also excellent and confirms the superiority of current mode over voltage mode when it comes to mode transition. Crossover frequencies between CCM and DCM are almost identical (8 and 5 kHz) unlike in voltage-mode control (5 kHz CCM and 190 Hz DCM).

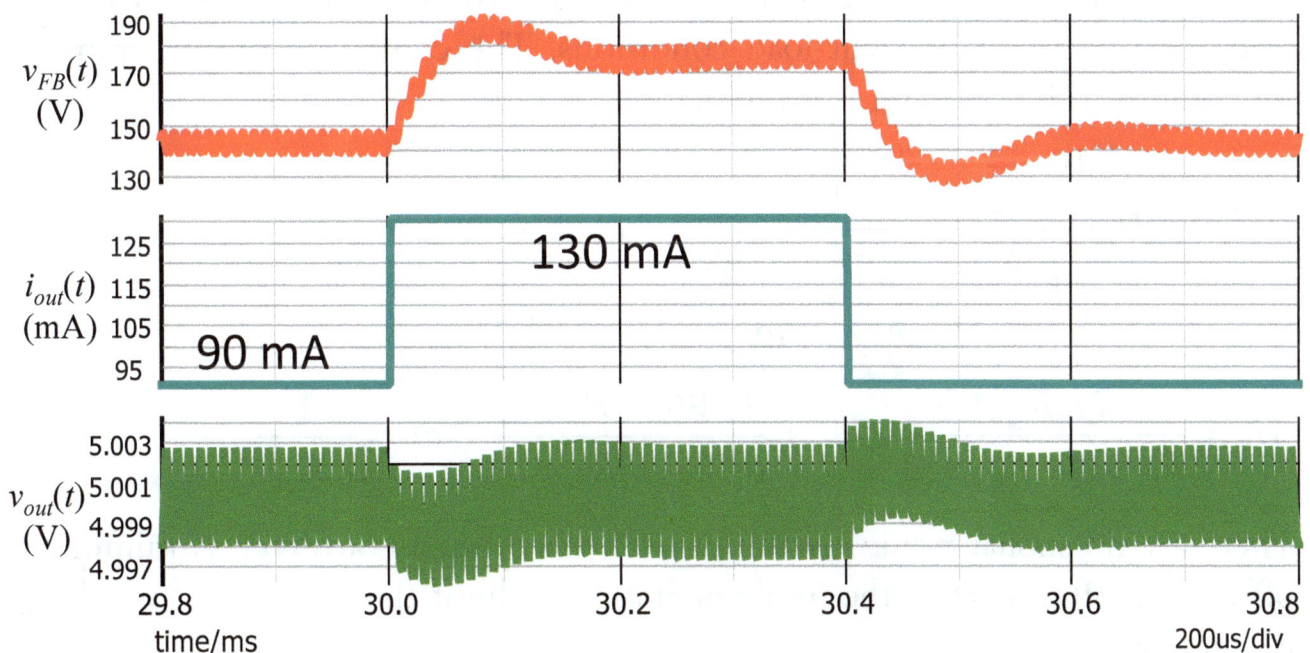

Buck Converter in COT

THE BUCK CONVERTER operated in constant on-time (COT) is a dc-dc structure used, for example, in a computer motherboard for point-of-load regulation: the di/dt absorbed by the microprocessor can be very large, requiring an extremely-fast response to limit the undershoot and recovery time.

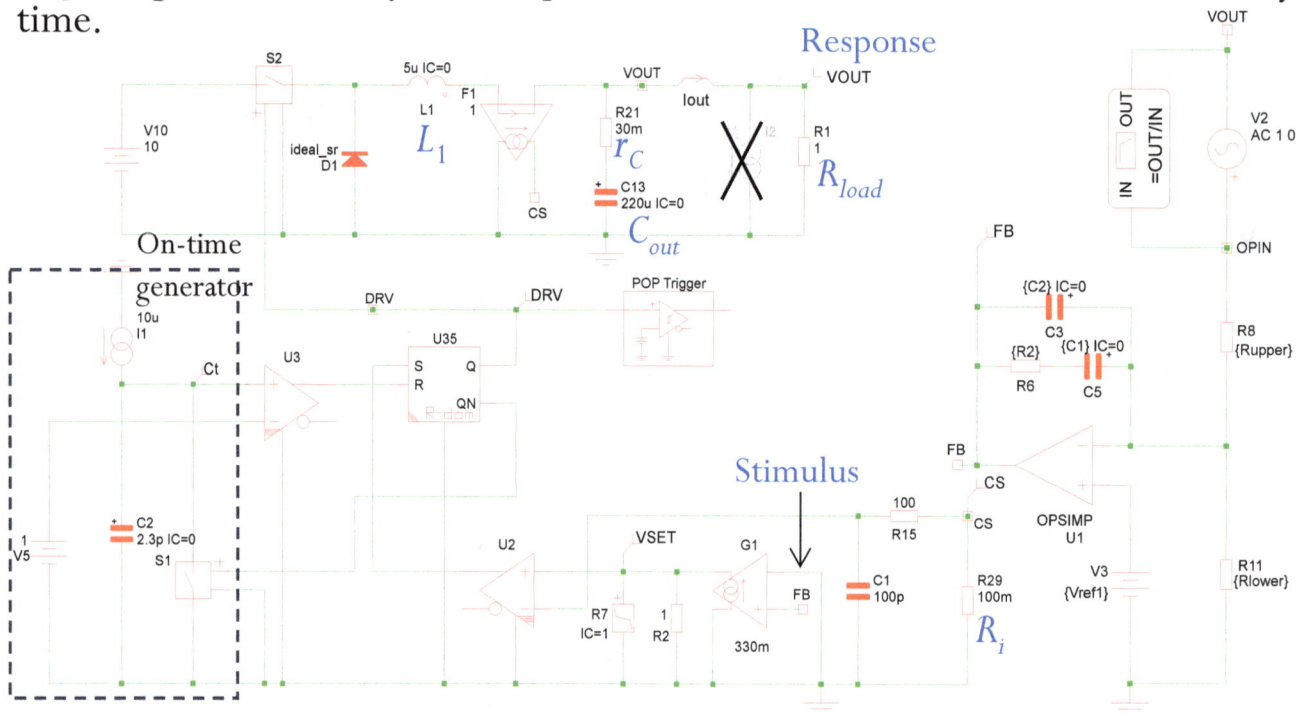

In a current-mode-controlled COT, the feedback voltage sets the inductor *valley* current at which the power switch turns back on again. The good thing with this type of converter is the natural standby mode in which switching frequency decreases in light-load conditions. The control-to-output transfer function for this converter is given below:

$$H(s) = H_0 \frac{1+\dfrac{s}{\omega_z}}{1+\dfrac{s}{\omega_p}} \frac{1}{1+\dfrac{s}{\omega_1 Q}+\left(\dfrac{s}{\omega_1}\right)^2} \qquad \omega_1 = \frac{\pi}{t_{on}} \qquad Q = \frac{2}{\pi}$$

$$H_0 = \frac{2L_1 R_{load}}{R_i\left(2L_1 + R_{load}t_{on}\right)} \qquad \omega_p = \frac{L_1 + 0.5 \cdot R_{load}t_{on}}{C_{out}L_1 R_{load}} \qquad \omega_z = \frac{1}{r_C C_{out}}$$

There are no subharmonic instability with this type of converter. A minimum off-time is often part of the implementation to limit the maximum switching frequency.

Power Stage and Compensation

IT IS INTERESTING to check the switching frequency when simulating this converter as it depends on the operating point. SIMPLIS® proposes dedicated probes from which you can extract a variety of data like duty ratio or switching frequency. The bias point confirms the 5-V output voltage and a 2.2-MHz frequency:

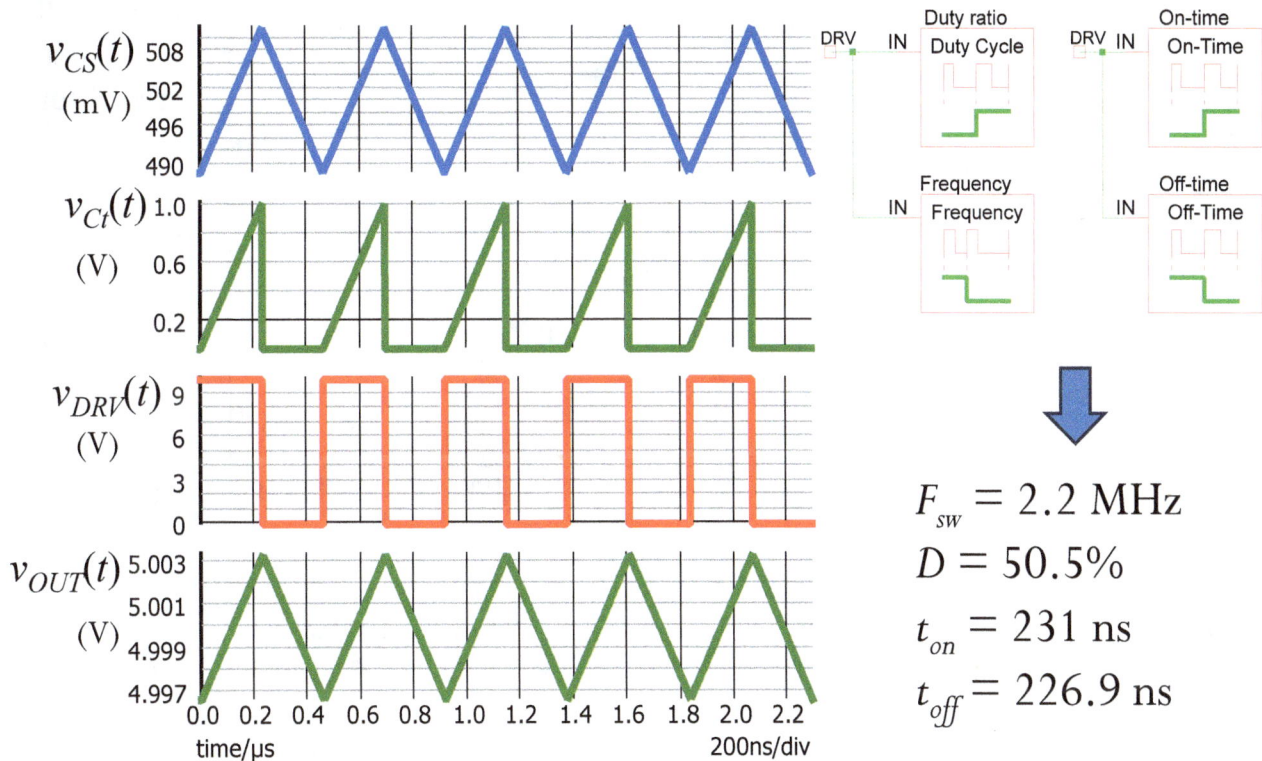

$$F_{sw} = 2.2 \text{ MHz}$$
$$D = 50.5\%$$
$$t_{on} = 231 \text{ ns}$$
$$t_{off} = 226.9 \text{ ns}$$

The ac response is obtained in a fraction of a second after the bias point.

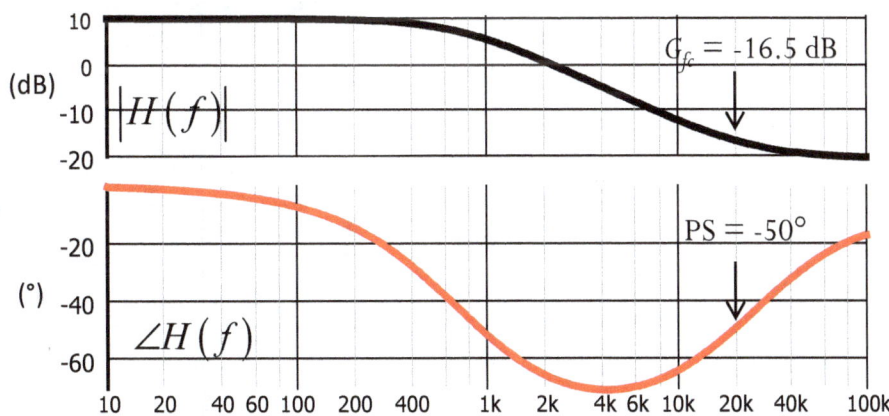

The power ac plot confirms the smooth response which lets you push crossover to 20 kHz considering the 2.2-MHz switching frequency. A type 2 compensator will do well for this purpose.

Loop Gain and Transient Response

THE MACRO PLACES poles and zeroes to bring a phase margin of 60°. Here, the 20-kHz crossover frequency target implies a high gain-bandwidth (GBW) op-amp. If you select a slower type, the ac response of the compensator will suffer and fail to provide gain and phase boost at 20 kHz.

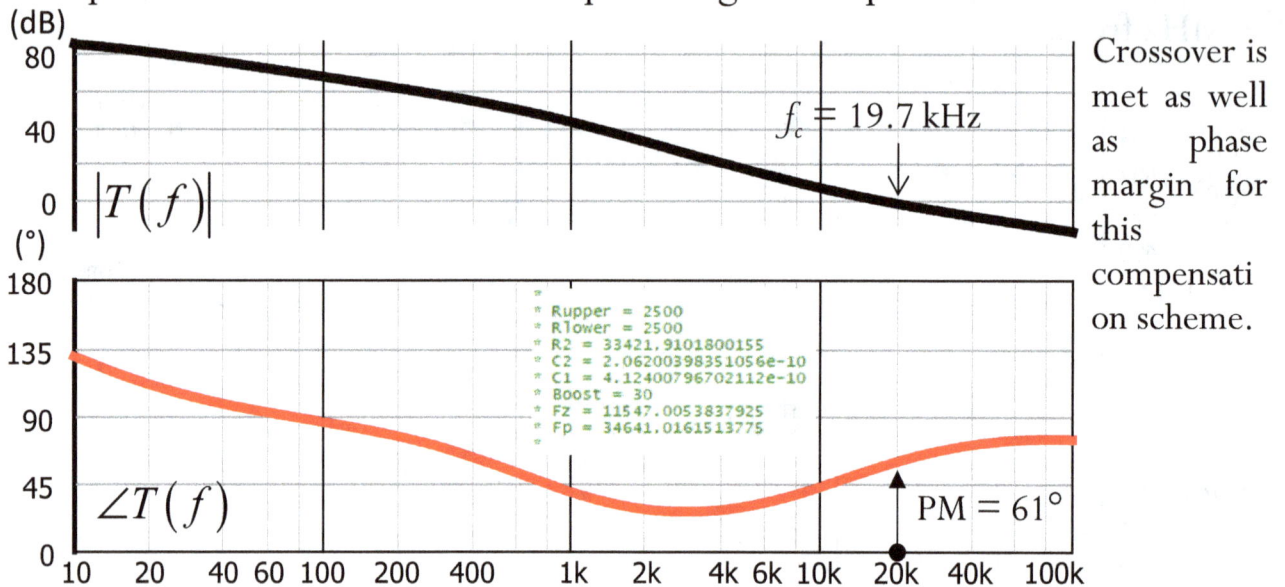

Crossover is met as well as phase margin for this compensation scheme.

$f_c = 19.7\,\text{kHz}$

$|T(f)|$

$\angle T(f)$

PM = 61°

```
* Rupper = 2500
* Rlower = 2500
* R2 = 33421.9101800155
* C2 = 2.06200398351056e-10
* C1 = 4.12400796702112e-10
* Boost = 30
* Fz = 11547.0053837925
* Fp = 34641.0161513775
*
```

The load step consists of a current stepped from 1 A to 5 A.

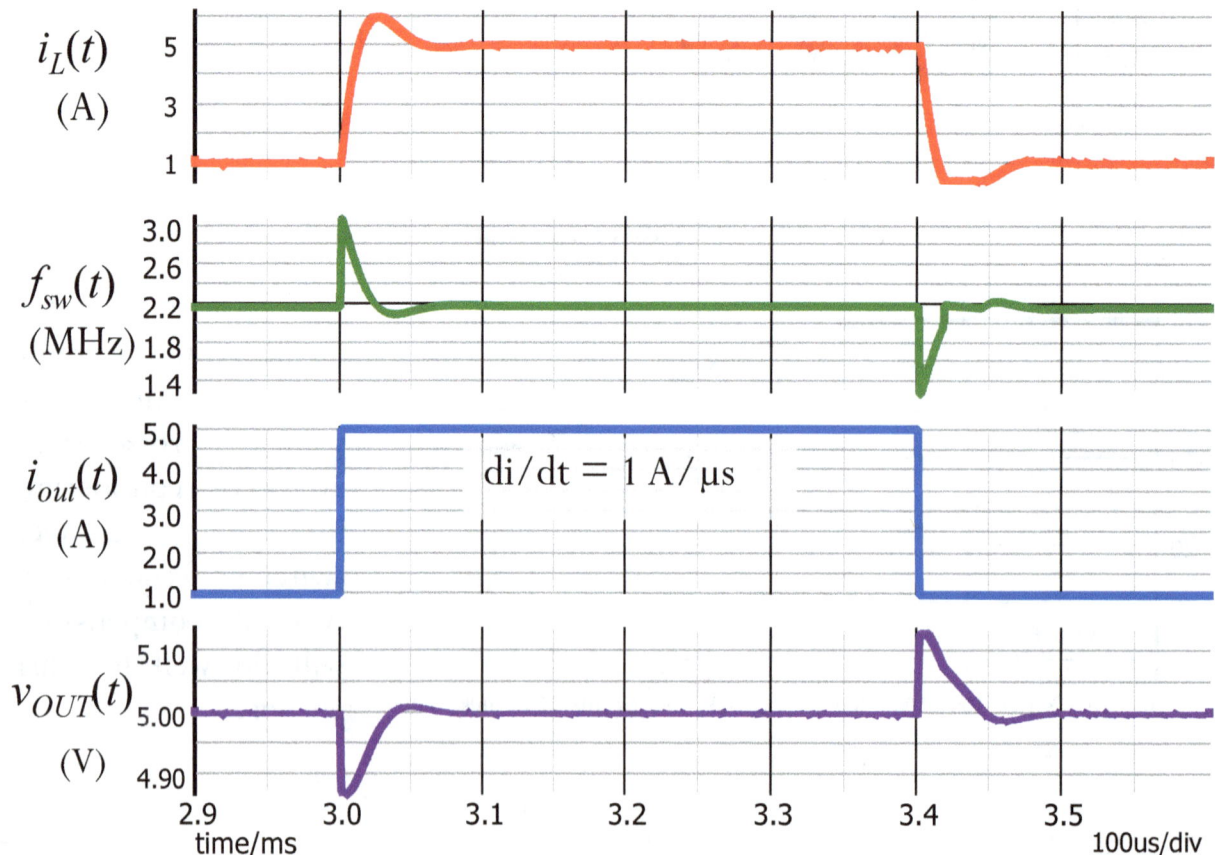

$i_L(t)$ (A)

$f_{sw}(t)$ (MHz)

$i_{out}(t)$ (A)

di/dt = 1 A/µs

$v_{OUT}(t)$ (V)

time/ms

100us/div

The Forward Converter in VM

THE FORWARD CONVERTER belongs to the buck-derived topology: a transformer scales the input voltage V_{in} up or down via its turns ratio N. The dc transfer characteristic thus becomes $V_{out} = NDV_{in}$. A demagnetization means is necessary to reset the transformer core, cycle-by-cycle, and a third winding can be used for that purpose. The below circuit shows a non-isolated single-switch voltage-mode forward converter. The compensation strategy is similar to that adopted for the buck converter.

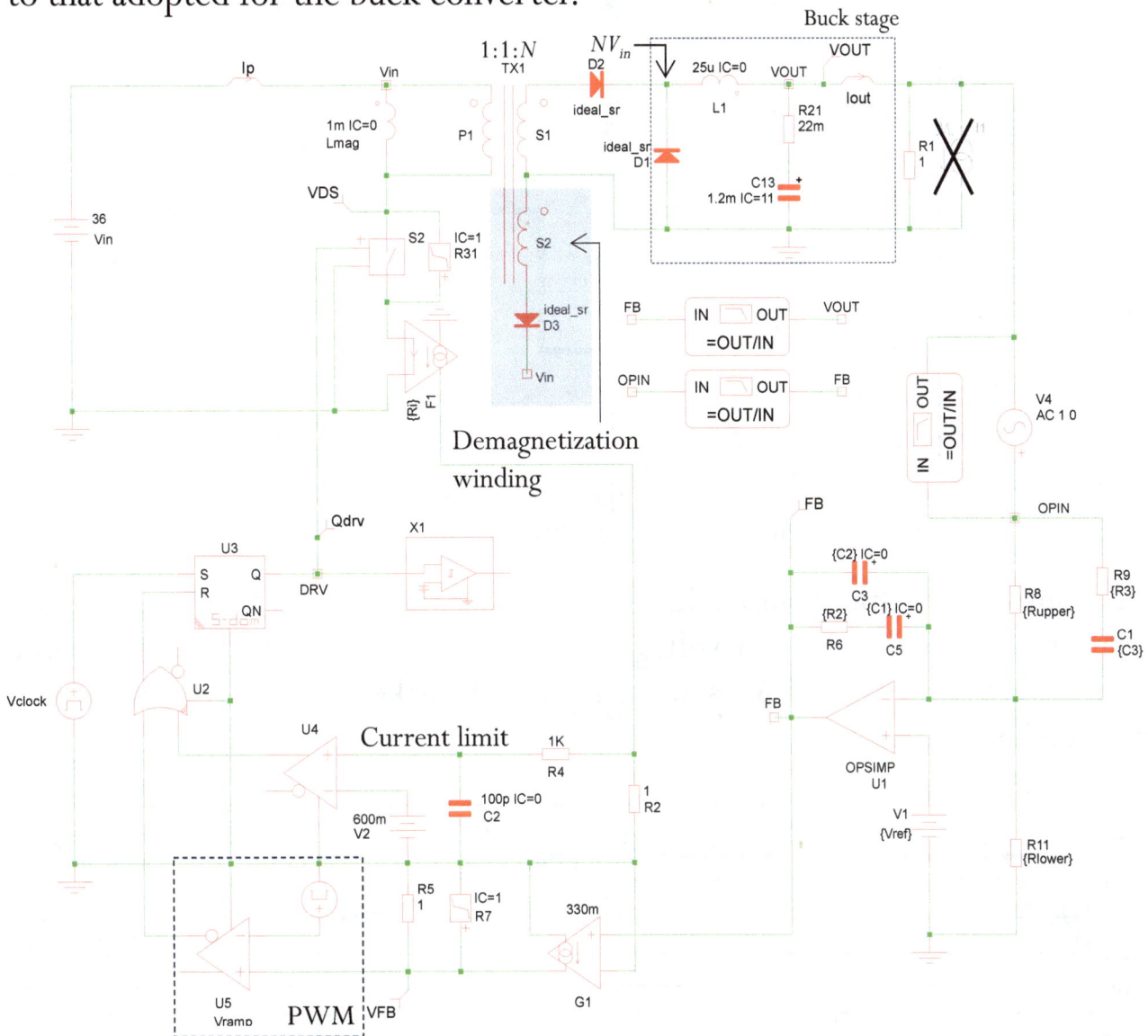

The duty ratio is limited below 50% to offer enough demagnetization time to the transformer. This circuit delivers 12 V / 12 A from a telecom dc bus ranging from 36 to 75 V. The control-to-output transfer function is similar to that of the VM buck converter except for the dc gain: $H_0 = NV_{in}/V_p$

Steady-State Operation

THE SIMULATION is fast and SIMPLIS® delivers the dc operating point together with the control-to-output transfer function in a few seconds. You see how the drain jumps to $2V_{in}$ during the off-time due to the third winding.

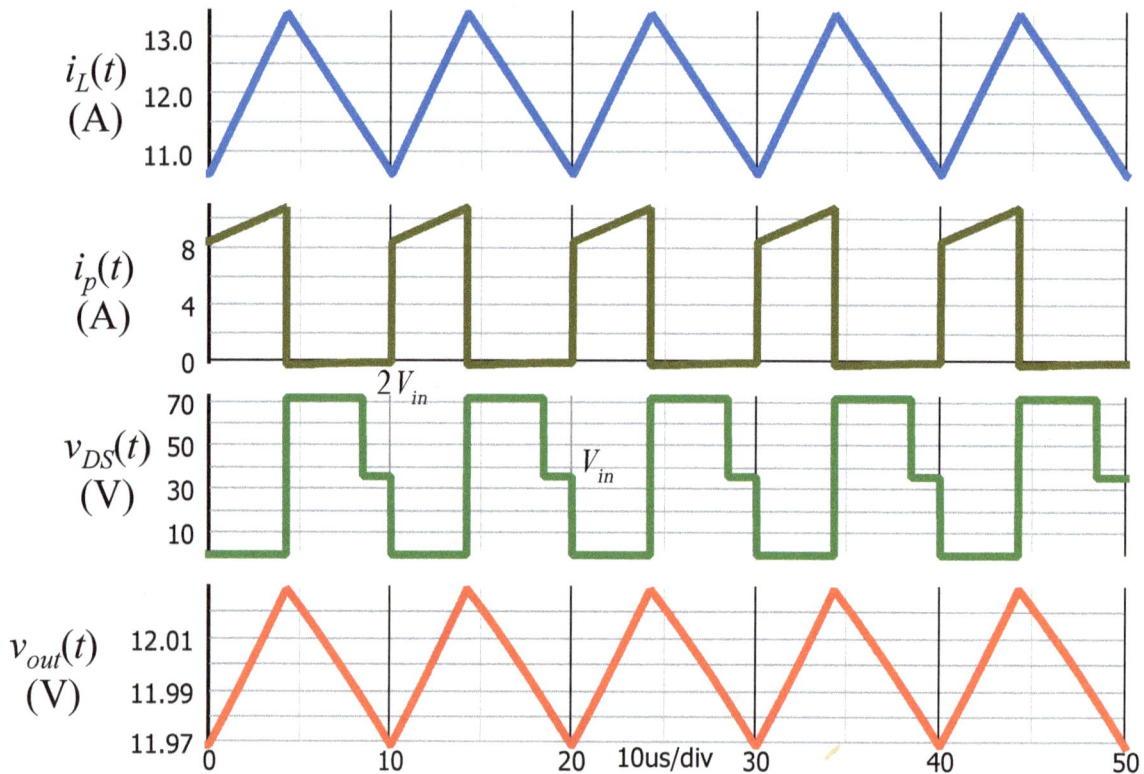

The control-to-output transfer function shows a peak at the LC resonance and a gain varying with the input voltage as with the buck in voltage mode. Some feedforward on the PWM sawtooth would make the power stage gain independent from V_{in}.

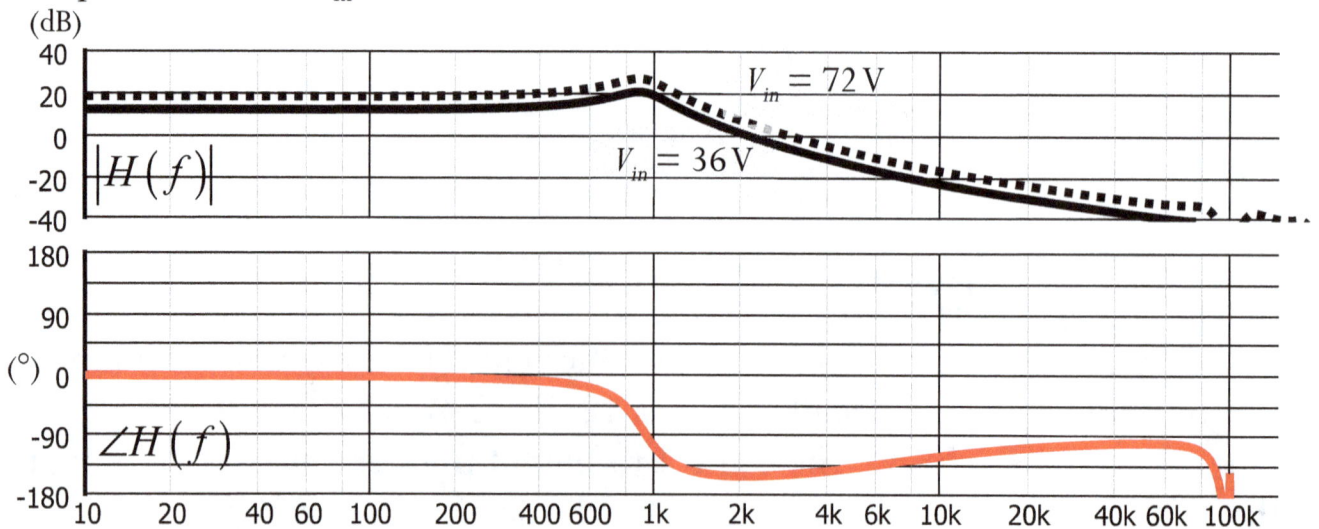

Compensated Gain - Transient Response

WITH A RESONANCE located at 900 Hz, I will place one zero at this frequency and another one at 600 Hz. I can adjust it later at a lower position if conditional stability is a problem when operating in DCM. A pole is installed at 50 kHz ($F_{sw}/2$) while the second is adjusted to meet phase margin at the selected 10-kHz crossover frequency. The macro gives the following values:

```
.VAR fz1=600        .VAR fc=10k * targeted crossover *
.VAR fz2=900        .VAR PM=60 * choose phase margin at crossover *
.VAR fp2=50k
*
* Do not edit the below lines *
.VAR boost=PM-PS-90
.VAR G=10^(-Gfc/20)
.VAR fp1=fc/tan(atan(fc/fz1)+atan(fc/fz2)-atan(fc/fp2)-boost*pi/180)
*
* adjust second pole for targetted boost *
*
.VAR a=sqrt((fc^2/fp1^2)+1)
.VAR b=sqrt((fc^2/fp2^2)+1)
.VAR c=sqrt((fz1^2/fc^2)+1)
.VAR d=sqrt((fc^2/fz2^2)+1)
.VAR R2=((a*b/(c*d))/(fp1-fz1))*Rupper*G*fp1
.VAR C1=1/(2*pi*fz1*R2)
.VAR C2=C1/(C1*R2*2*pi*fp1-1)
.VAR C3=(fp2-fz2)/(2*pi*Rupper*fp2*fz2)
.VAR R3=Rupper*fz2/(fp2-fz2)
.VAR G0=((R2*C1)/(Rupper*(C1+C2)))*c*d/(a*b) * Gain at fc sanity check *
*
```

$V_{out} = 12\,\text{V}$

```
* Rupper = 9500
* Rlower = 2500
* R2 = 38462.7917485801
* R3 = 174.134419551935
* C1 = 6.89648947534561e-09
* C2 = 1.37141196820791e-09
* C3 = 1.82795501890341e-08
* Boost = 90
* Fz1 = 600
* Fz2 = 900
* Fp1 = 3617.25067385445
* Fp2 = 50000
*
```

The loop is compensated for a 10-kHz crossover with a phase margin of 51°: we wanted 60° in the first place but the op-amp GBW affects the compensator response. Crossover must either be reduced or a faster op-amp be selected.

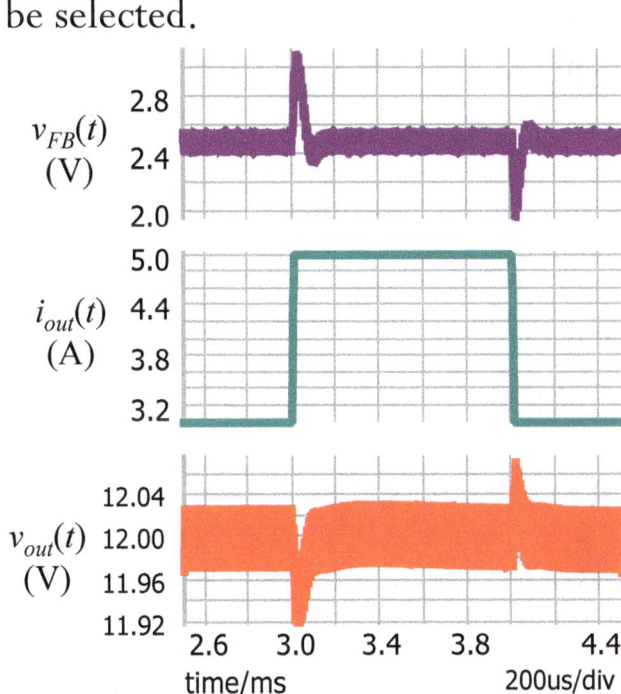

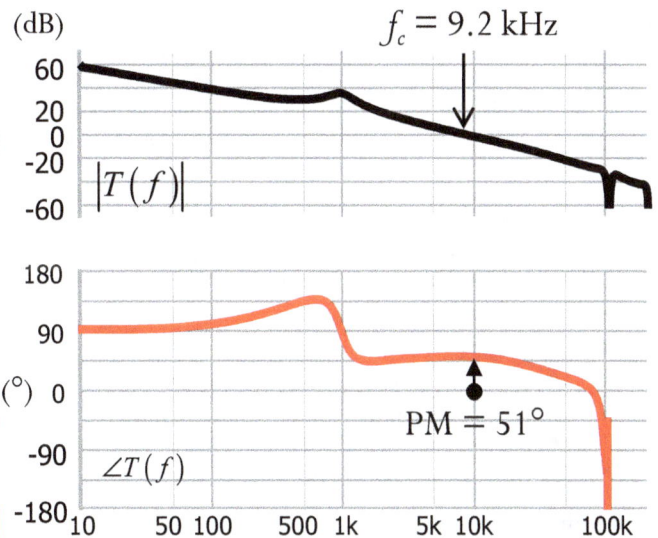

The transient response does not show any oscillations and the converter recovers quickly.

The Forward Converter in CM

WHEN OPERATED IN CURRENT-MODE, the circuit must monitor the primary-side current with a sense resistance or a current-sense transformer. The primary-side current circulating in the power switch at turn on, is made of the reflected secondary-side inductor current to which you add the transformer magnetizing current. This current acts as a stabilizing ramp and you must size its contribution before applying any slope compensation.

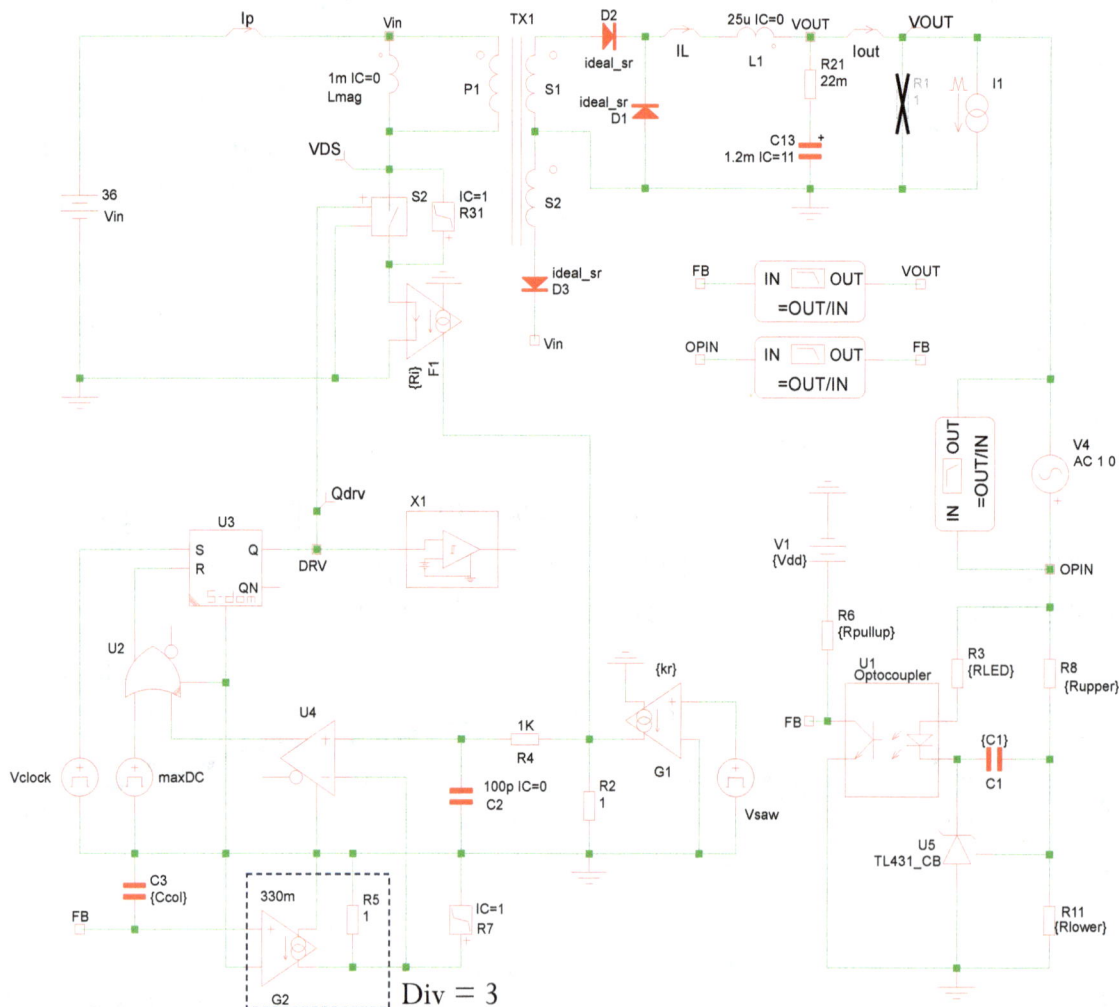

Let's see if this free compensation ramp does help:

$$V_{in} := 36V \quad R_i := 0.04\Omega \quad L_p := 1mH \quad N_1 := 0.8 \quad V_{out} := 12V \quad L_1 := 25\mu H$$

Magnetizing ramp:

$$S_e := \frac{V_{in}}{L_p} \cdot R_i = 1.44\frac{mV}{\mu s} \qquad S_n := \frac{N_1 \cdot V_{in} - V_{out}}{L_1} \cdot N_1 \cdot R_i = 21.504\frac{mV}{\mu s} \qquad m_c := 1 + \frac{S_e}{S_n} = 1.067$$

$$Q := \frac{1}{\pi \cdot \left[m_c \cdot (1 - D) - 0.5 \right]} = 2.601 \quad Q \text{ is still greater than 1 and damping is necessary.}$$

Power Stage AC Response

WITH THE LOW magnetizing current, an extra stabilizing ramp is necessary to reduce the quality factor of the double poles to 1 or below:

$$m_c = \frac{0.818}{1-D} = 1.4 \quad \Longrightarrow \quad S_e = S_n\left(\frac{0.818}{1-D} - 1\right) = 8.6 \text{ mV/μs} \quad \Longrightarrow \quad S_e = 8.6m - 1.44m = 7.2 \text{ mV/μs}$$

Subtract mag. ramp

Compensation ramp

```
.VAR Sn={((((N*Vin-Vout)/L)*N)*Ri}
.VAR Sramp={1/Ts}
*.VAR mc=1 * set this value for ramp comp 1 = no ramp *
*.VAR Se={(mc-1)*Sn}
.VAR Se=7.2k
.VAR kr={Se/Sramp}
*
```

The macro will adjust coefficient k_r to inject the right amount of compensation ramp.

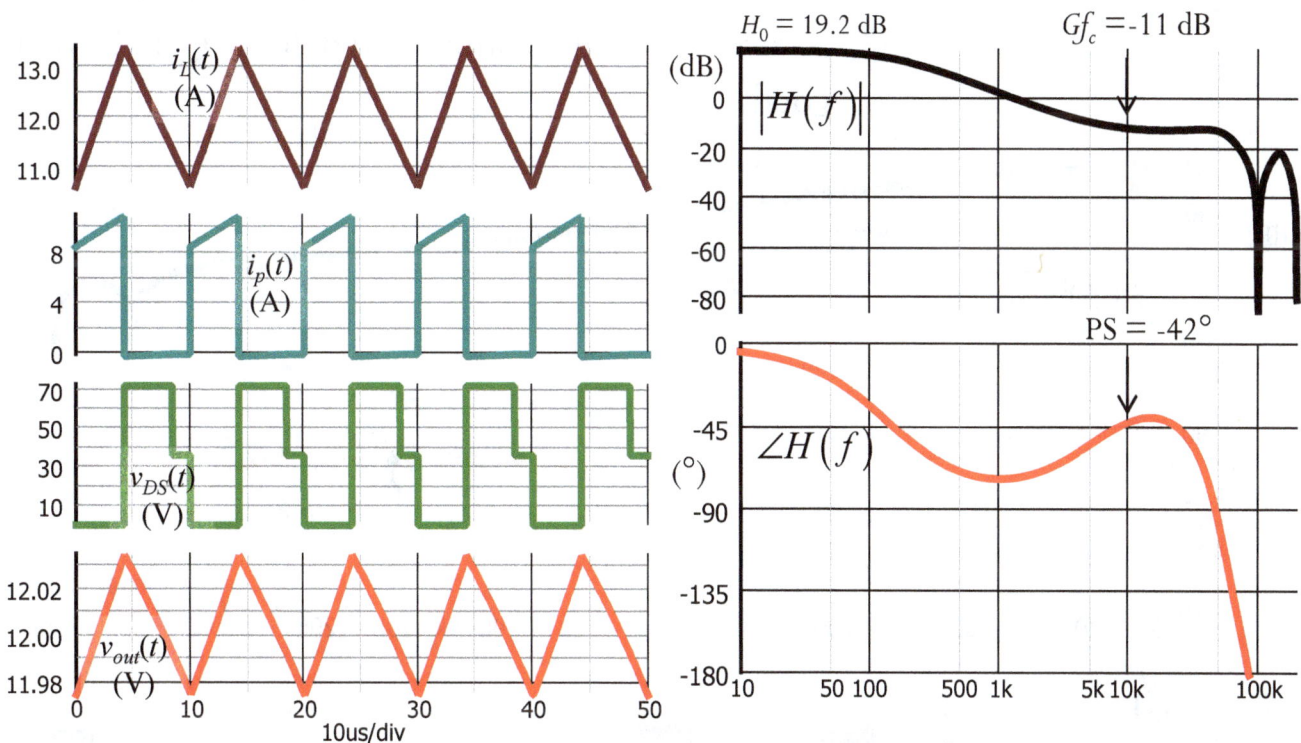

$H_0 = 19.2$ dB $Gf_c = -11$ dB

$\|H(f)\|$

PS $= -42°$

$\angle H(f)$

10μs/div

The control-to-output transfer function is similar to that of the CM buck converter except for the dc gain H_0 which accounts for the turns ratio N and the divider installed between the op-amp and the peak current setpoint:

$$H_0 := \frac{R_{load}}{Div \cdot R_i \cdot N_1} \cdot \frac{1}{1 + \dfrac{R_{load} \cdot T_{sw}}{L_1} \cdot \left[m_c \cdot (1-D) - 0.5\right]} = 9.241 \qquad 20 \cdot \log(H_0) = 19.315$$

The power stage ac plot reveals well-behaved magnitude and phase responses. The double poles at $F_{sw}/2$ are well damped and there is no peaking.

Compensated Gain

CONSIDERING THE ROLE played by the output capacitor ESR in the power stage response, you must sweep its minimum and maximum values then check phase margin in all extreme conditions. Fortunately, there are no gain variations with the input voltage owing to current-mode operation provided CCM is kept. We will select a crossover of 10 kHz for this 100-kHz converter.

```
* Rupper = 9500
* Rlower = 2500
* RLED = 2511.88643150958
* RMAX = 22978.7234042553
* C1 = 2.39259804989311e-09
* C2 = 1.11441490896282e-09
* Gcal = 3.98107170553497
* Boost = 20
* Fz = 7002.0753820971
* Fp = 14281.4800674211
* Sn = 21504
* Se = 7200
* kr = 0.072
*
```

The needed phase boost is modest considering the phase lag of the power stage below 45° and the 60° of phase margin target. The macro automates the process and places the pole and the zero accordingly. The compensated loop gain is very good.

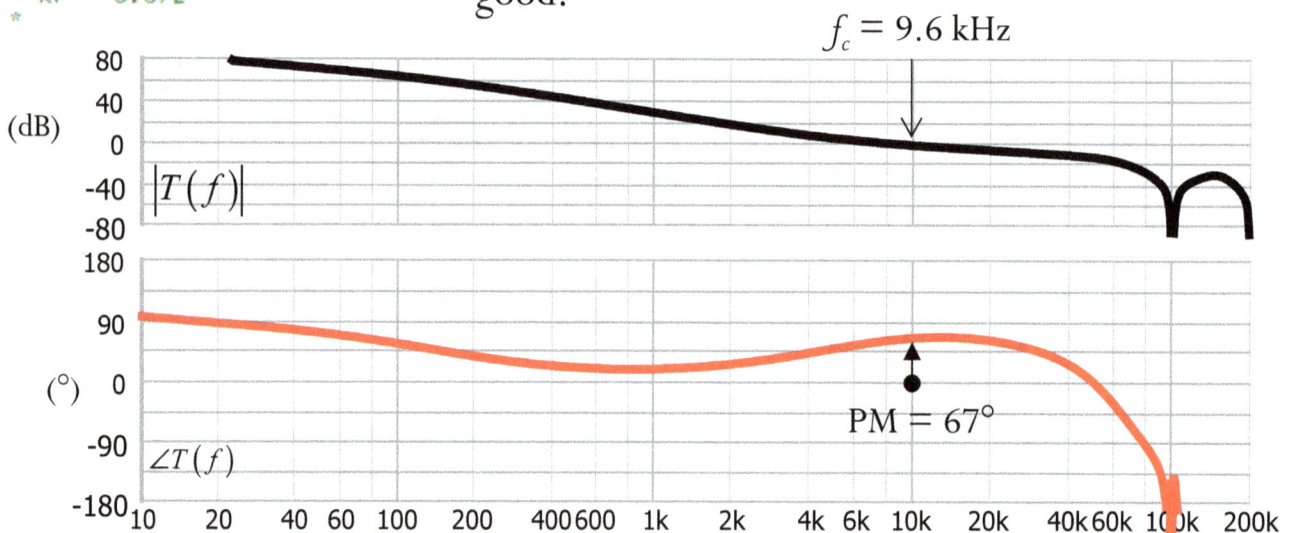

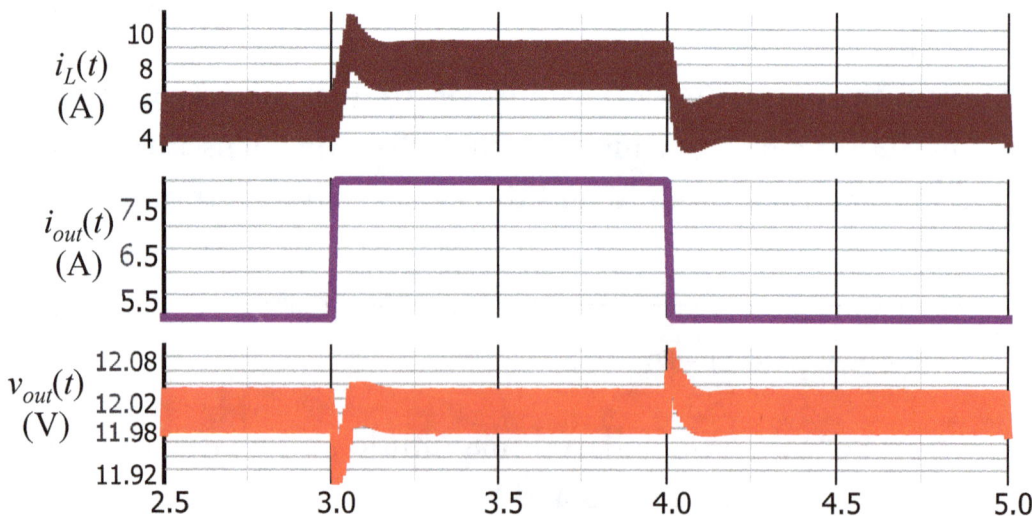

$f_c = 9.6$ kHz

The transient response is fast and stable with a drop of less than 100 mV for a 3-A load step (1 A/µs).

Forward Active Clamp in VM

THE FORWARD CONVERTER operated in active-clamp mode, offers interesting characteristics such as an extended duty ratio range and zero or quasi-zero-voltage switching (ZVS) in certain conditions. It is a popular structure in dc-dc bricks for telecom applications where the input voltage ranges from 36 to 75 V dc.

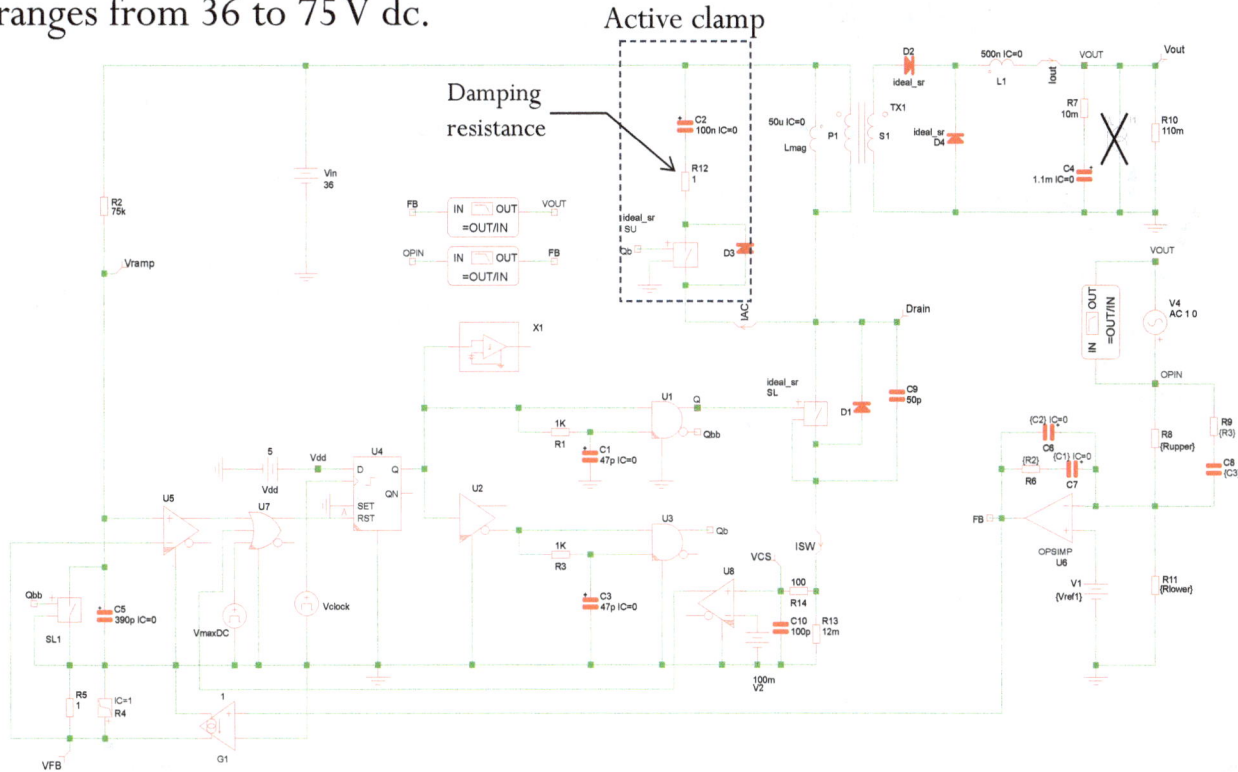

In the above circuit, an extra switch is used to demagnetize the core while pushing the transformer into the third quadrant. This switch can either be a N-channel high-side type (as here) or a ground-referenced type where a P-channel is selected for breakdown levels before 200 V.

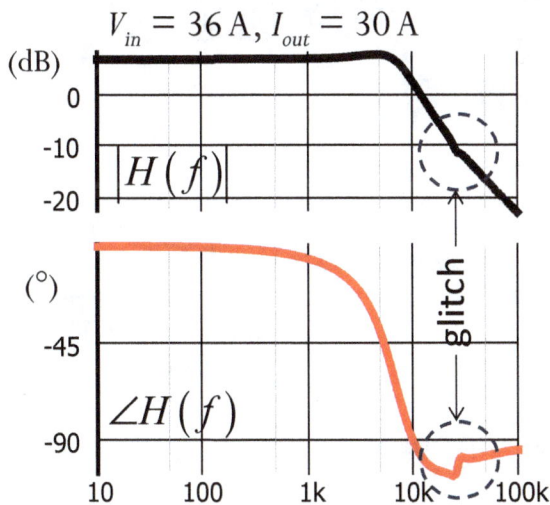

The active-clamp capacitor resonates with the magnetizing inductance. At resonance, you can observe a glitch in the magnitude and phase response. By adding a resistance in series with the clamp capacitor, the phase distortion is minimized and it is possible to select a crossover frequency past this point.

Compensated Gain

FOR THIS DESIGN, we have selected a 20-kHz crossover frequency, considering the *LC* output filter resonating peak at 6.8 kHz. A type 3 compensator is used for achieving a decent phase margin. Needless to say that a good op-amp must be selected with such high crossover goal. I remember successfully testing the LM8261 which offers a GBW of 15 MHz (minimum across temperature).

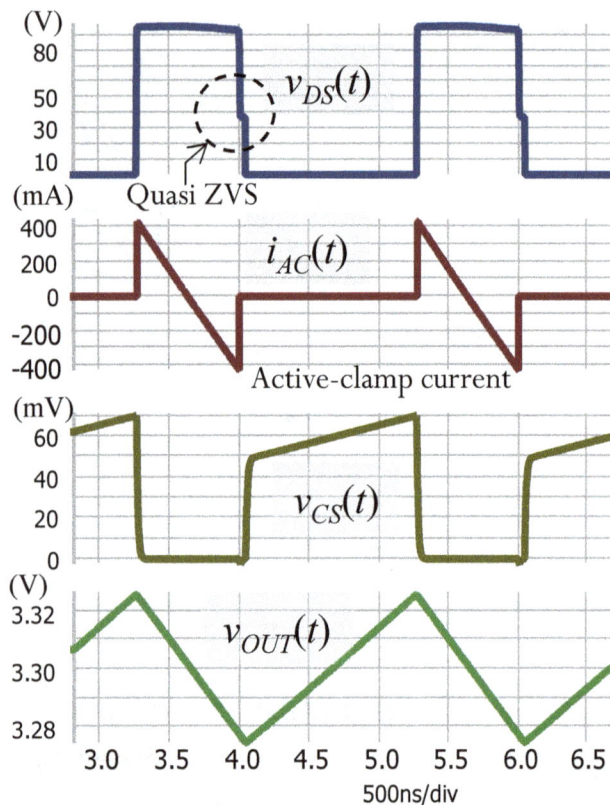

$V_{in} = 36\,\text{A},\ I_{out} = 30\,\text{A}$

$f_c = 20\,\text{kHz}$

$|T(f)|$

$\angle T(f)$

PM = 61°

```
*
.VAR Vin=36
.VAR Vout=3.3
.VAR L=500n
.VAR Ri=160m
*

*
.VAR fz1=6k
.VAR fz2=6k
.VAR fp2=250k
*
* Do not edit the below lines *
.VAR boost=PM-PS-90
.VAR G=10^(-Gfc/20)
.VAR fp1=fc/tan(atan(fc/fz1)+atan(fc/fz2)-atan(fc/fp2)-boost*pi/180)
*
```

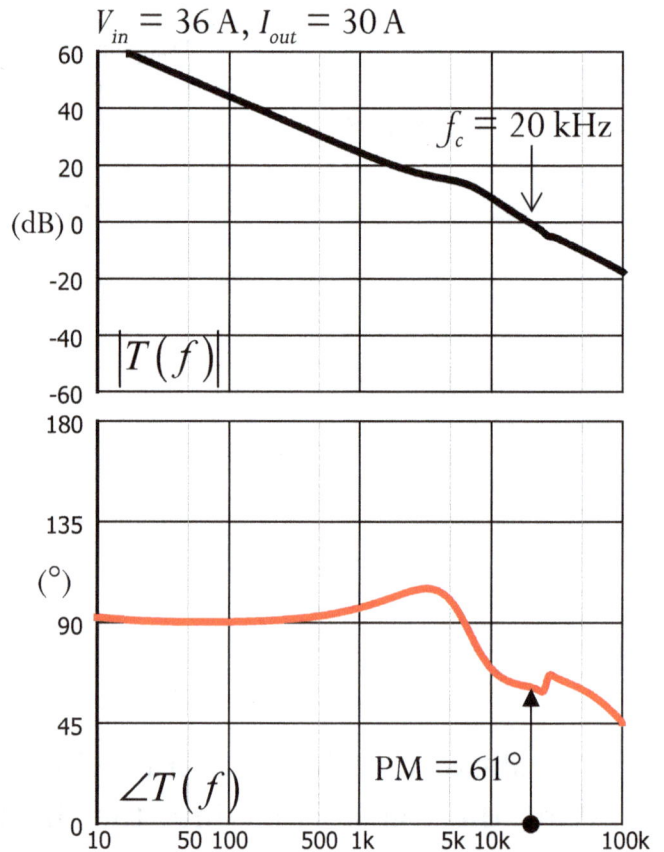

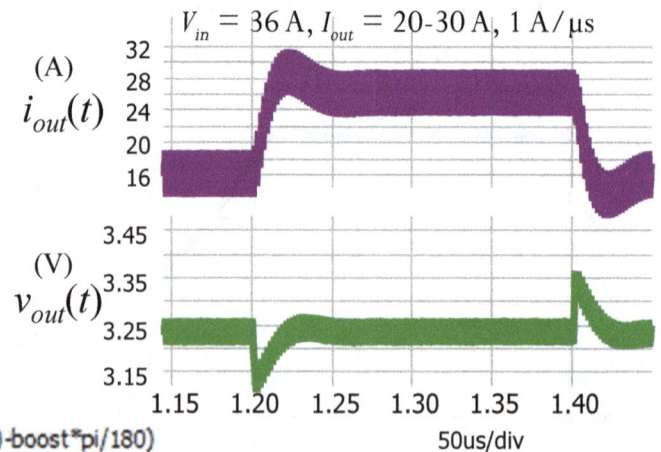

```
* Boost = 80
* Fz1  = 6000
* Fz2  = 6000
* Fp1  = 10621.8381956715
* Fp2  = 250000
```

$V_{in} = 36\,\text{A},\ I_{out} = 20\text{-}30\,\text{A},\ 1\,\text{A}/\mu\text{s}$

$i_{out}(t)$

$v_{out}(t)$

The transient response is excellent when the output current is stepped from 20 to 30 A at a 36- or 72-V input voltage.

Full-Bridge Converter in CM

ALSO BELONGING to the buck-derived family, this converter lets you deliver power levels beyond the kW. The system implements two half-bridges and requires adequate drivers for the high-side transistors. A drive transformer or a bootstrapped structure can be used for this purpose. Below is a current-mode version delivering 5 V / 100 A from a 24-V source.

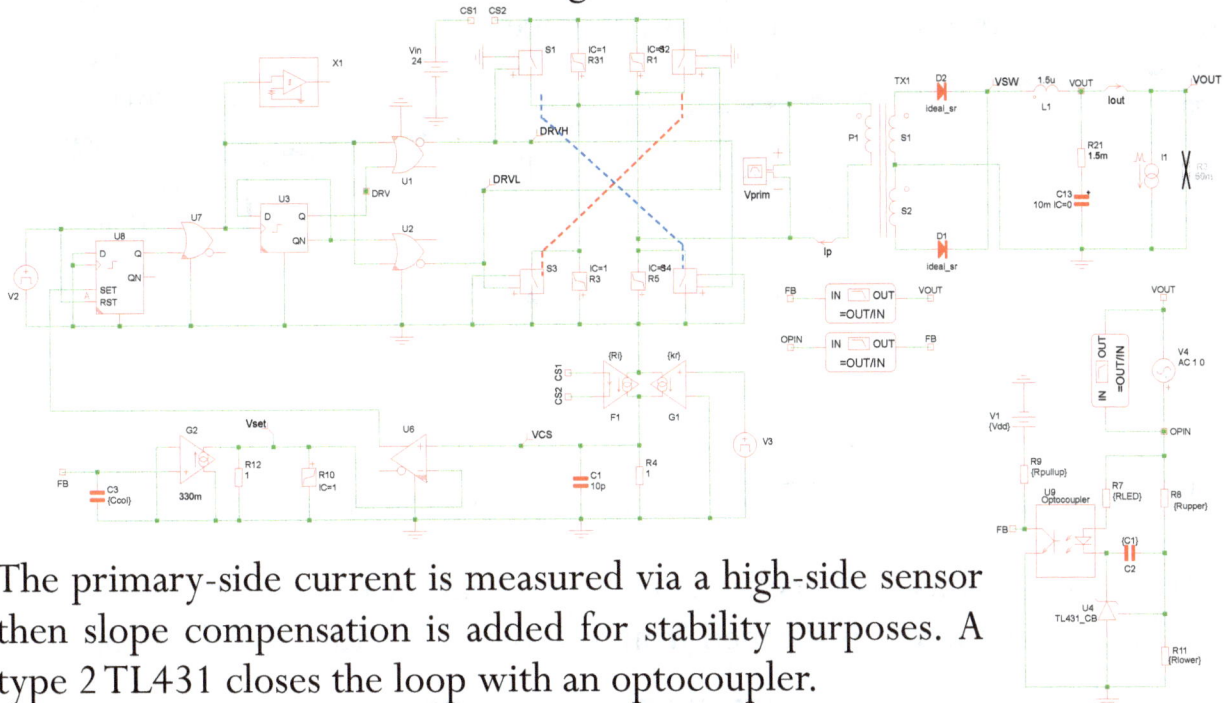

The primary-side current is measured via a high-side sensor then slope compensation is added for stability purposes. A type 2 TL431 closes the loop with an optocoupler.

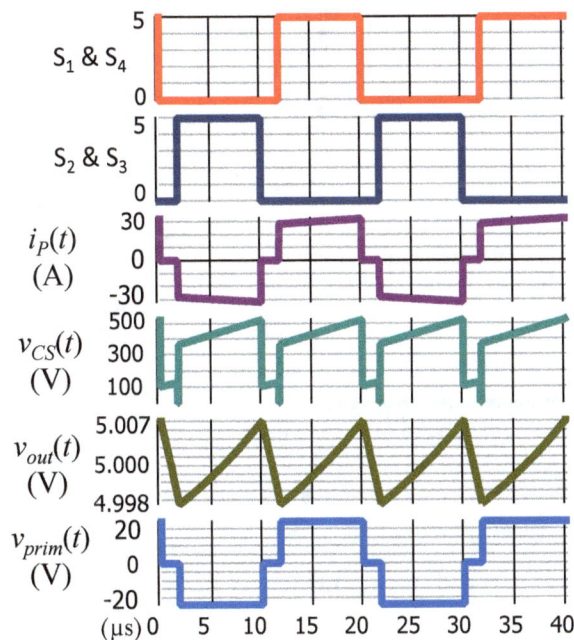

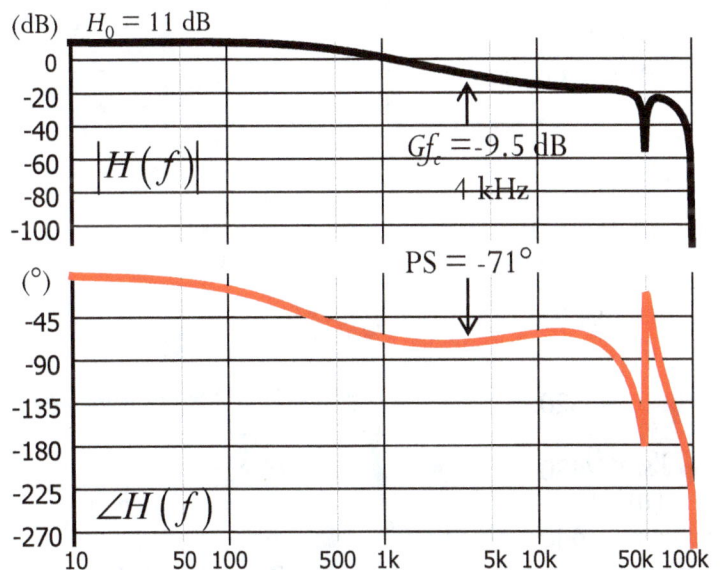

The control-to-output transfer function is well behaved and eases the compensation process.

Compensation and Transient Response

THE AC RESPONSE of the full-bridge converter is very similar to that of the buck converter. The compensation strategy is simple here and we will place a zero and pole on both sides of the crossover frequency (4 kHz). I did not evaluate it in this example, but the magnetizing current also circulates and damps the subharmonic poles. You could thus fine-tune the amount of injected compensation ramp to account for its presence.

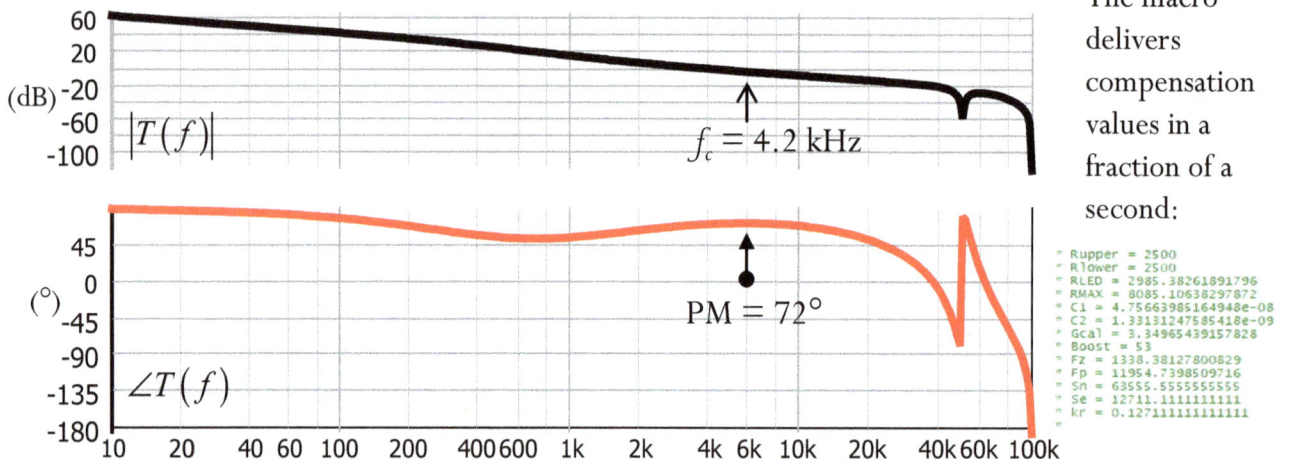

The macro delivers compensation values in a fraction of a second:

```
* Rupper = 2500
* Rlower = 2500
* RLED = 2985.38261891796
* RMAX = 8085.10638297872
* C1 = 4.75663985164948e-08
* C2 = 1.33131247585418e-09
* Gcal = 3.34965439157828
* Boost = 53
* Fz = 1338.38127800829
* Fp = 11954.7398509716
* Sn = 63555.5555555555
* Se = 12711.1111111111
* kr = 0.127111111111111
*
```

The graph shows $|T(f)|$ in dB and $\angle T(f)$ in degrees versus frequency, with $f_c = 4.2$ kHz and $PM = 72°$.

The transient response is excellent and the deviation remains at a minimum when the output current is stepped from 80 to 100 A with a 1-A/µs slope.

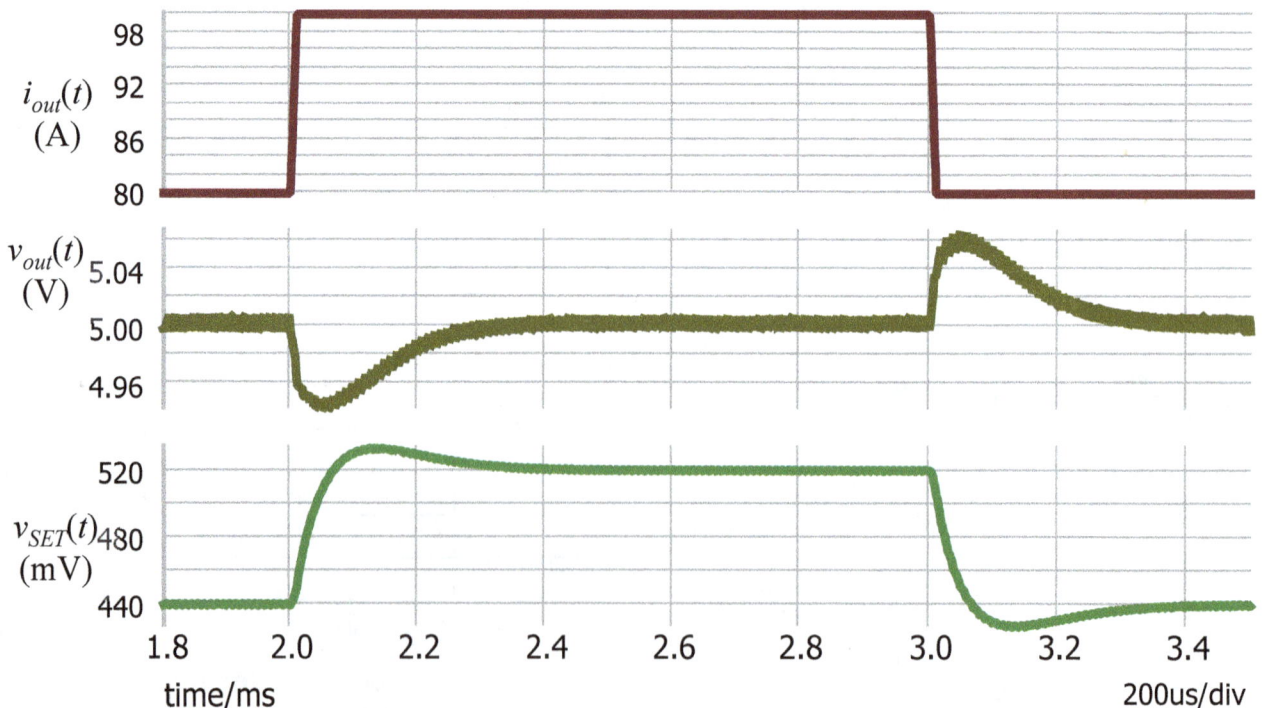

The graph shows $i_{out}(t)$ (A), $v_{out}(t)$ (V), and $v_{SET}(t)$ (mV) versus time/ms at 200us/div.

Phase-Shifted Full-Bridge in VM

THE PSFB converter operates in a full-bridge configuration where each leg is operated at 50% duty ratio and delivers a square-wave at its middle point. When the two waveforms entirely overlap, the differential voltage applied at the transformer is zero. It is when the two signals start shifting apart that power is transmitted to the secondary side. The series inductor helps meeting zero-voltage switching (ZVS) when sufficient energy is stored cycle-by-cycle.

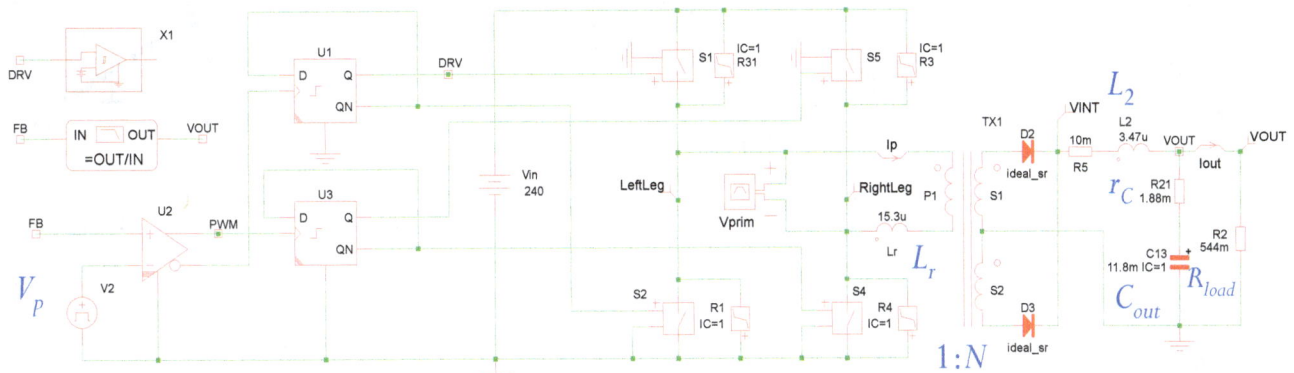

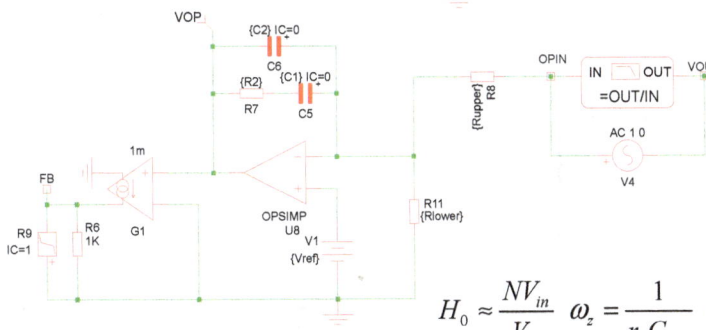

$$H(s) = H_0 \frac{1 + \dfrac{s}{\omega_z}}{1 + \dfrac{s}{\omega_0 Q} + \left(\dfrac{s}{\omega_0}\right)^2} e^{-s\Delta t} \quad \text{Delay}$$

$$H_0 \approx \frac{N V_{in}}{V_p} \quad \omega_z = \frac{1}{r_C C_{out}} \quad Q = \frac{(R_{load} + r_d)\sqrt{\dfrac{C_{out} L_2 R_{load}}{R_{load} + r_d}}}{L_2 + C_{out} R_{load} r_d} \quad r_d = 2N^2 L_r F_{sw}$$

$$\omega_0 = \frac{1}{\sqrt{\dfrac{C_{out} L_2 R_{load}}{R_{load} + r_d}}}$$

The voltage that appears at the cathodes junction in the secondary side is reduced in duration because of the primary-side inductance presence. The inductance helps obtaining ZVS but it affects the *effective* duty ratio transmitting the power to the load. It has the effect of delaying the information by an amount Δt, included in the exponential term.

$$\Delta D := \frac{N_1 \left[2 \cdot I_{out} - \dfrac{1}{F_{sw}} \cdot \dfrac{V_{out}}{L_1} \cdot (1 - d_1) \right]}{\dfrac{V_{in}}{L_{lk}} \cdot \dfrac{1}{F_{sw}}} \qquad \Delta t := \Delta D \cdot T_{sw}$$

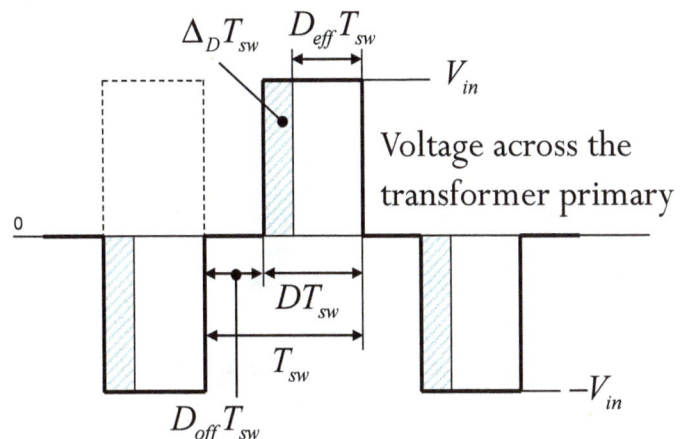

Voltage across the transformer primary

Operating Point and AC Response

AFTER SIMULATION is complete, you see the typical primary-side current signature on top with the magnetization and demagnetization cycles of the series inductance in the primary side. The effect of the delay is clearly visible as highlighted in the graph. The output voltage is perfectly regulated to 12 V.

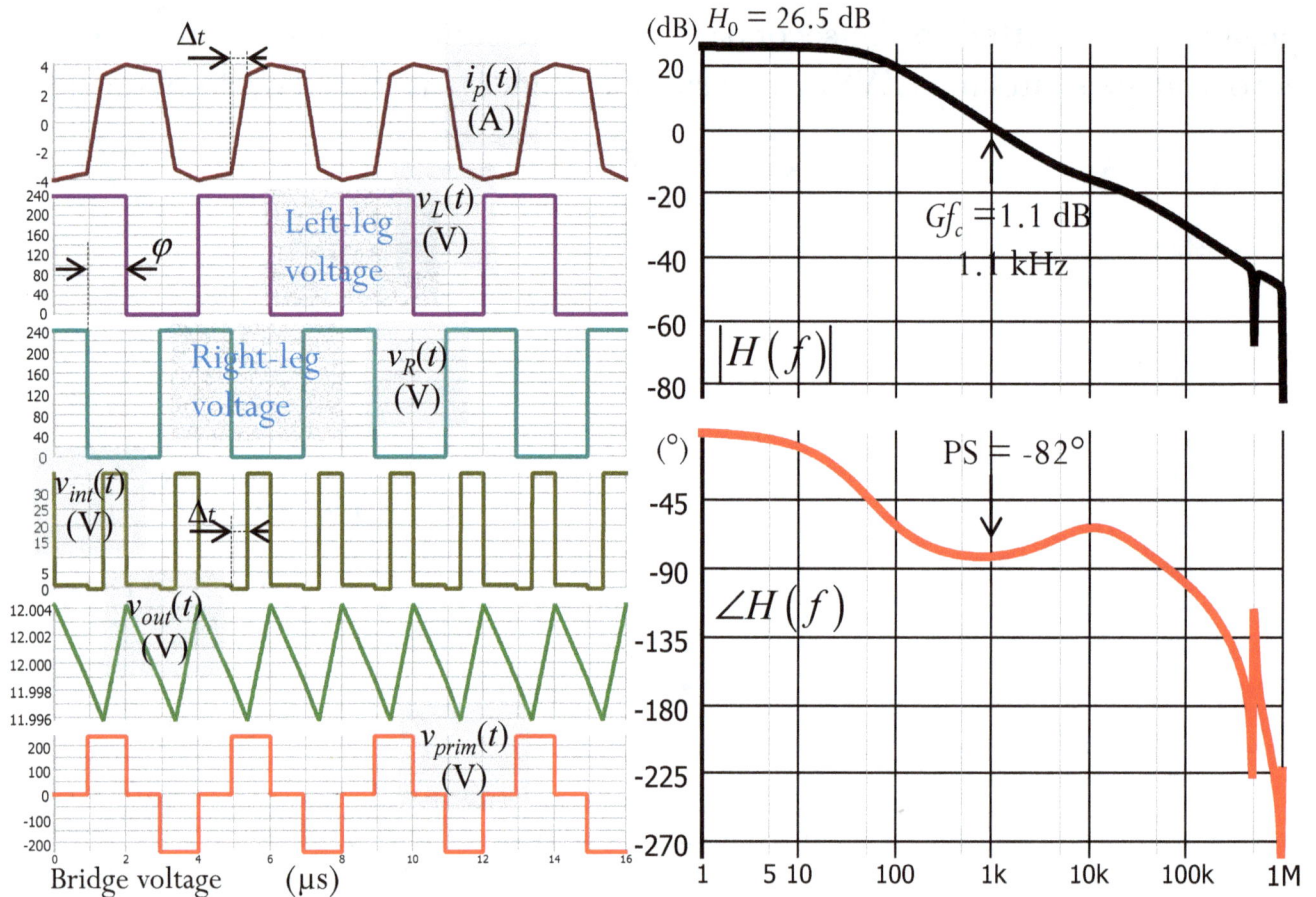

The transfer function is actually very well behaved and you can notice the absence of peaking despite a voltage-mode control. This is because the series inductance provides additional damping through the r_d term. There are no particular difficulties in closing the loop. At 1 kHz, the phase shift is 82° and a type 2 compensator will be perfectly suited for this dc-dc converter:

```
*
.VAR Gfc=1.1 * magnitude at crossover *
.VAR PS=-82 * phase lag at crossover *
*
* Enter Design Goals Information Here *
*
.VAR fc=1k * targetted crossover *
.VAR PM=60 * choose phase margin at crossover *
*
```

```
* Rupper = 9500
* Rlower = 2500
* k = 2.90421087767582
* R2 = 9495.80072081567
* C2 = 6.5473925307118e-09
* C1 = 4.86762023080509e-08
* Boost = 52
* Fz = 344.327613289665
* Fp = 2904.21087767582
* ROL = 316227766.016838
*
```

Compensation and Transient Response

THE COMPENSATED LOOP gain is shown below and confirms the selected crossover frequency with a robust phase margin. The transient response is also very good and does not show any excessive overshoot after the step load release. The selected example implements an op-amp but an isolated version featuring an optocoupler is of course possible. The PSFB finds application in high-power converters.

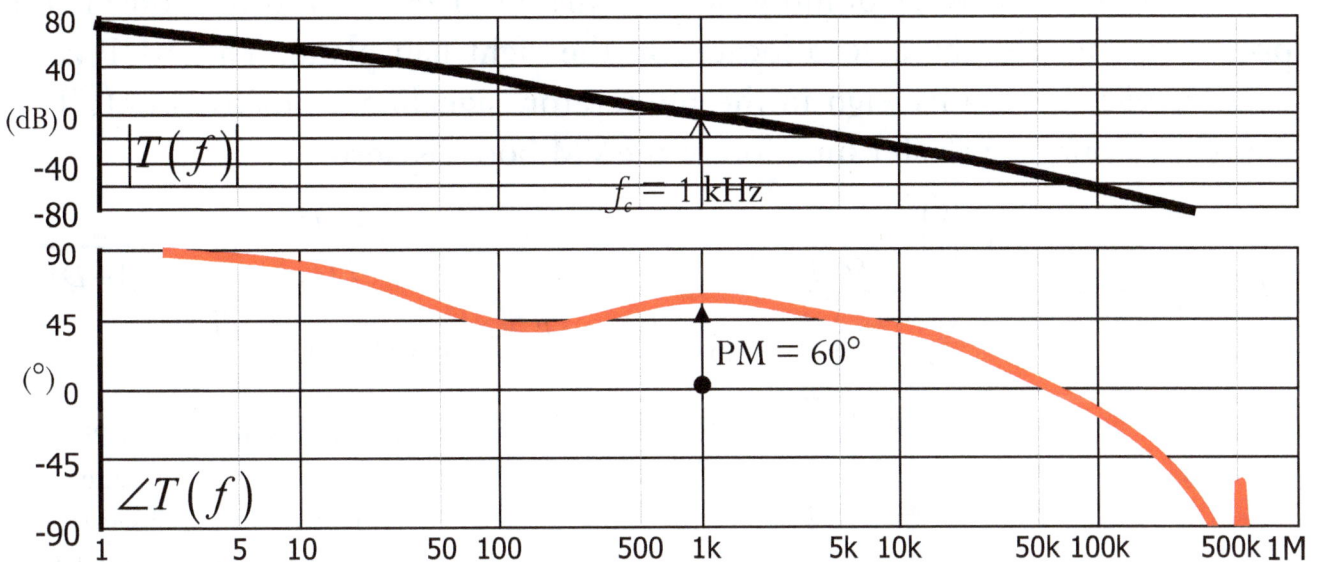

Transient response to a 7-A load step with a 1-A/μs slope.

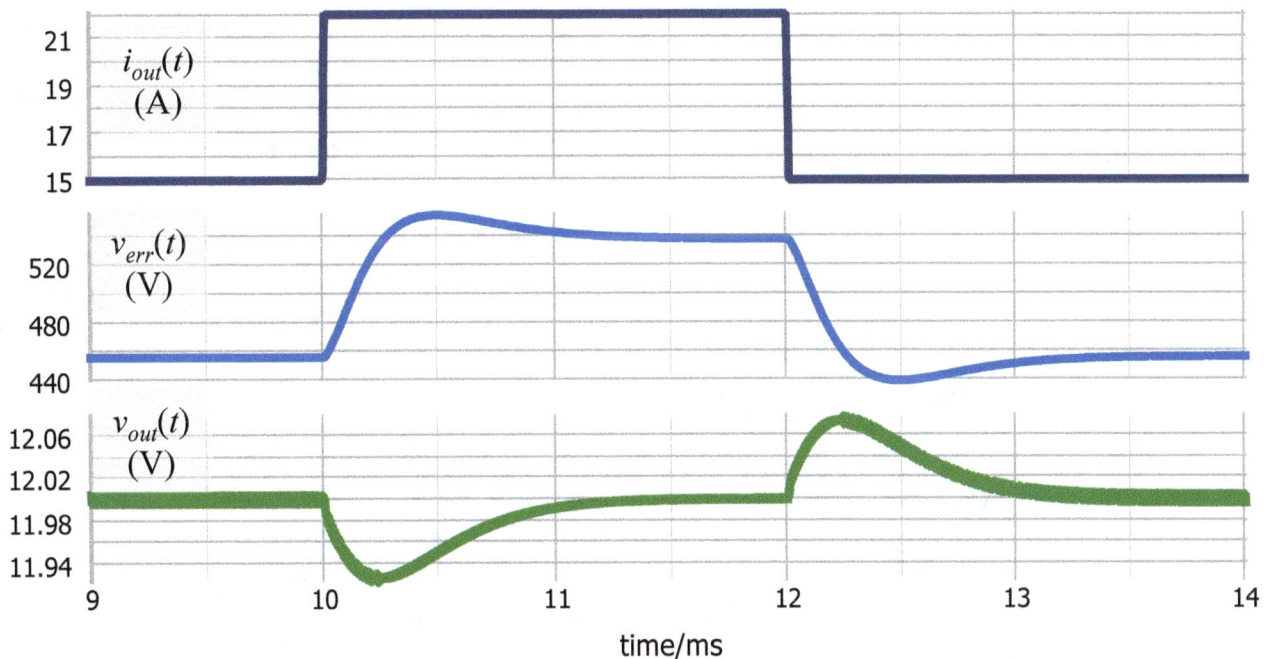

The Boost Converter in VM

A BOOST CONVERTER elevates the input voltage by first storing energy in the inductor placed in series with the source. It then releases it to the load during the demagnetization phase. Any sudden output power increase cannot thus be instantaneously answered, as more energy needs first to be stored in this two-step conversion process. The inductance value and the available volt-seconds severely limit the speed at which the current in the inductor can vary and it delays the delivery of more output current. This delay in the conversion process is modeled by a zero located in the right half-plane. This RHP zero appears with a negative sign in the numerator, signaling a positive root. The control-to-output transfer function of the VM boost is here:

$$H(s) = \frac{V_{out}(s)}{V_{err}(s)} = H_0 \frac{\left(1 + \dfrac{s}{\omega_{z_1}}\right)\left(1 - \dfrac{s}{\omega_{z_2}}\right)}{1 + \dfrac{s}{\omega_0 Q} + \left(\dfrac{s}{\omega_0}\right)^2}$$

$$\omega_{z_1} = \frac{1}{r_C C_{out}} \qquad \omega_0 \approx \frac{1-D}{\sqrt{L_1 C_{out}}} \qquad H_0 \approx \frac{V_{in}}{V_p (1-D)^2}$$

$$\omega_{z_2} = \frac{(1-D)^2 R_{load}}{L_1} \qquad Q \approx (1-D) R_{load} \sqrt{\frac{C_{out}}{L_1}}$$

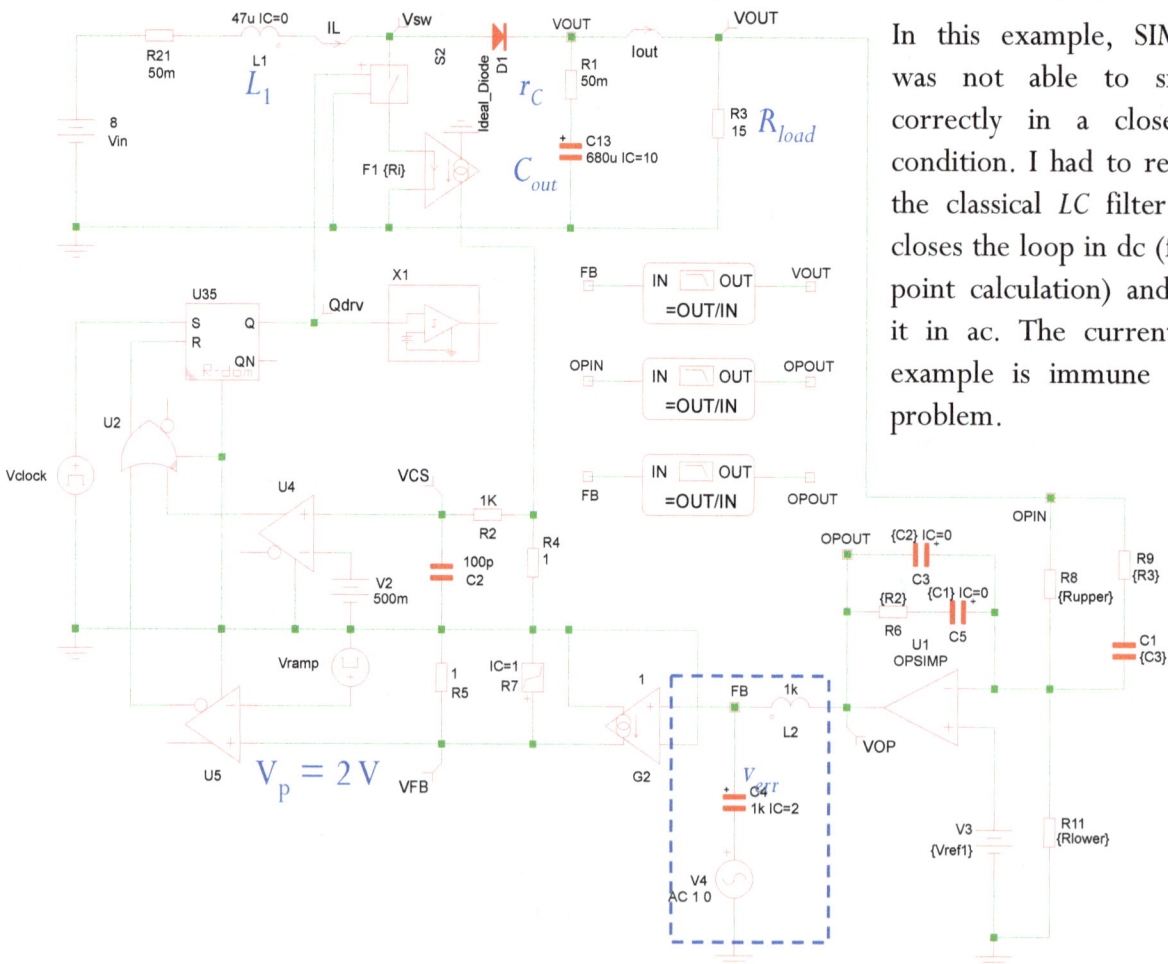

In this example, SIMPLIS® was not able to simulate correctly in a closed-loop condition. I had to resort to the classical *LC* filter which closes the loop in dc (for bias point calculation) and opens it in ac. The current-mode example is immune to this problem.

Compensating the Boost VM

THE BOOST CONVERTER operated in voltage-mode control exhibits a peaky 2nd-order response whose resonant frequency f_o varies with the operating conditions. Added to the RHP zero phase lag, it sets two limits for selecting a crossover frequency. Here, at the lowest input voltage of our example (8 V), the resonance occurs at 466 Hz while the RHPZ is located at 14 kHz:

$$r_L := 0.05\Omega \qquad r_C := 0.05\Omega \qquad C_2 := 680\mu F \qquad L_1 := 47\mu H \qquad V_{out} := V_{in} \cdot \frac{1}{1 - D_0} = 15.267V$$

$$V_{in} := 8V \qquad V_p := 1V \qquad D_0 := 47.6\% \qquad R_L := 15\Omega \qquad H_0 := \frac{V_{in}}{V_p \cdot (1 - D_0)^2} = 29.136 \xrightarrow{\text{dc gain (dB)}}$$

$$20 \cdot \log(H_0) = 29.289$$

$$\omega_{z1} := \frac{1}{r_C \cdot C_2} \qquad f_{z1} := \frac{\omega_{z1}}{2 \cdot \pi} = 4.681 \cdot kHz \qquad \boxed{\omega_{z1} := \frac{1}{r_C \cdot C_2}} \qquad f_{z1} \approx 4.7 \text{ kHz} \quad \text{LHP zero}$$

$$\omega_{z2} := \frac{R_L \cdot (1 - D_0)^2}{L_1} \qquad f_{z2} := \frac{\omega_{z2}}{2 \cdot \pi} = 13.947 kHz \qquad \boxed{\omega_{z2} := \frac{R_L \cdot (1 - D_0)^2}{L_1}} \implies \boxed{f_{z2} \approx 14 \text{ kHz}} \quad \text{RHP zero}$$

$$Q := (1 - D_0) \cdot R_L \cdot \sqrt{\frac{C_2}{L_1}} = 29.897 \qquad \omega_0 := \frac{1 - D_0}{\sqrt{L_1 \cdot C_2}} \qquad \boxed{f_0 := \frac{\omega_0}{2 \cdot \pi} = 466.496 Hz}$$

Slow-down the converter to allow the inductor current to build up.

$$\implies \boxed{3 \cdot f_0 < f_c < 0.2 \cdot f_{z_2} \to f_c = 2 \text{ kHz}}$$

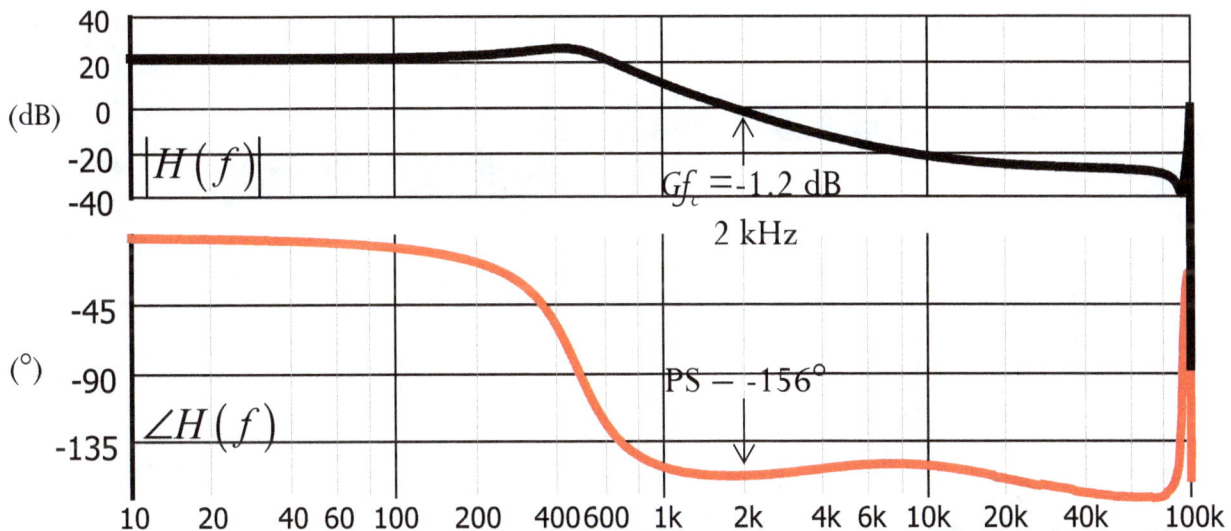

In this compensation exercise, the phase lag is quite pronounced at 2 kHz and a type 3 compensator is necessary. This control-to-output response includes the PWM gain which is -6 dB in this example (2-V peak).

Operating Point and Compensated Gain

THE OPERATING POINT confirms a duty ratio at 47% and an output voltage regulated at 15 V with a 1-A current as expected.

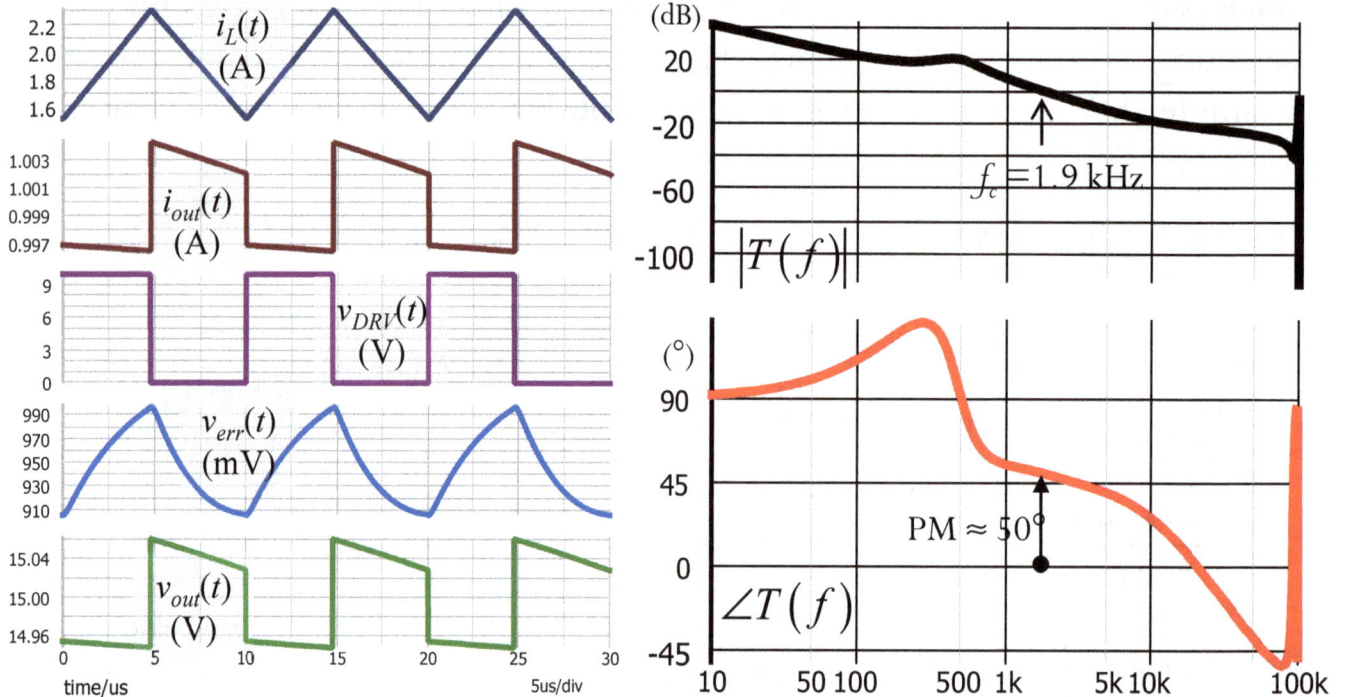

The adopted compensation values lead to ≈2-kHz crossover value with a 50° phase margin. The below graph shows the transient response with a 0.5-A step load in 1 A/µs.

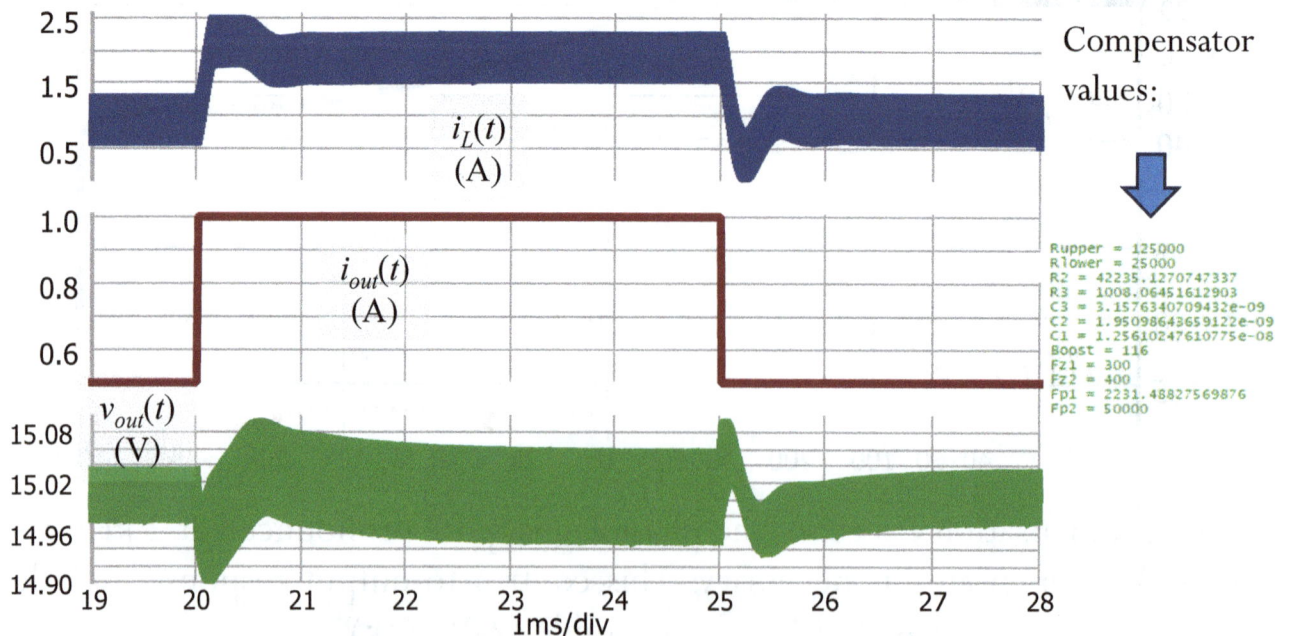

Compensator values:

```
Rupper = 125000
Rlower = 25000
R2 = 42235.1270747337
R3 = 1008.06451612903
C3 = 3.15763407094320e-09
C2 = 1.95098643659122e-09
C1 = 1.25610247610775e-08
Boost = 116
Fz1 = 300
FZ2 = 400
Fp1 = 2231.48827569876
Fp2 = 50000
```

The Boost Converter in CM

WHEN OPERATED in current-mode control, the boost converter ac response loses its second-order peak but the RHP zero remains. As a result, you cannot push the crossover too far. The transfer function is now described by a 3rd-order polynomial and hosts two subharmonic poles at $F_{sw}/2$.

$$H(s) = H_0 \frac{\left(1 + \dfrac{s}{\omega_{z_1}}\right)\left(1 - \dfrac{s}{\omega_{z_2}}\right)}{1 + \dfrac{s}{\omega_{p_1}}} \frac{1}{1 + \dfrac{s}{\omega_n Q_p} + \left(\dfrac{s}{\omega_n}\right)^2} \qquad H_0 \approx \frac{R_{load}}{R_i} \frac{1}{2M + \dfrac{1}{\tau_L M^2}\left(\dfrac{1}{2} + \dfrac{S_e}{S_n}\right)}$$

$$\omega_{z_1} = \frac{1}{r_C C_{out}}$$

$$\omega_{z_2} \approx \frac{(1-D)^2 R_{load}}{L_1}$$

$$\omega_{p_1} \approx \frac{\dfrac{2}{R_{load}} + \dfrac{T_{sw}}{L_1 M^3}\left(1 + \dfrac{S_e}{S_n}\right)}{C_{out}} \qquad Q_p = \frac{1}{\pi(m_c D' - 0.5)} \qquad \omega_n = \frac{\pi}{T_{sw}} \qquad m_c = 1 + \frac{S_e}{S_n} \Leftarrow S_n = \frac{V_{in}}{L_1} R_i$$

External slope [V]/[s]

$$\tau_L = \frac{L_1}{R_{load} T_{sw}} \qquad M = \frac{V_{out}}{V_{in}}$$

The current sense could classically be done with a resistance, but I preferred to use a current source instead as it is easier to combining with the external compensation ramp S_e in SIMPLIS®.

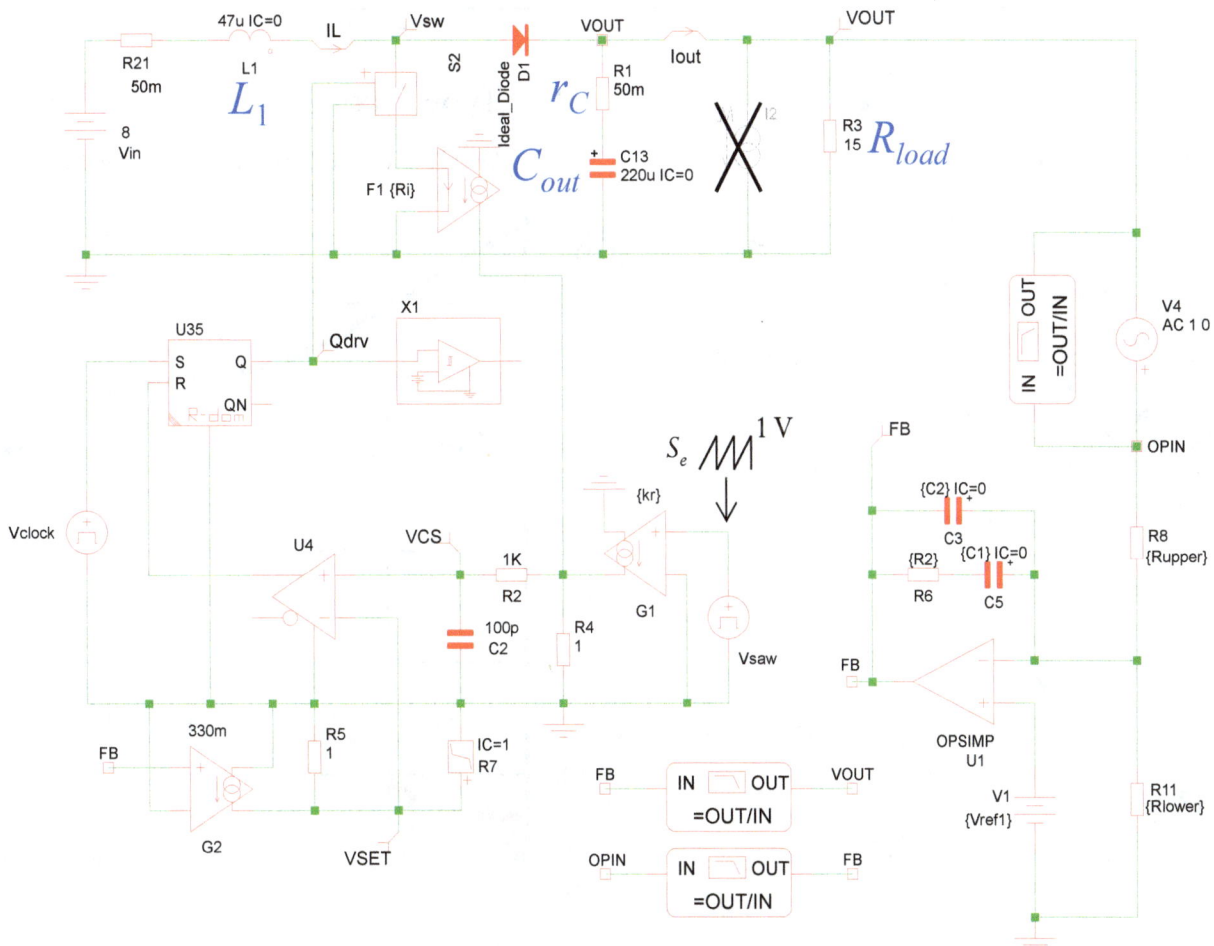

Power Stage AC Response

COMPENSATING THE BOOST converter in CM is simpler than in VM, as you no longer have the variable peaking in the control-to-output transfer function. Despite the disappearance of the peak, the crossover frequency is still limited below 20% of the lowest RHP zero which implies a 2-kHz f_c.

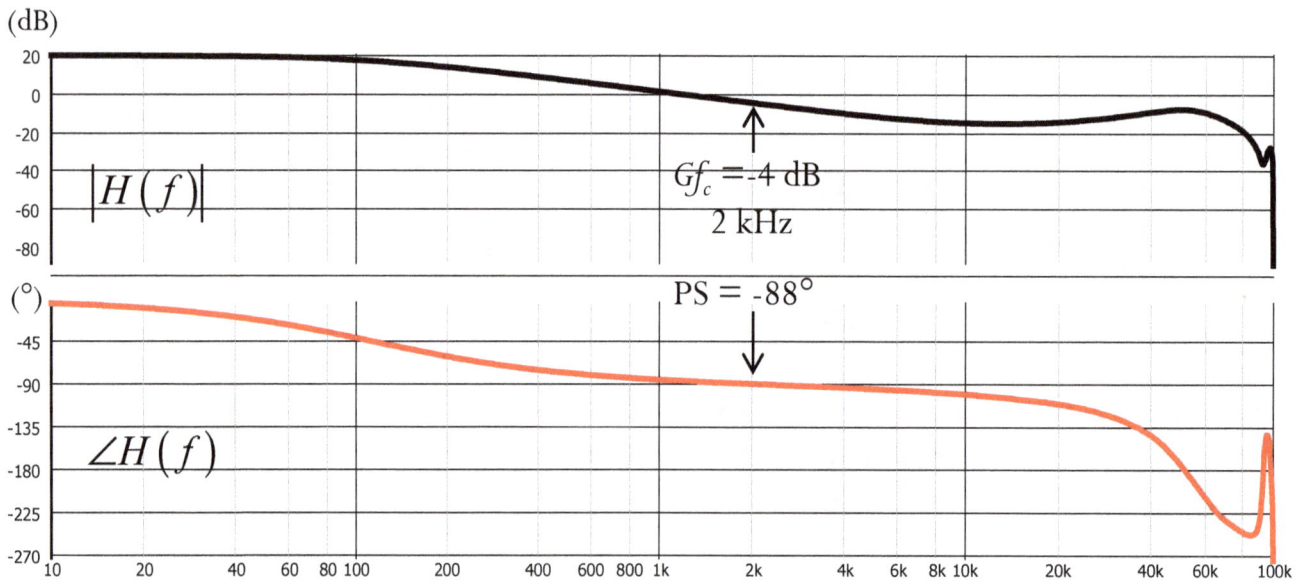

```
.VAR Gfc=-4 * magnitude at crossover *
.VAR PS=-88 * phase lag at crossover *
*
* Enter Design Goals Information Here *
*
.VAR fc=2k * targeted crossover *
.VAR PM=60 * choose phase margin at crossover *
*
.VAR boost=PM-PS-90
.VAR G=10^(-Gfc/20)
.VAR fp=(tan(boost*pi/180)+sqrt((tan(boost*pi/180))^2+1))*fc
.VAR fz=fc^2/fp
.VAR a=sqrt((fc^2/fp^2)+1)
.VAR b=sqrt((fz^2/fc^2)+1)
.VAR R2=((a/b)*G*Rupper*fp)/(fp-fz)
.VAR C1=1/(2*pi*R2*fz)
.VAR C2=C1/(C1*R2*2*pi*fp-1)
*
```

Calculated values:
```
*
* Rupper = 12500
* Rlower = 2500
* R2 = 21586.0313540429
* C2 = 1.15179863903482e-09
* C1 = 1.2856444944508e-08
* Boost = 58
* Fz = 573.490771517616
* Fp = 6974.82888768182
* Sn = 17021.2765957447
* Se = 6808.51063829787
* kr = 0.068085106382978
*
```

Type 2 compensator ac response with the automated values:

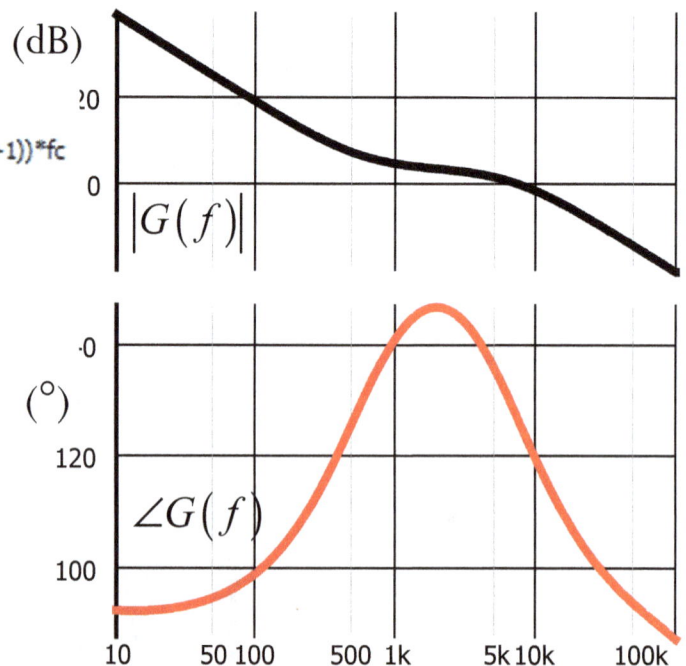

Closed-Loop Transient Response

THE OPERATING POINT of the boost converter operated in CM is the same as in the voltage-mode case. However, the current comparator is active, cycle-by-cycle, while it remained silent in voltage-mode control. After compensation, the loop gain confirms a good crossover with an improved phase margin of 58°.

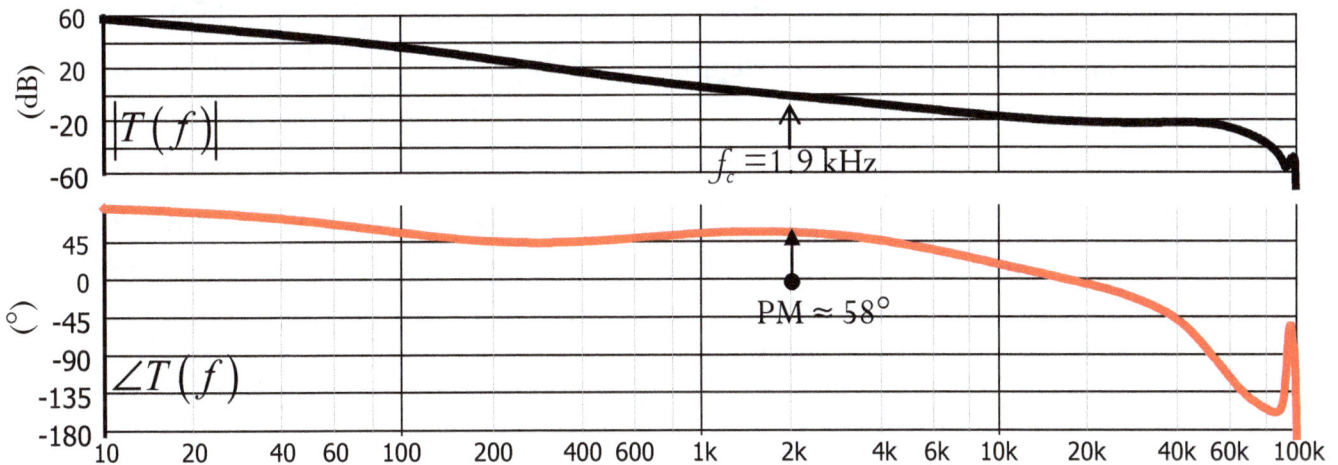

The transient response is very stable and shows a small 20-mV undershoot.

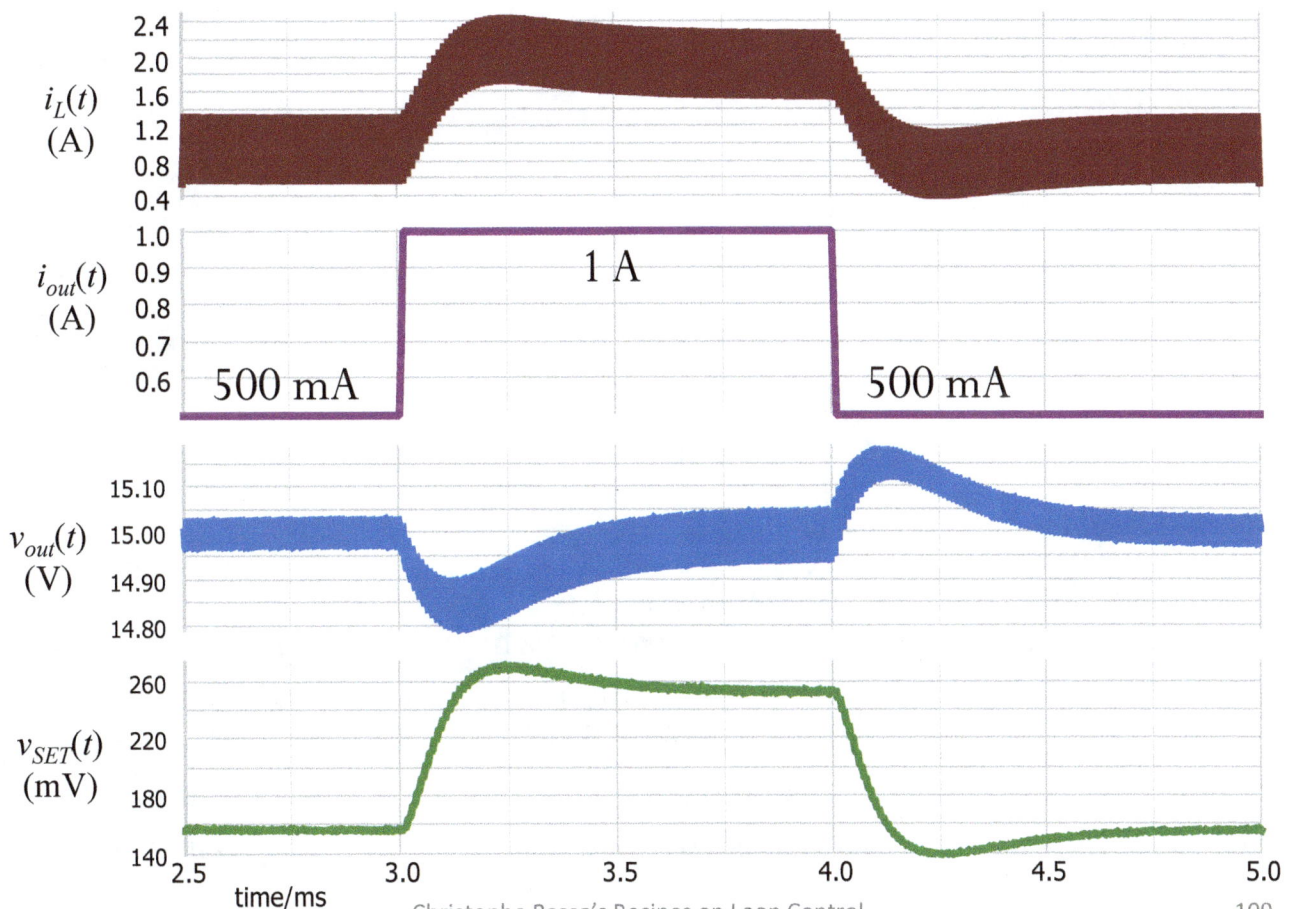

Power Factor Correction - BCM

POWER FACTOR CORRECTION is a typical application in which the boost converter occupies the largest part of the ac-dc adapters market for power levels exceeding 75 W. By forcing a sinusoidal current absorption and boosting the output voltage up to 380 or 400 V, the boost converter plays the role of a pre-converter, feeding a downstream high-voltage dc-dc converter such as a flyback converter for instance. For low power levels, below ≈150 W for a single-switch approach, this boost stage can be operated in constant-on-time type of control, borderline conduction mode (BCM), without sensing the input voltage. For simulating the circuit and extracting the control-to-output transfer function, you can supply the converter with a dc source whose value corresponds to the rms source the PFC will operate from. For instance, if you want to check stability at 100 V rms, just bias the converter from a 100-V dc source:

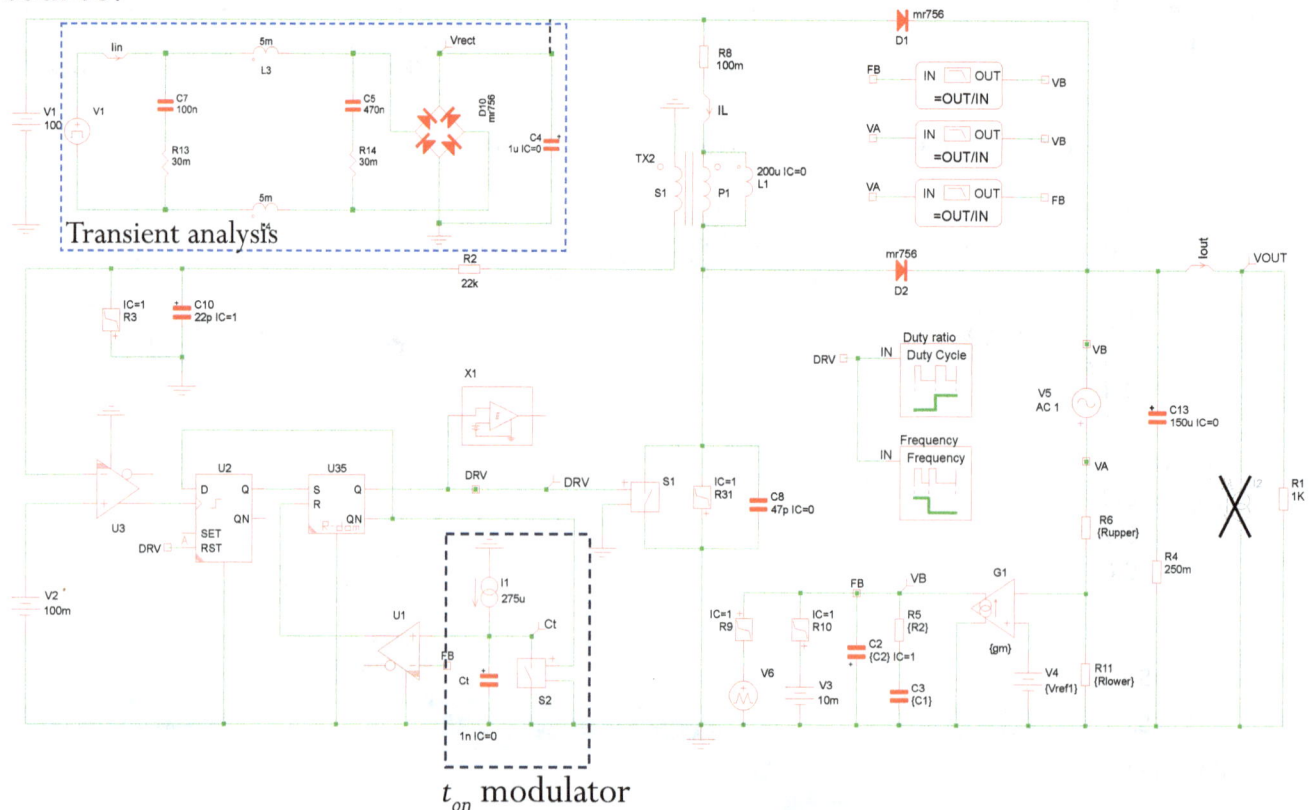

There is no clock in this type of converter which is self-relaxing: the power switch turns on for a programmed duration $- t_{on} -$ and when the switch turns off, the magnetic activity in the inductor is monitored by an auxiliary winding. When the core is fully demagnetized, a new cycle is initiated after a short deadtime for quasi-zero-voltage-switching: it is a free-running structure.

Selecting the Crossover Frequency

IN A POWER FACTOR CORRECTION stage, we can show that the gain depends on the input voltage squared. Therefore, if you choose a crossover frequency f_c at 100 V rms, then at a 230-V, this crossover will approximately be shifted by a factor $(230/100)^2 \approx 5$. In lack of a feedforward term, we should select a crossover frequency of 20 Hz at 230 V which will drop to around 4-5 Hz at 100 V. Pushing crossover higher certainly makes the loop react faster but more ripple in V_{FB} is detrimental to harmonic distortion.

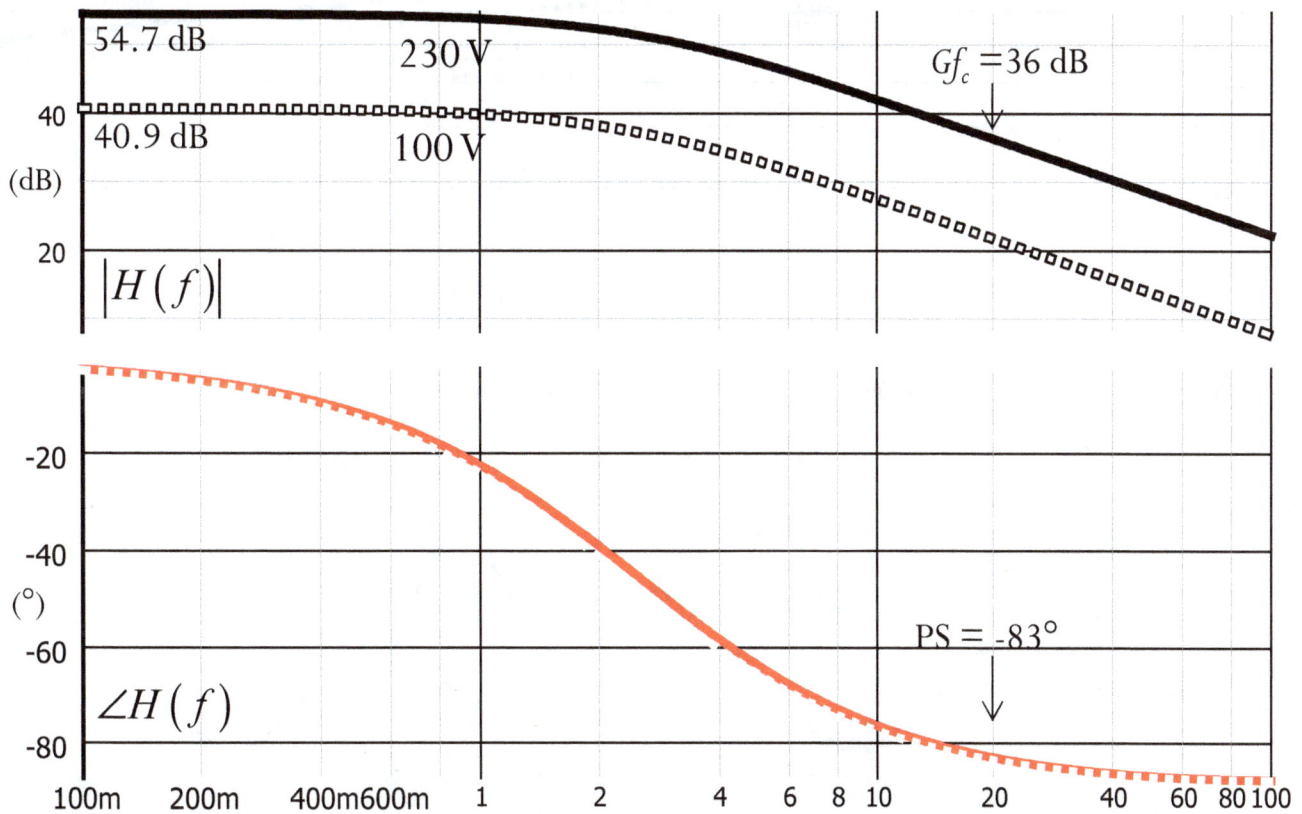

The OTA circuit produces a type 2 compensator whose components values are automated by the macro:

```
* Choose OTA characteristics *
*
.VAR gm=110u * transconductance in Siemens *
*
*
* Do not edit the below lines *
.VAR boost=PM-PS-90
.VAR G=10^(-Gfc/20)
.VAR fp=(tan(boost*pi/180)+sqrt((tan(boost*pi/180))^2+1))*fc
.VAR fz=fc^2/fp
.VAR a=sqrt((fc^2/fp^2)+1)
.VAR b=sqrt((fz^2/fc^2)+1)

.VAR R2=(a/b)*(fp*G)*(Rlower+Rupper)/((fp-fz)*Rlower*gm)
.VAR C1=1/(2*pi*R2*fz)
.VAR C2=(Rlower*gm/(2*pi*fp*G*(Rlower+Rupper)))(b/a)
*
```

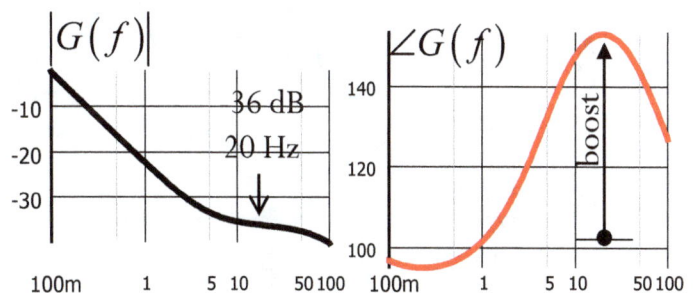

```
*
* Rupper = 2996031.74603175
* Rlower = 19841.2698412698
* C1 = 1.42627525677878e-06
* C2 = 8.7235442996365e-08
* R2 = 23239.8355338028
*
```

Transient Performance

ONCE COMPENSATED, the PFC exhibits the 20-Hz crossover at 230 V input voltage and it drops to 5 Hz at a 100-V input. Some PFC controllers sense the input voltage and adjust the compensator mid-band gain to compensate the gain variation with input line and keep a constant f_c. Phase margin is good at both crossover points:

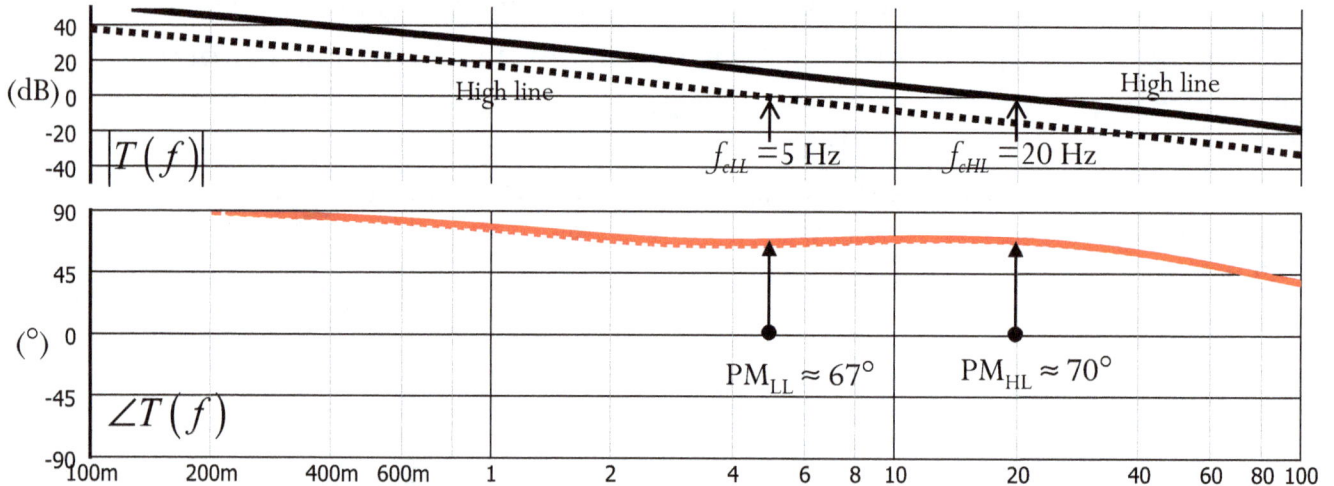

The simulation is fast, giving a THD of 6.8% at a 100-V rms input.

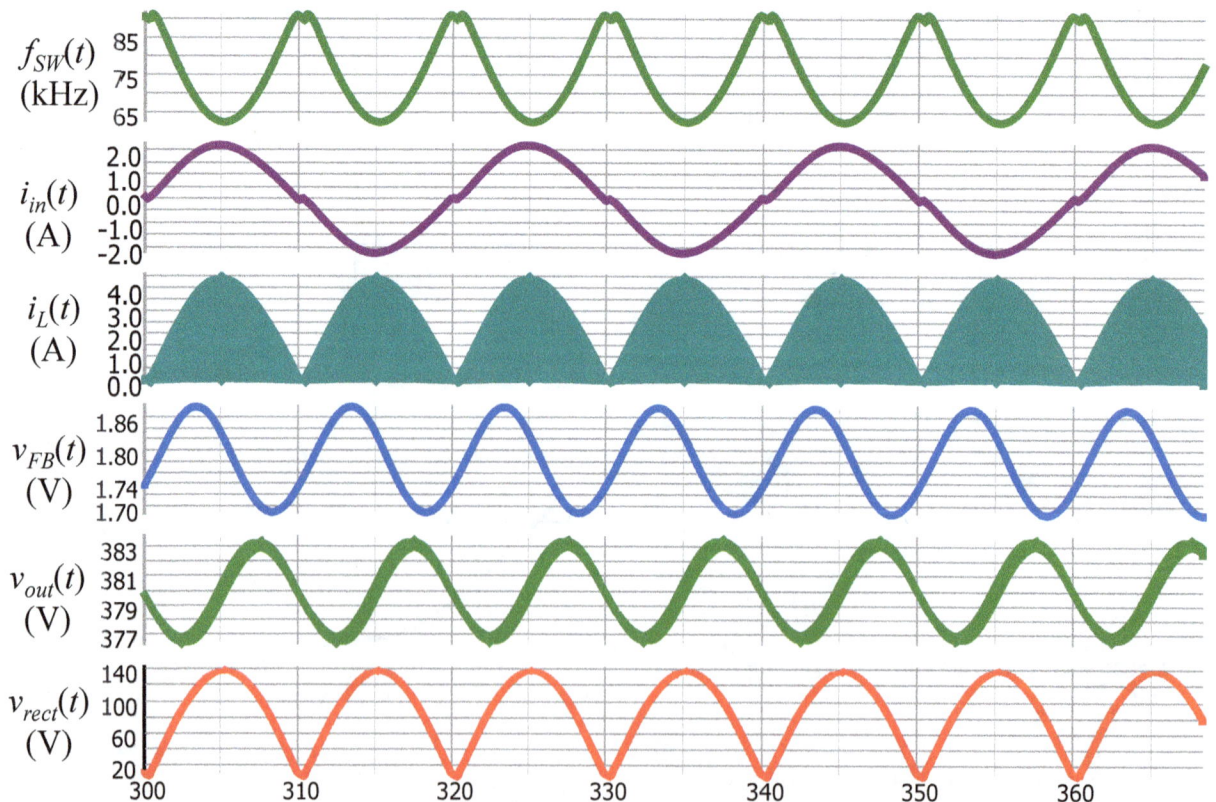

Power Factor Correction - CCM

HIGHER POWER LEVELS can be obtained using CCM. Numerous control laws exist to obtain a good input current harmonic distortion. A classical approach consists of using a multiplier sensing the rectified high-voltage input to force the inductor current to follow a sinusoidal shape. In these circuits, there are two loops: one with a sufficiently-high bandwidth shapes the current via average mode control and a second, slower, classically regulates the output voltage. The below template is a way to realize this operating scheme:

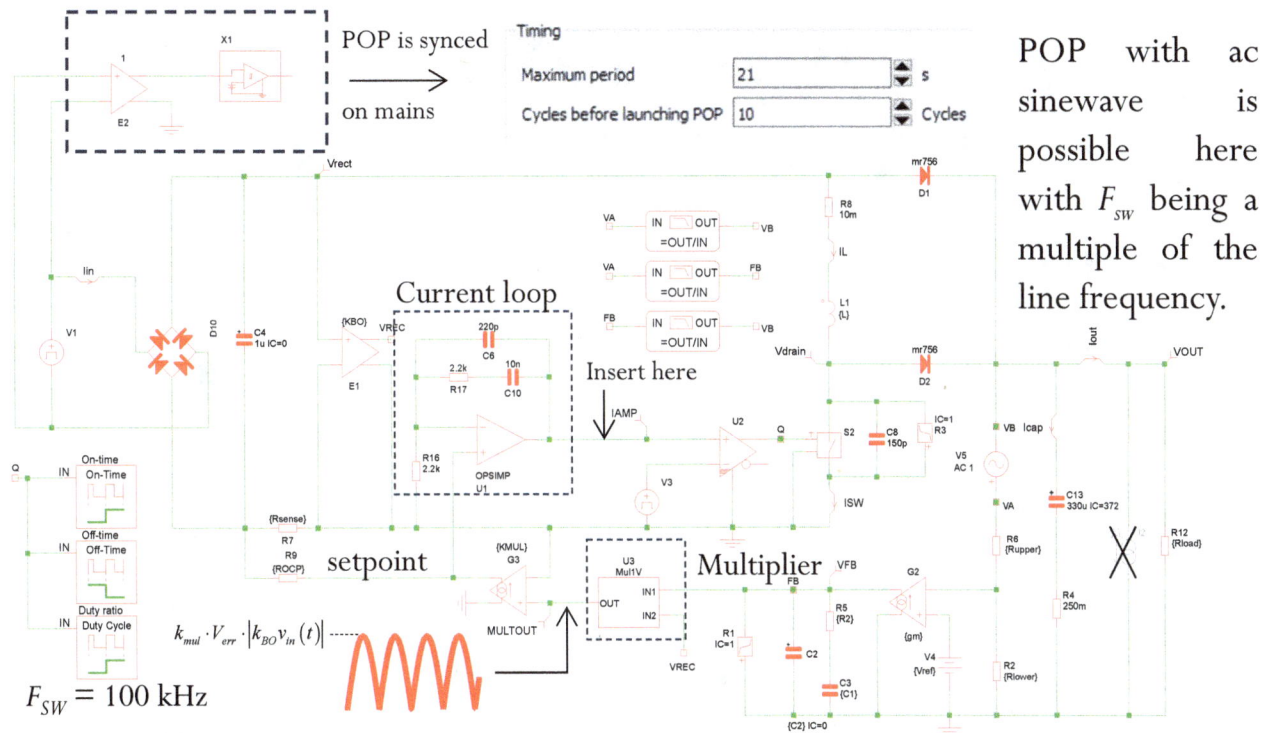

For stabilizing a two-loop system, you first need to ensure the inner loop is stable with sufficient margins before considering the second one. It is possible to ac-sweep this path by inserting the ac stimulus at the point indicated in the schematic diagram:

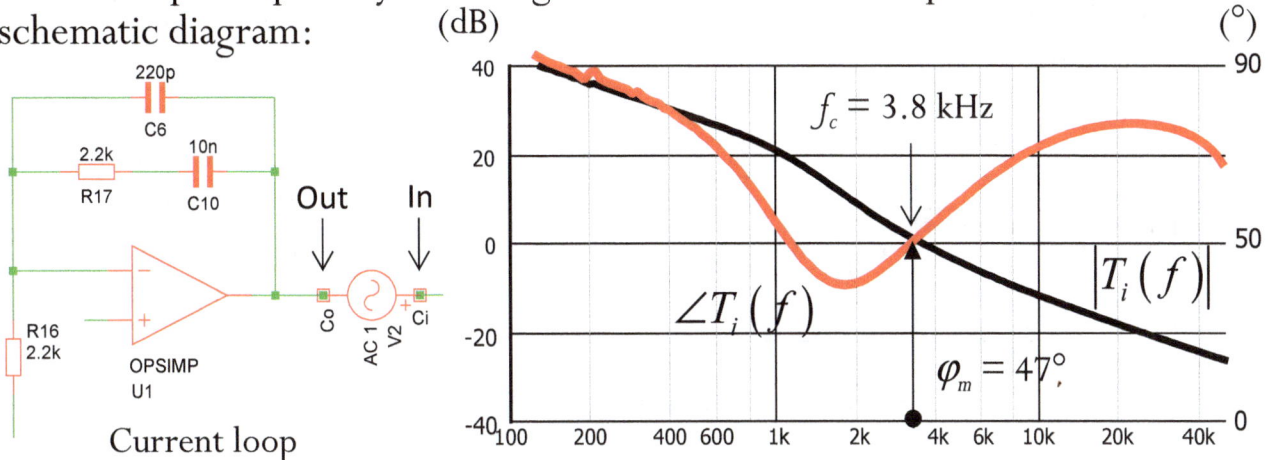

Total Harmonic Distortion

ONCE THE CURRENT LOOP is stabilized, you can run a periodic operating point (POP) by synchronizing the trigger block directly from the mains input. The operating point is quickly obtained and reveals a low THD of 2.7% at a 100-V input voltage and a 1.3-kW output power:

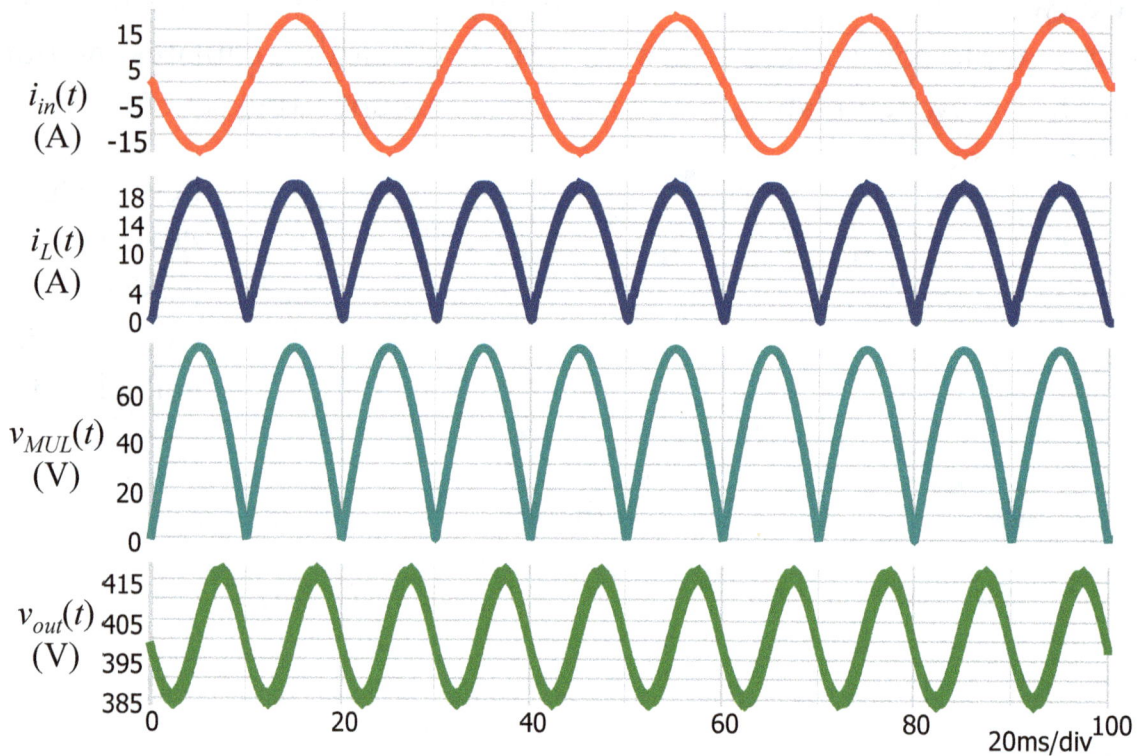

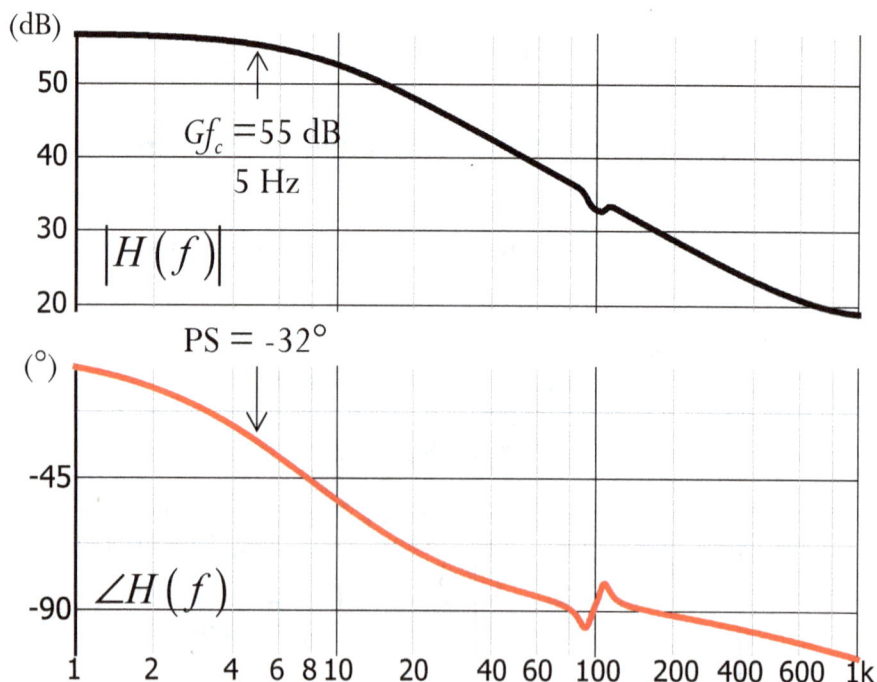

The control-to-output transfer function of the voltage loop, shows a gain of 55 dB at 5 Hz. In an integrated circuit like the UC1854, feedforward is provided by dividing the multiplier output by the input voltage square. Not in this example as I wanted to keep the circuit simple. It is thus optimized for a 100-V input.

The Compensated CCM PFC

ONCE THE TYPE 2 COMPENSATOR is designed, we can check the compensated loop gain at a 100-V rms input. The PFC will regulate at 230 V but the proposed circuit is not optimized at this level as previously mentioned:

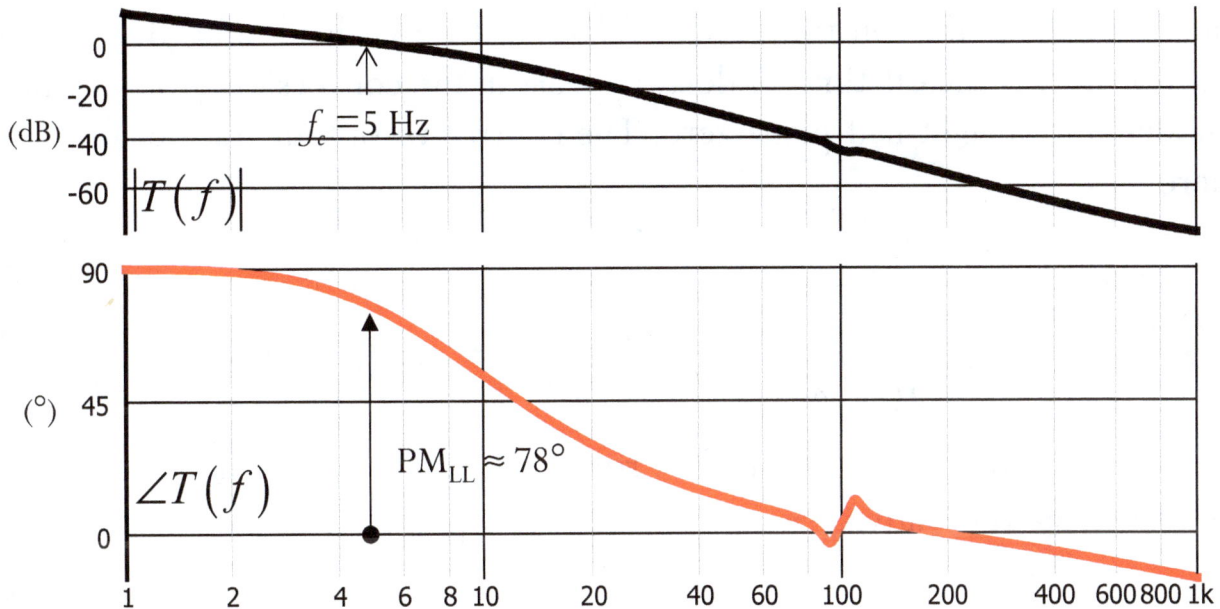

The transient response is stable when the load is changed from 1 kW to 1.5 kW in 10 µs. You can see how the multiplier input immediately follows the setpoint delivered by the error amplifier:

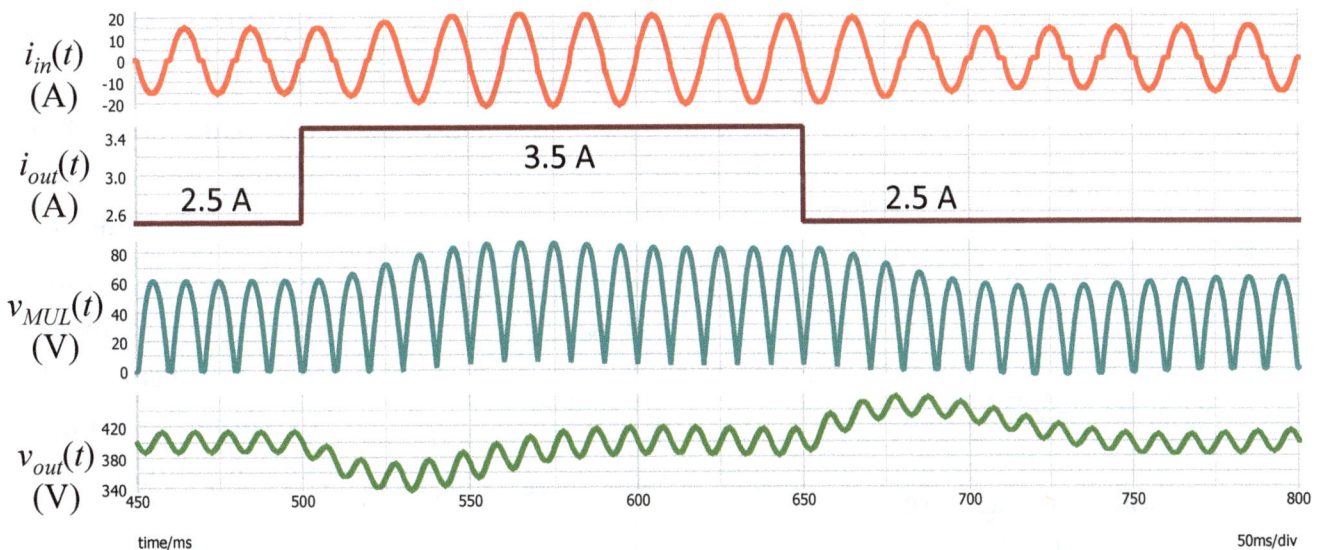

There are plenty of different techniques for implementing a CCM PFC, including the control of operating modes (multi-mode PFCs, BCM-DCM-CCM) which improve THD and efficiency in wide load range.

Buck-Boost Converter in VM

THE BUCK-BOOST CONVERTER offers the possibility to increase or decrease the input voltage in relationship with the duty ratio D. In this single-switch converter, the output voltage is negative. Like the boost converter, the buck-boost transfers energy in two steps: the energy is first stored in the inductor during the on-time then released to the load and the output capacitor during the off-time. A delay appears in the conversion process which is modeled by a right-half-plane zero. The control-to-output transfer function is here:

$$H(s) = \frac{V_{out}(s)}{V_{err}(s)} = H_0 \frac{\left(1 + \dfrac{s}{\omega_{z_1}}\right)\left(1 - \dfrac{s}{\omega_{z_2}}\right)}{1 + \dfrac{s}{\omega_0 Q} + \left(\dfrac{s}{\omega_0}\right)^2} \qquad \omega_{z_1} = \frac{1}{r_C C_{out}} \qquad \omega_0 \approx \frac{1-D}{\sqrt{L_1 C_{out}}} \quad H_0 \approx -\frac{V_{in}}{V_p (1-D)^2}$$

$$\omega_{z_2} = \frac{(1-D)^2 R_{load}}{D L_1} \qquad Q \approx (1-D) R_{load} \sqrt{\frac{C_{out}}{L_1}}$$

An inversion block is inserted before routing V_{out} to the error amplifier as the output voltage is negative.

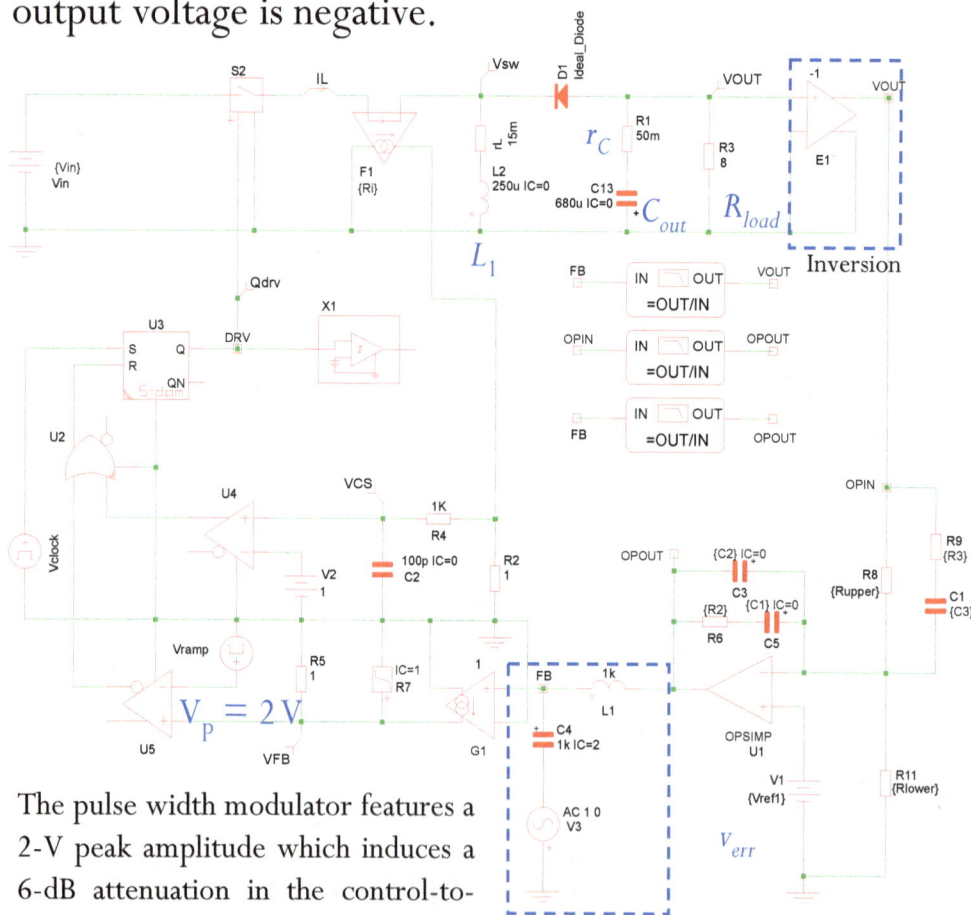

As with the boost example, SIMPLIS® was not able to simulate correctly in closed-loop condition. I had to resort to the classical *LC* filter which closes the loop in dc (for bias point calculation) and opens it in ac. The current-mode example is immune to that problem.

The pulse width modulator features a 2-V peak amplitude which induces a 6-dB attenuation in the control-to-output transfer function.

Compensating the VM Buck-Boost

THE BUCK-BOOST CONVERTER operated in voltage-mode control, like the boost converter, exhibits a peaky 2nd-order response whose resonant frequency f_o varies with the operating conditions. Added to the RHP zero phase lag, it sets two limits for selecting a crossover frequency. Here, at the selected input voltage of our example (40 V), the resonance occurs at 296 Hz while the RHPZ is located at 13 kHz. A 2-kHz crossover frequency is chosen.

$$r_L := 0.05\Omega \quad r_C := 0.05\Omega \quad C_2 := 680\mu F \quad L_1 := 250\mu H \quad R_L := 8\Omega$$

$$V_{in} := 40V \quad V_p := 2V \quad D_0 := 23.2\% \quad V_{out2} := -\left(V_{in} \cdot \frac{D_0}{1 - D_0}\right) = -12.083\,V$$

$$H_0 := -\frac{V_{in}}{(1 - D_0)^2} \cdot \frac{1}{V_p} \qquad 20 \cdot \log(|H_0|) = 30.606 \quad \text{dc gain (dB)}$$

$$\omega_{z2} := \frac{(1 - D_0)^2 \cdot R_L}{D_0 \cdot L_1} \qquad f_{z2} := \frac{\omega_{z2}}{2\pi} = 12.948 \cdot kHz \qquad \Rightarrow \quad f_{z1} \approx 4.7\,kHz \quad \text{LHP zero}$$

$$\omega_{z1} := \frac{1}{r_C \cdot C_2} \qquad f_{z1} := \frac{\omega_{z1}}{2\pi} = 4.681 \cdot kHz \qquad \qquad f_{z2} \approx 13\,kHz \quad \text{RHP zero}$$

$$Q := (1 - D_0) \cdot R_L \sqrt{\frac{C_2}{L_1}} = 10.133 \quad \omega_0 := \frac{1 - D_0}{\sqrt{L_1 \cdot C_2}} \quad f_0 := \frac{\omega_0}{2\pi} = 296.454 \cdot Hz$$

Slow-down the converter to allow the inductor current to build up.

$$\Rightarrow \quad 3 \cdot f_0 < f_c < 0.2 \cdot f_{z_2} \rightarrow f_c = 2\,kHz$$

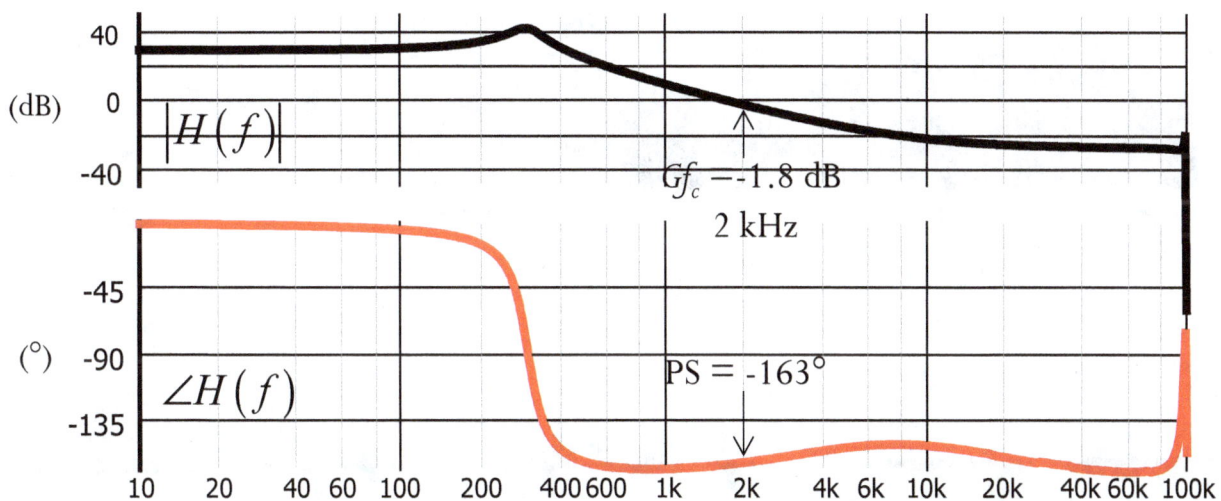

The power stage response of the buck-boost converter is very close to that of the boost converter. It requires a type 3 compensator to meet the phase margin requirements in voltage-mode control.

Operating Point and Compensated Gain

THE OPERATING POINT confirms a duty ratio at 23% and an output voltage regulated at -12 V with a 1.5-A current as expected.

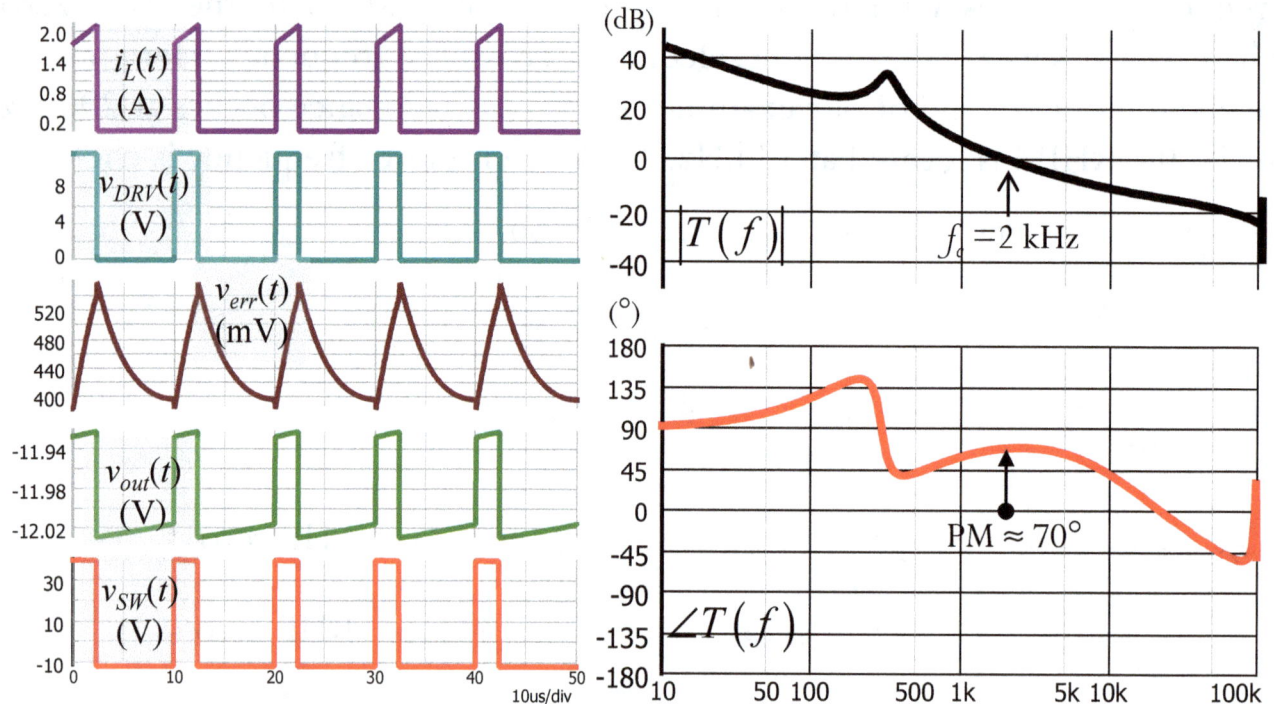

The adopted compensation values lead to ≈2-kHz crossover with a 70° phase margin. The below graph shows the transient response with a 0.5-A step load in 1 A/μs.

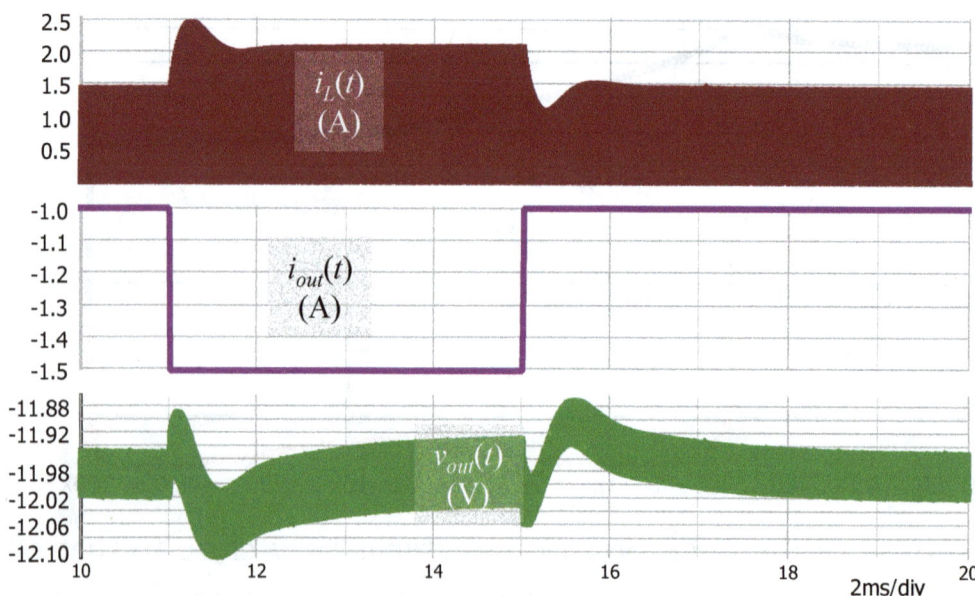

Compensator values:

```
Rupper = 9500
Rlower = 2500
R2 = 1887.79659560479
R3 = 57.3440643863179
C3 = 5.55087766432786e-08
C2 = 1.43237730527306e-08
C1 = 3.37229007537028e-07
Boost = 144
Fz1 = 250
Fz2 = 300
Fp1 = 6135.82711928582
Fp2 = 50000
```

Buck-Boost Converter in CM

WHEN OPERATED in current-mode control, the buck-boost converter ac response no longer peaks since the low-frequency resonance has gone. The RHP zero, however, remains and occupies a similar location. Again, you cannot push the crossover too far. The transfer function is now described by a 3rd-order polynomial and classically hosts two subharmonic poles at $F_{sw}/2$. These poles are damped by the addition of a compensation ramp S_e.

$$H(s) = H_0 \frac{\left(1 + \dfrac{s}{\omega_{z_1}}\right)\left(1 - \dfrac{s}{\omega_{z_2}}\right)}{1 + \dfrac{s}{\omega_{p_1}}} \cdot \frac{1}{1 + \dfrac{s}{\omega_n Q_p} + \left(\dfrac{s}{\omega_n}\right)^2}$$

$$H_0 \approx -\frac{R_{load}}{R_i} \frac{1}{\dfrac{(1-D)^2}{2\tau_L}\left(1 + 2\dfrac{S_e}{S_n}\right) + 2M + 1}$$

External slope [V]/[s]

$$\omega_{z_2} \approx \frac{(1-D)^2 R_{load}}{D L_1}$$

$$\omega_{z_1} = \frac{1}{r_C C_{out}}$$

$$\tau_L = \frac{L_1}{R_{load} T_{sw}}$$

$$\omega_{p_1} \approx \frac{\dfrac{(1-D)^3}{2\tau_L}\left(1 + 2\dfrac{S_e}{S_n}\right) + 1 + D}{R_{load} C_{out}}$$

$$Q_p = \frac{1}{\pi(m_c D' - 0.5)}$$

$$\omega_n = \frac{\pi}{T_{sw}}$$

$$m_c = 1 + \frac{S_e}{S_n} \leftarrow S_n = \frac{V_{in}}{L_1} R_i$$

$$M = \left|\frac{V_{out}}{V_{in}}\right|$$

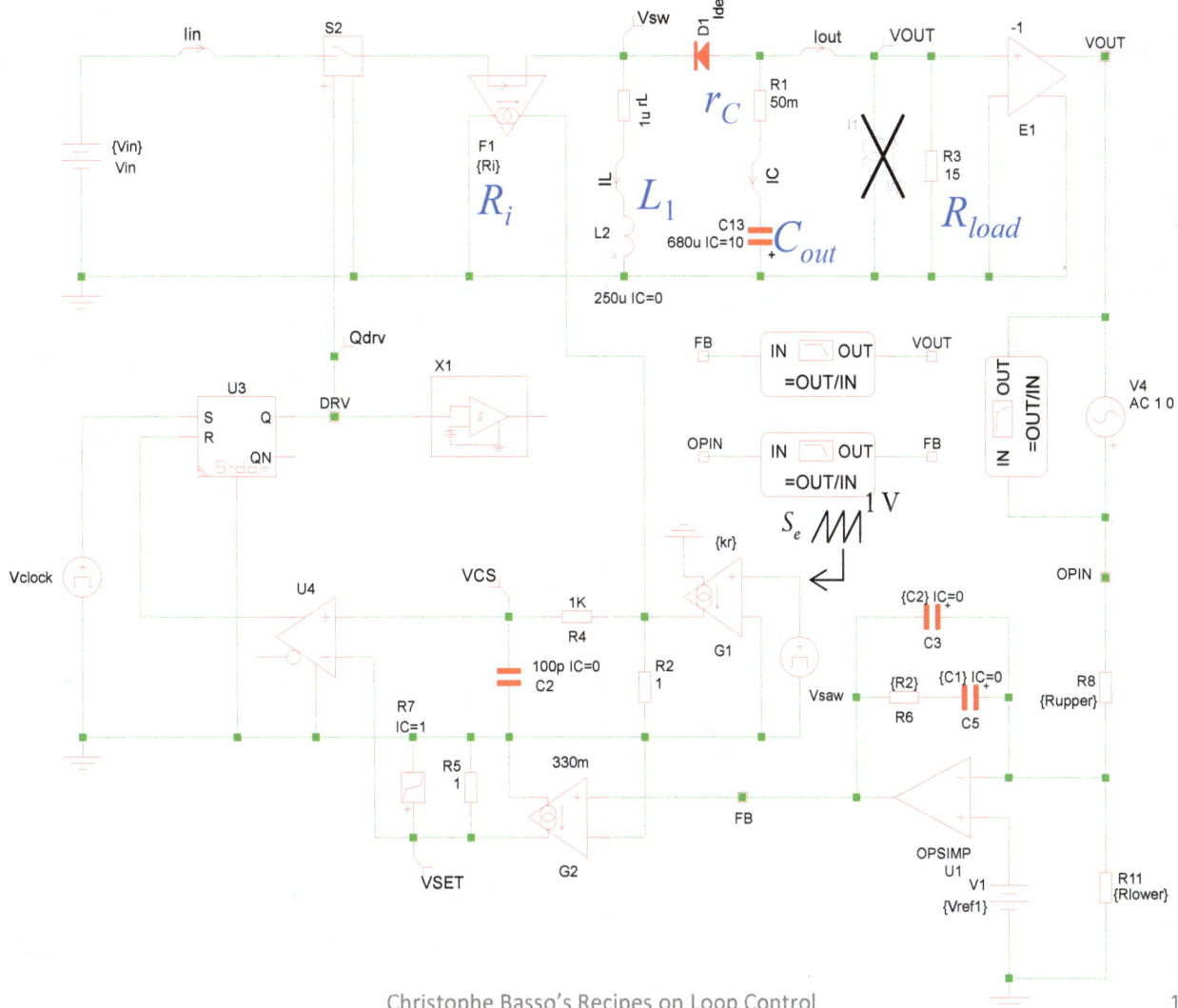

Power Stage AC Response

THE CURRENT-MODE version no longer restricts you for the lower limit of the crossover frequency but, as the RHPZ remains, you are still stuck with an upper limit set at 20% of its lowest position (minimum V_{in} and maximum I_{out}). In this example, it implies a 2-kHz f_c.

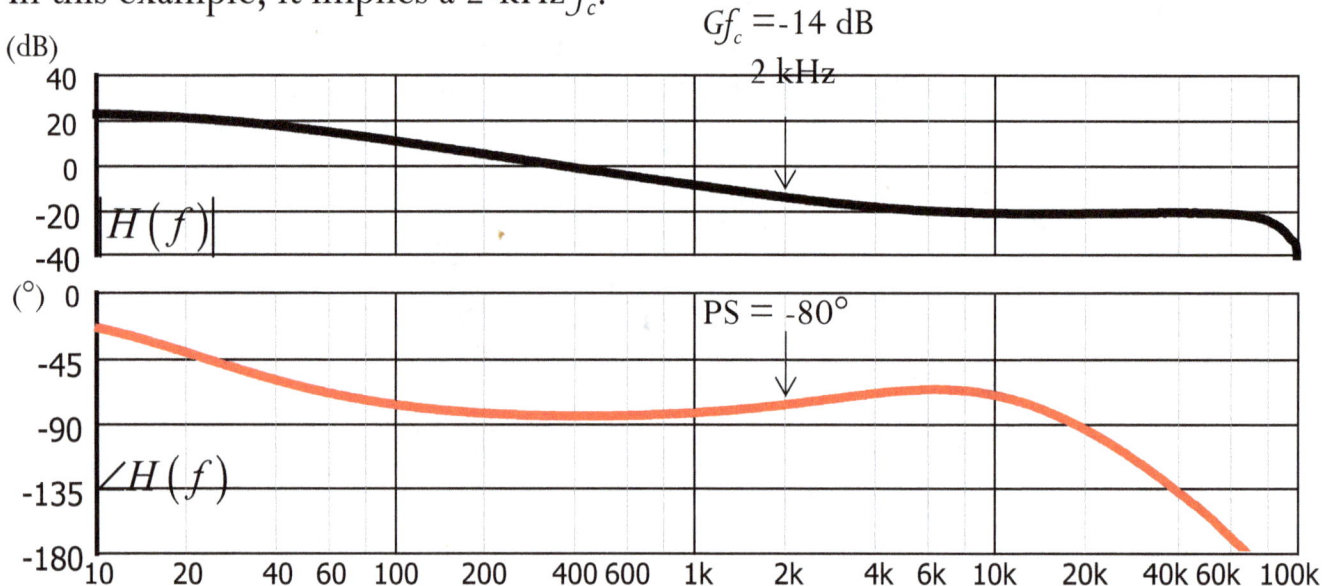

$Gf_c = -14$ dB

2 kHz

$|H(f)|$

PS = $-80°$

$\angle H(f)$

```
*
.VAR Vin=40
.VAR Vout=12
.VAR L=250u
.VAR Ri=150m
.VAR Ts=10u * please update clock and ramp generators *
*
.VAR Gfc=-14 * magnitude at crossover *
.VAR PS=-80 * phase lag at crossover *
*
* Enter Design Goals Information Here *
*
.VAR fc=2k * targeted crossover *
.VAR PM=60 * choose phase margin at crossover *
*|
.VAR Sn={((Vin)/L)*Ri}
.VAR Sramp={1/Ts}
.VAR mc=1.5 * set this value for ramp comp *
.VAR Se={(mc-1)*Sn}
.VAR kr={Se/Sramp}
*
```

Calculated values:

```
* Rupper = 9500
* Rlower = 2500
* R2 = 54883.4333319182
* C2 = 6.08320425404147e-10
* C1 = 3.98366670772094e-09
* Boost = 50
* Fz = 727.940468532405
* Fp = 5494.95483890924
* Sn = 24000
* Se = 12000
* kr = 0.12
```

Type 2 compensator ac response with the automated values:

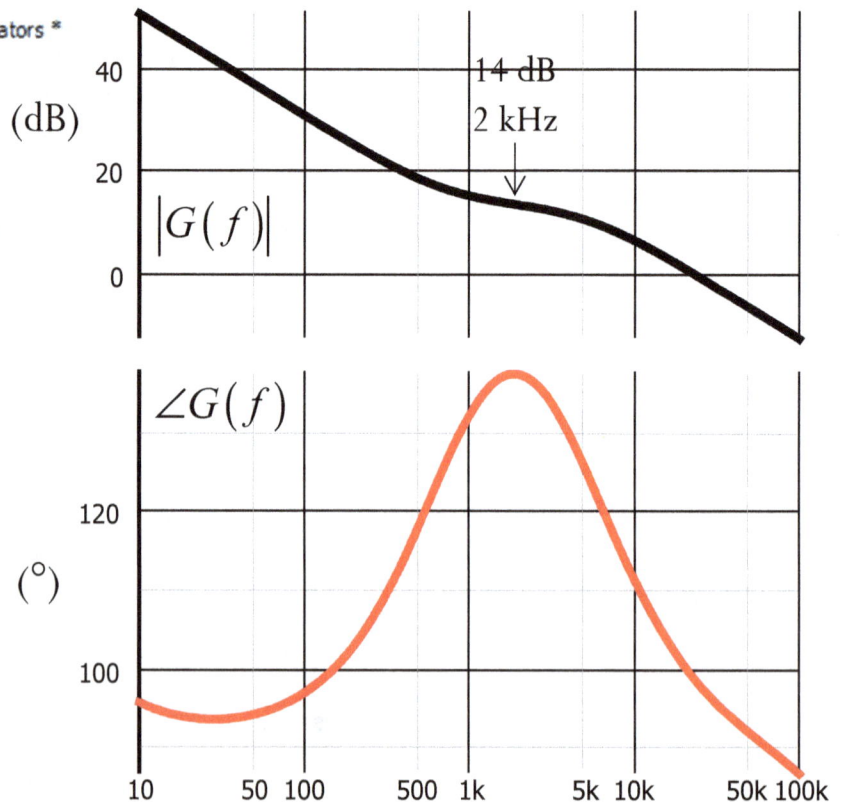

14 dB

2 kHz

$|G(f)|$

$\angle G(f)$

Closed-Loop Transient Response

THE OPERATING POINT of the buck-boost converter operated in CM does not differ from the voltage-mode case except that the current comparator is now active, cycle-by-cycle. After compensation, the loop gain confirms a 2-kHz crossover frequency with a comfortable phase margin of 62°.

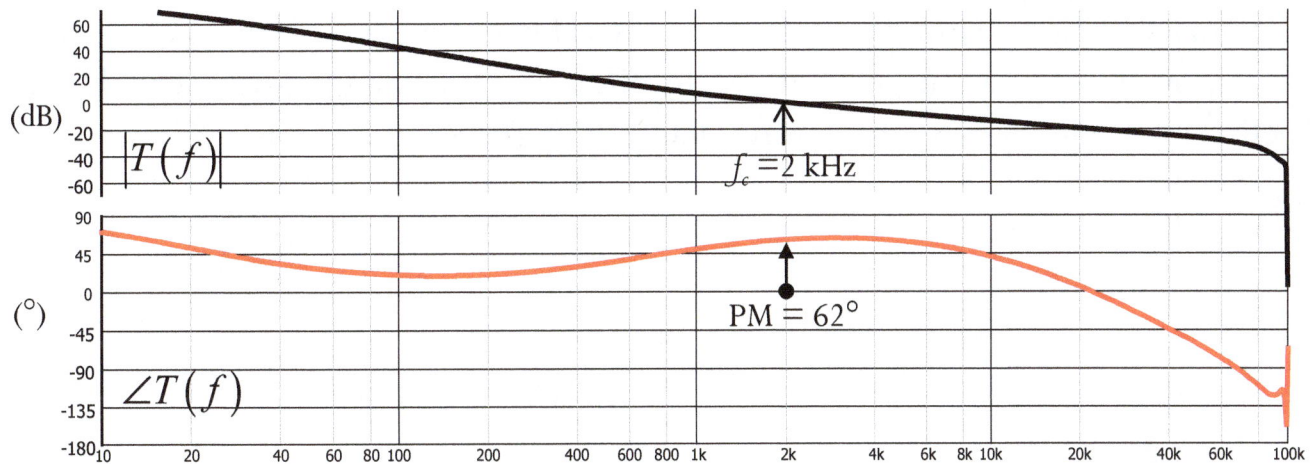

The transient response is very stable and shows a small 20-mV undershoot. The output current is swept with a 1-A/µs slope. Always mention this value.

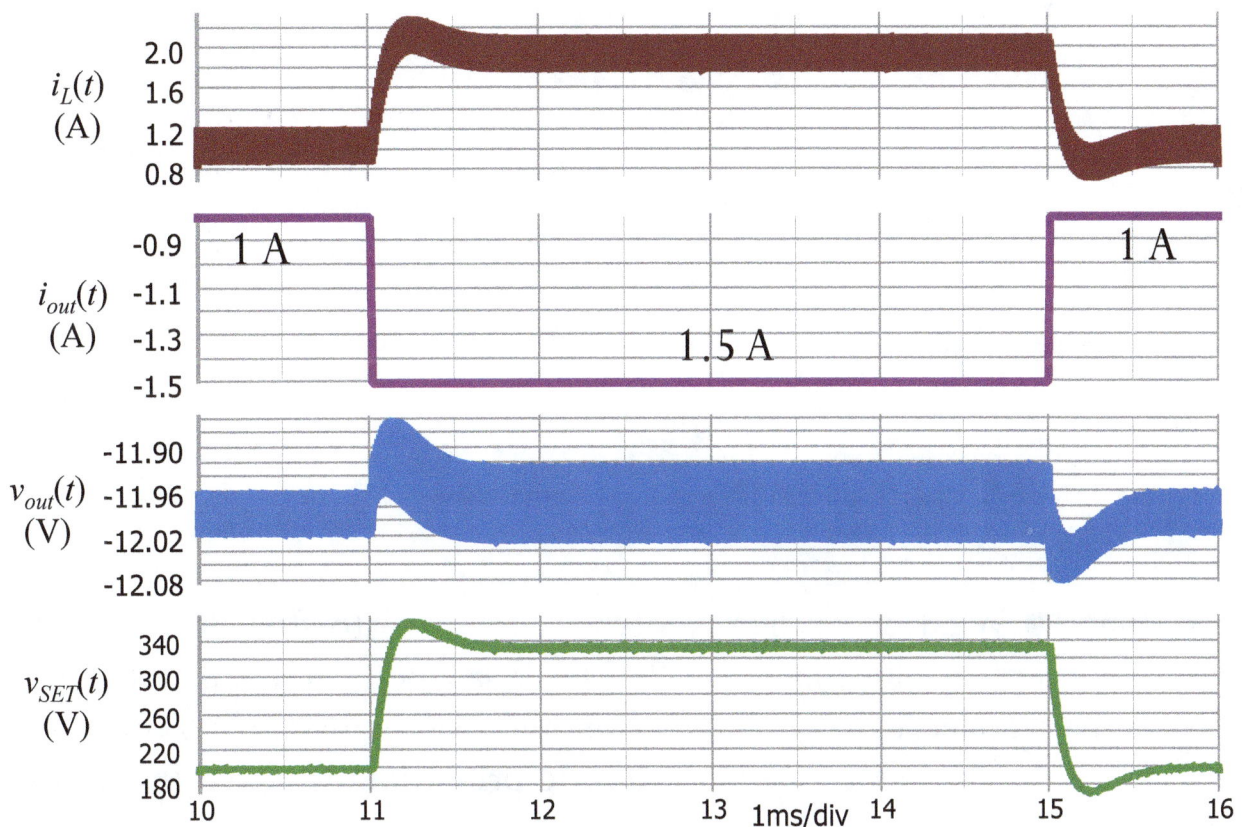

The Flyback Converter in VM

THE FLYBACK CONVERTER represents the isolated version of the buck-boost converter. Through the added transformer, it offers the option to deliver single or multiple output voltages, of positive or negative polarities. The flyback converter is probably the most popular ac-dc topology in consumer and industrial applications. If current-mode control is implemented in the vast majority of cases, voltage-mode control has been the control scheme of choice for Power Integrations with its wide offer of high-voltage switchers. Unlike current-mode control, there is no need for slope compensation in VM.

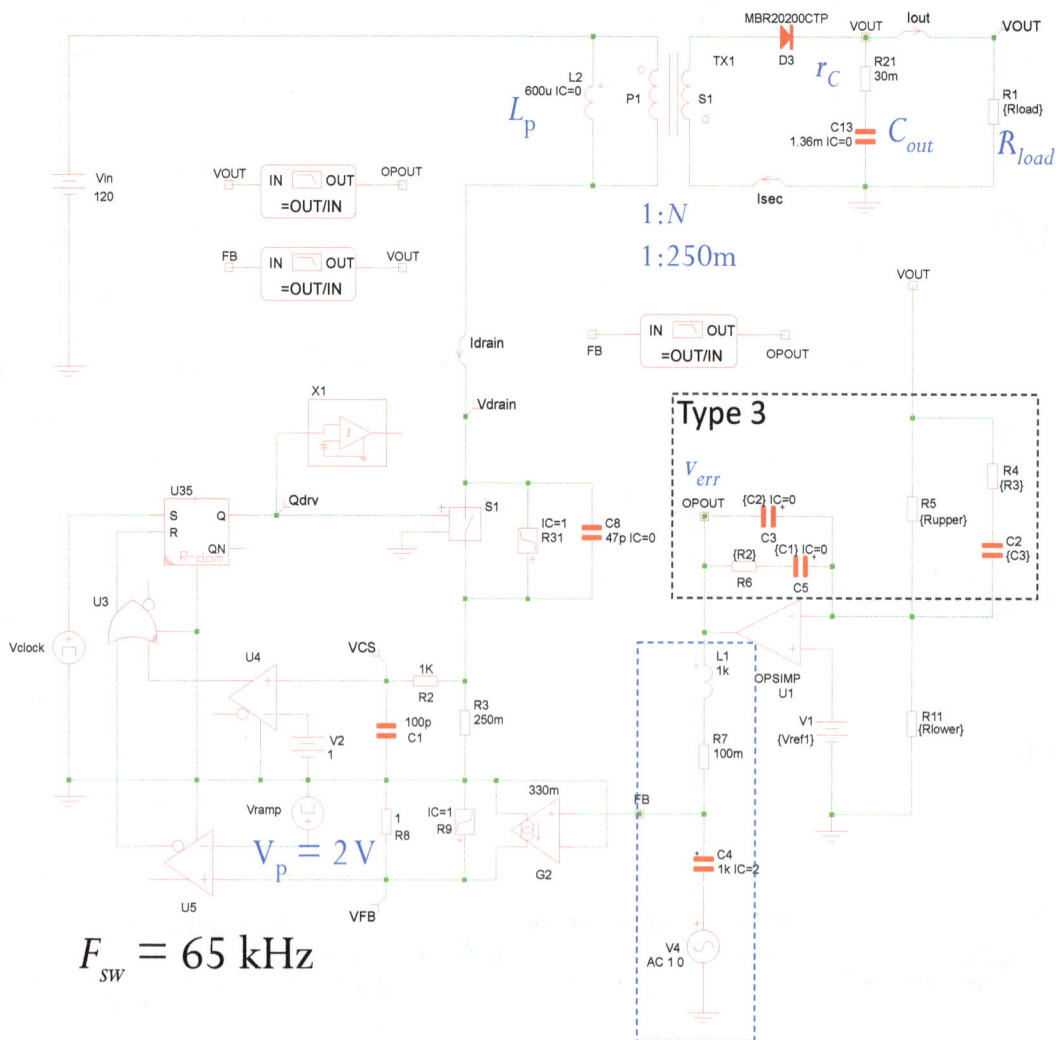

This 19-V/60-W adapter requires a type 3 compensator for closing the loop. The pulse-width modulator uses a 2-V peak sawtooth in a classical way. The transformer model is very simple for the best possible convergence. It associates a dc subcircuit together with the primary inductance L_p for storing and releasing energy cycle-by-cycle.

Power Stage Response in VM

DESPITE THE PRESENCE of a transformer, the control-to-output transfer function of the flyback converter resembles that of the buck-boost converter. The inductor at play in this isolated version, is the primary inductance of the flyback transformer, often labeled L_p.

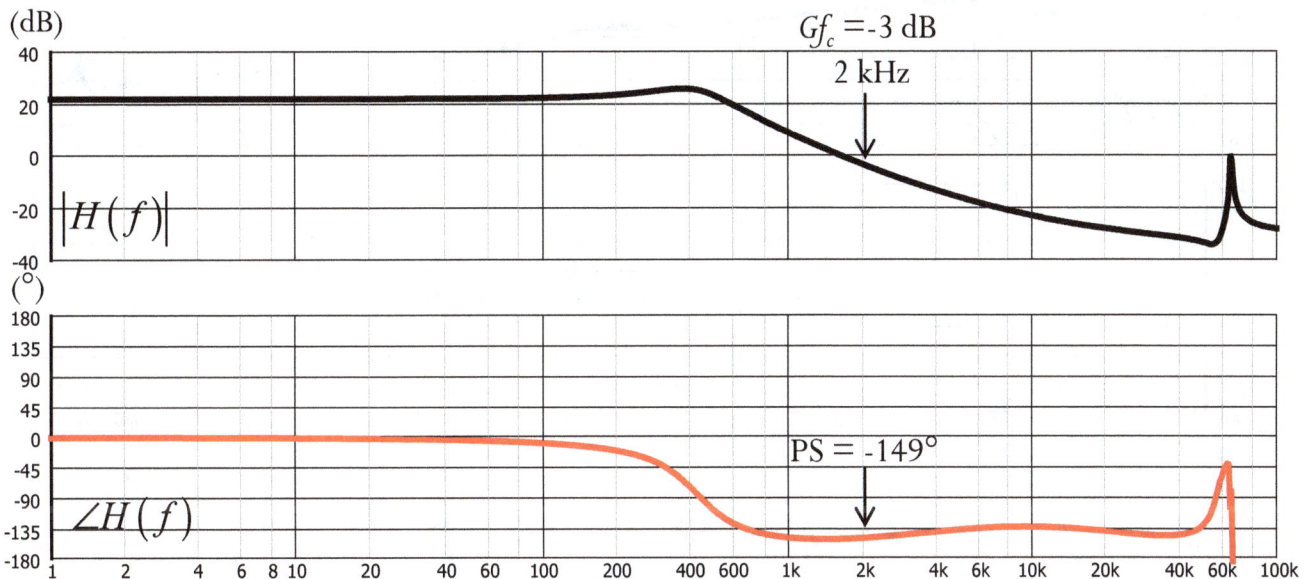

For this exercise, the control-to-output transfer function is extracted at the lowest input voltage and the maximum current. If this is a high-voltage power supply operated from the rectified mains, select a dc bias for V_{in} equal to the valley voltage of the rectified bulk with some margin, e.g. 70 V when plugged from a 85-V ac outlet. First, determine the resonant frequency and the RHP zero position in this worst-case:

$$f_0 = \frac{1-D}{N\sqrt{L_p C_{out}}} \frac{1}{2\pi} \approx 430 \text{ Hz} \qquad f_{RHPZ} = \frac{(1-D)^2 R_{load}}{D L_p N^2} \frac{1}{2\pi} \approx 24.6 \text{ kHz}$$

We will select a 2-kHz crossover frequency.

```
*
.VAR Vin=120
.VAR Vout=19
.VAR Lp=600u
.VAR Cout=1.36m
.VAR Ri=250m
.VAR N=250m
.VAR Rload=6
.VAR Ts=15u * please update clock and ramp generators
*
.VAR D=Vout/(Vout+N*Vin) * duty ratio calculation *
.VAR fRHPZ={((1-D)^2*Rload/(D*Lp*N^2))/(2*pi)}
.VAR fcMAX=0.2*fRHPZ
.VAR fo=((1-D)/(N*sqrt(Lp*Cout)))/(6.28)
*
```

```
*
*  Rupper = 66000
*  Rlower = 10000
*  R2 = 36691.9518364894
*  R3 = 1808.21917808219
*  C3 = 2.93391688022938e-09
*  C2 = 5.80072137767065e-10
*  C1 = 5.42199771087195e-09
*  Boost = 119
*  Fz1 = 800
*  Fz2 = 800
*  Fp1 = 8277.68749141228
*  Fp2 = 30000
*  FRHPZ = 24616.8762677045
*  FCMAX = 4923.37525354091
*  fo = 431.69840384445
*
```

The macro calculates the RHPZ position as well as the resonant frequency. Crossover frequency is selected at 20% of the RHP zero and well above the resonant peak.

Compensated Loop Gain

THE COMPENSATION is done for a 2-kHz crossover and the phase margin approaches 60°, which is excellent. A quick transient test at both input levels will tell us if the design is acceptable or not.

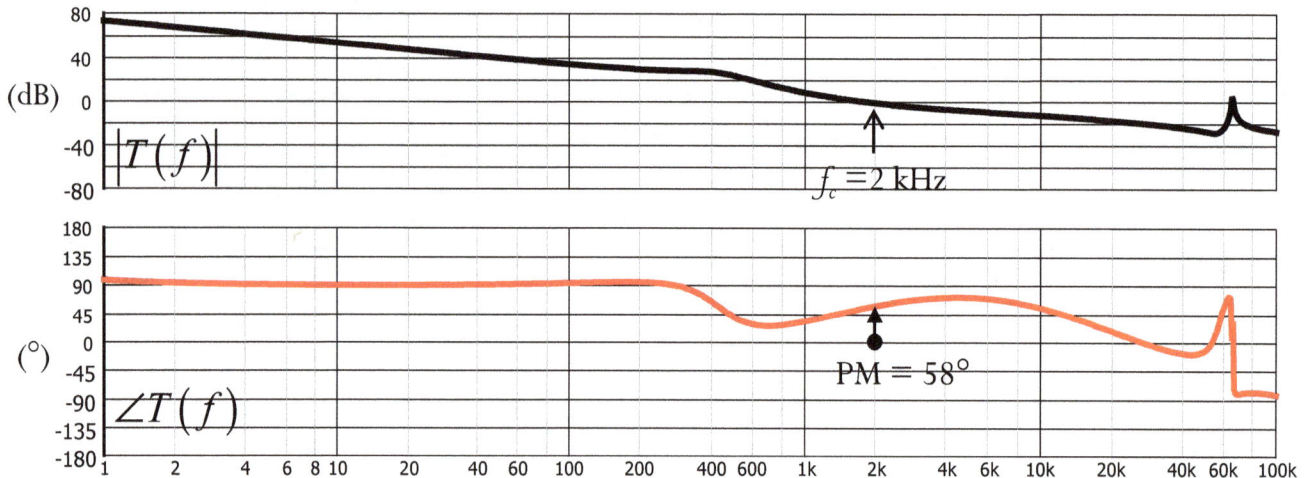

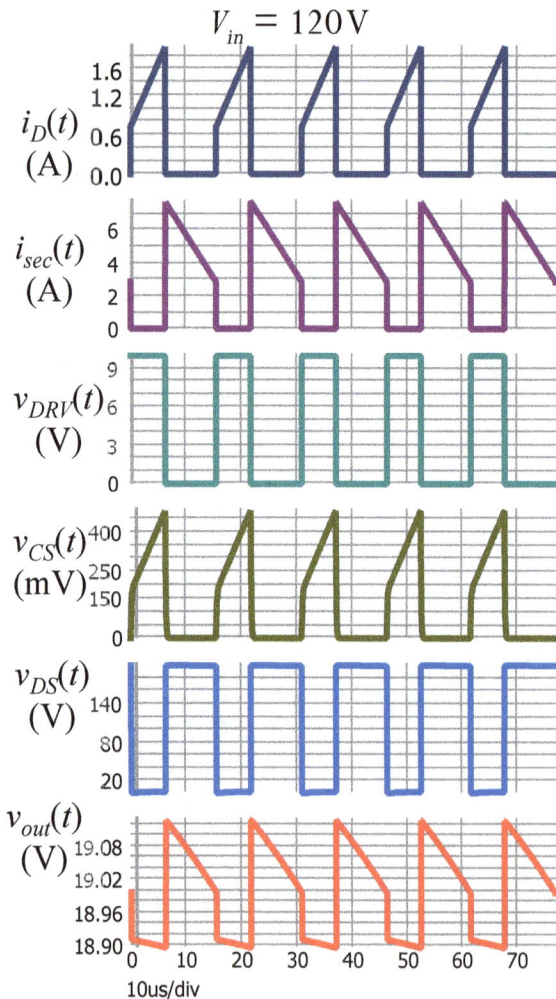

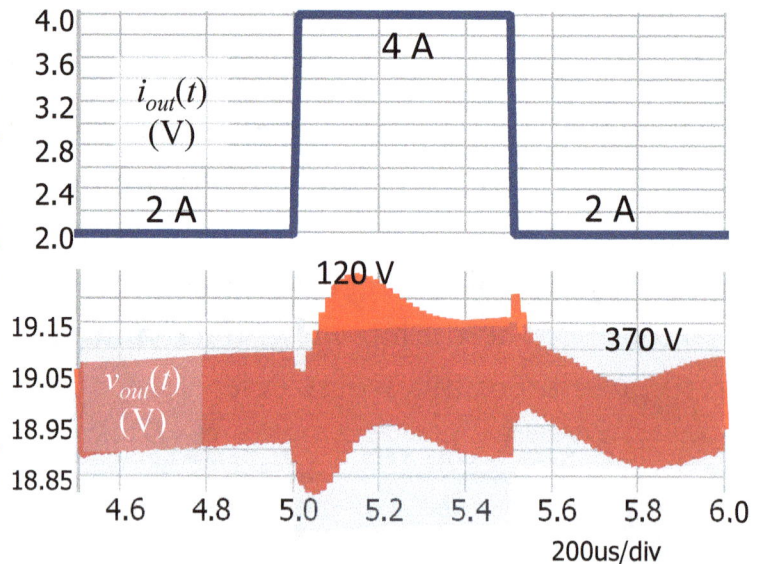

The load step confirms the stable response at both input voltage extremes. From the dc operating point (left) obtained at a low input voltage, you can extract rms and average values for the selection of components such as power capacitors like C_{out}, for instance.

The Flyback Converter in CM

THE CURRENT-MODE-CONTROLLED flyback converter offers a better input line rejection than its VM counterpart in ac-dc applications. The 100- or 120-Hz ripple needs to be vigorously rejected and current-mode is naturally immune to input voltage changes. The sense resistance and the maximum peak current in the transformer must be sized to cope with the lowest input voltage, e.g. the valley of the rectified bulk voltage. The compensation implies a proper level of slope compensation and a type 2 compensator, often built around a TL431 and an optocoupler in offline applications.

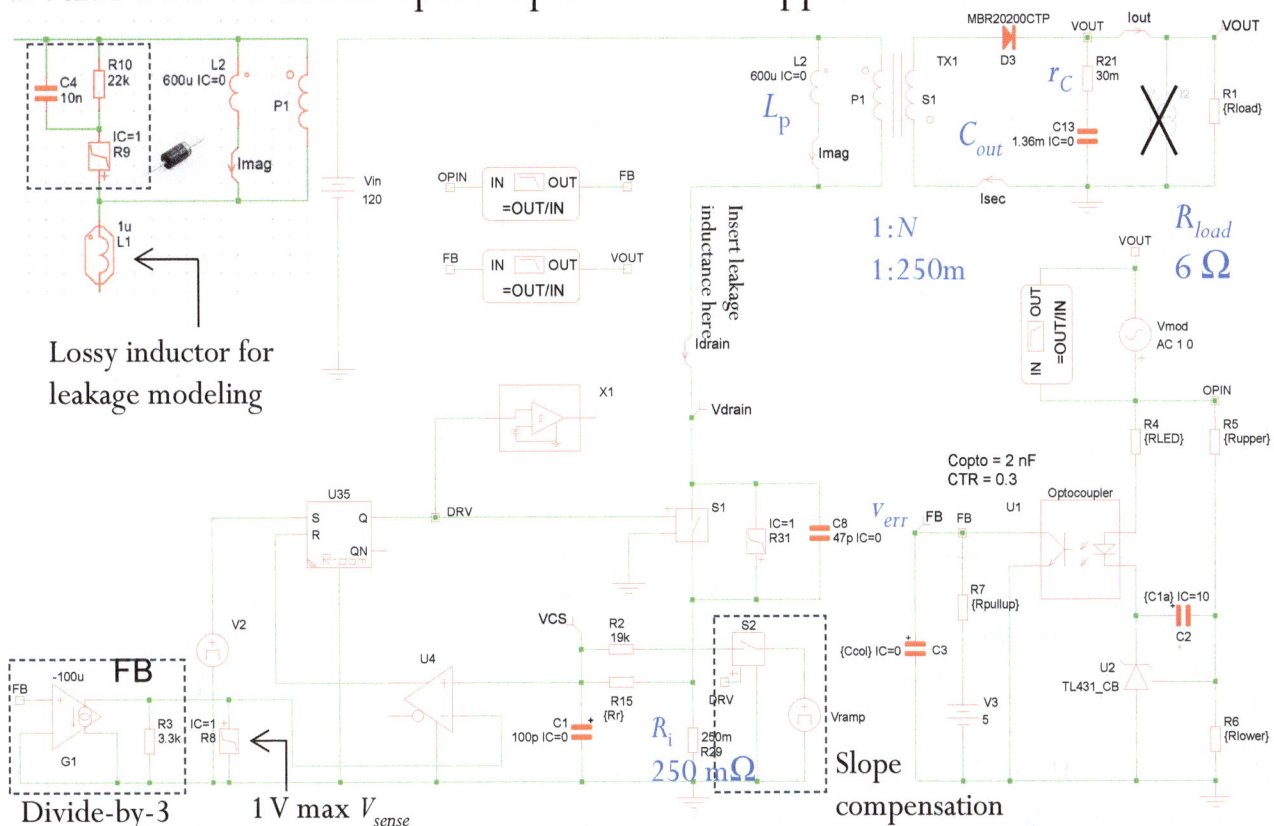

Lossy inductor for leakage modeling

Divide-by-3   1 V max V_{sense} Slope compensation

You can see the absence of a *RCD* clamping network in this setup. Its presence can bother the POP algorithm. If you want to add it, choose the lossy inductor model for the leakage inductance, it converges better than a simple *L* component. An example is shown in the upper left corner of the schematic diagram. The divide-by-3 is a classic in these controllers: assume a 1-V maximum sense voltage, then you do not want the compensator output to be swinging between ≈0 and 1 V in normal operation. By inserting this block, the output of the op-amp or the optocoupler will swing between ≈0 and 3 V, offering better dynamics and noise immunity.

A First-Order Response

OPERATING IN CURRENT-MODE CONTROL brings a 3rd-order type of response which includes a dominant pole at low frequency and two subharmonic poles at $F_{sw}/2$. Once these subharmonic poles are properly damped with an adequate level of slope compensation, you can consider the ac response of the power stage as that of a 1st-order system for low frequency. As such, a type 2 compensator is well suited for the compensation job as the peak observed in VM has disappeared but the RHP zero remains at a similar position. The power stage response is shown below.

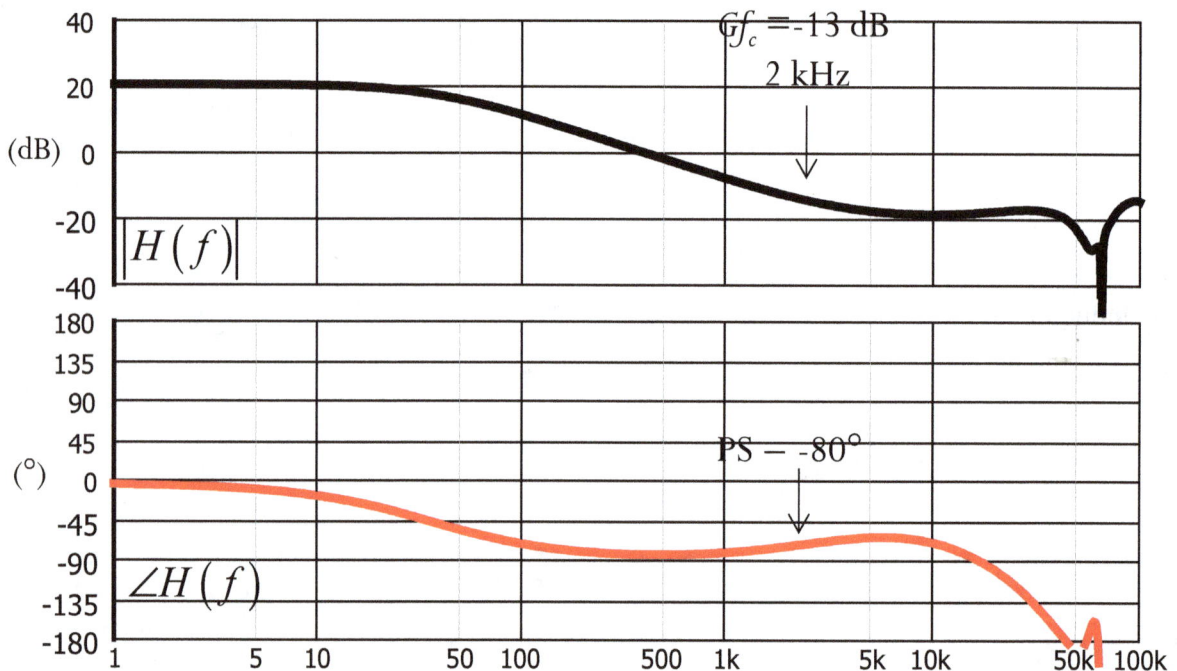

The control-to-output transfer function in CCM depends on the primary inductance L_p but also involves the turns ratio N and sense resistance R_i:

$$H(s) = H_0 \frac{\left(1+\dfrac{s}{\omega_{z_1}}\right)\left(1-\dfrac{s}{\omega_{z_2}}\right)}{1+\dfrac{s}{\omega_{p_1}}} \frac{1}{1+\dfrac{s}{\omega_n Q_p}+\left(\dfrac{s}{\omega_n}\right)^2} \qquad H_0 \approx \frac{R_{load}}{NR_i} \frac{1}{\dfrac{(1-D)^2}{2\tau_L}\left(1+2\dfrac{S_e}{S_n}\right)+2M+1}$$

$$\omega_{p_1} \approx \frac{\dfrac{(1-D)^3}{2\tau_L}\left(1+2\dfrac{S_e}{S_n}\right)+1+D}{R_{load}C_{out}} \qquad Q_p = \frac{1}{\pi(m_c D'-0.5)} \qquad \omega_n = \frac{\pi}{T_{sw}} \qquad m_c = 1+\frac{S_e}{S_n} \overset{\text{External slope [V]/[s]}}{\Longleftarrow} S_n = \frac{V_{in}}{L_p}R_i$$

$$\omega_{z_1} = \frac{1}{r_C C_{out}} \qquad \omega_{z_2} \approx \frac{(1-D)^2 R_{load}}{DL_p N^2} \qquad \tau_L = \frac{L_p N^2}{R_{load}T_{sw}} \qquad M = \frac{V_{out}}{NV_{in}}$$

Designing the Compensator

THE RHPZ POSITION is the same in VM and CM control, naturally limiting the crossover frequency selection. We will also choose 2 kHz in this example. The loop is closed via a TL431 and an optocoupler that we have duly characterized (to unveil its pole position and CTR). The macro accounts for the parasitic capacitor the component exhibits between its collector and emitter connections. If not properly factored in, it may significantly affect the phase margin. In this example, the pole has been measured at 6 kHz and should not be a problem. If the final capacitor value calculated by the macro is negative, it means you have to either push the optocoupler pole higher (by reducing the pull-up resistance if you can) or adopt a lower crossover frequency. Regarding the type 2 compensator with the TL431, make sure to verify that the LED series resistance, which sets the mid-band gain, is below the computed maximum value. It should not be a problem with a 19-V output voltage though.

This capacitor will be installed between the feedback and ground pins, very close to the controller. A negative value indicates either too high a f_c or too low an opto pole.

```
*
.VAR Vin=120
.VAR Vout=19
.VAR Lp=600u
.VAR Ri=250m
.VAR N=250m
.VAR Rload=6
.VAR Ts=15u * please update clock and ramp generators *
*
.VAR D=Vout/(Vout+N*Vin) * duty ratio calculation *
.VAR mc=0.818/(1-D) * recommended compensation value for a Q of 1 *
.VAR Sn={(Vin/Lp)*Ri}
.VAR Sramp={2.5/Ts} * 2.5 V over Ts - check your IC specs *
.VAR mc=1.5 * set this value for ramp comp *
.VAR Se={(mc-1)*Sn}
.VAR Rr={(Se/Sramp)*19k+1m}
.VAR fRHPZ={((1-D)^2*Rload/(D*Lp*N^2))/(2*pi)}
.VAR fcMAX=0.3*fRHPZ
*
* Enter values extracted from the plant Bode plot *
*
.VAR Gfc=-13 * magnitude at crossover *
.VAR PS=-80 * phase lag at crossover *
*
* Enter Design Goals Information Here *
*
.VAR fc=2k * targeted crossover *
.VAR PM=60 * choose phase margin at crossover *
*
```

ok

```
* Rupper = 66000
* Rlower = 10000
* C2 = 1.44819154804453e-09
* C1 = 3.31268645711794e-09
* Ccol = 1.2190035561207e-10
* Boost = 50
* Fz = 727.940468532405
* Fp = 5494.95483890924
* Sn = 50000
* Se = 25000
* D = 0.387755102040816
* Mc = 1.5
* Rramp = 2850.001
* RLED = 1477.5559514551
* Rmax = 24995.7446808511
* FRHPZ = 24616.8762677045
* FcMAX = 7385.06288031136
```

The TL431 and the optocoupler nicely build a type 2 compensator without resorting to extra components.

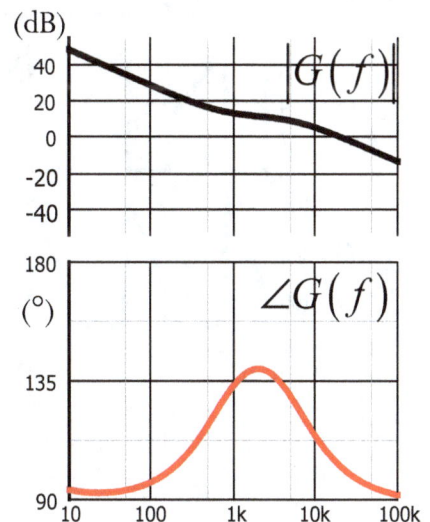

Compensated Loop Gain

WE CAN NOW RUN the CM flyback at different input levels and verify that crossover and margins meet the values we have targeted.

$V_{in} = 120\,\text{V dc}$

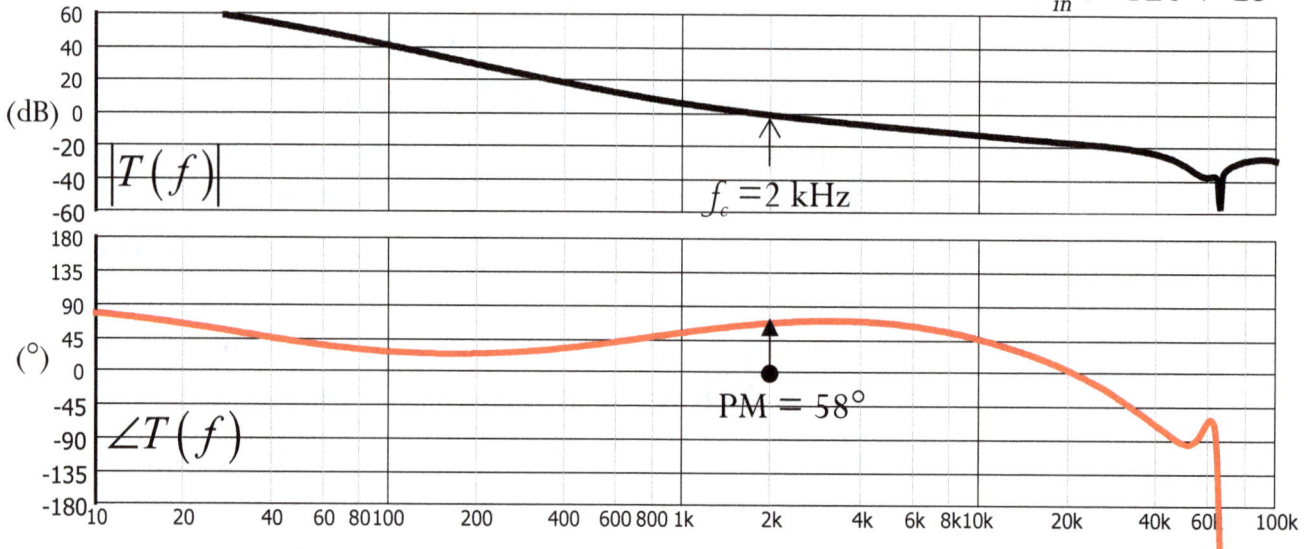

$|T(f)|$

$f_c = 2\,\text{kHz}$

$\angle T(f)$

$\text{PM} = 58°$

$V_{in} = 120\,\text{V dc}$
Gain Crossover Frequency   1.9857756kHz
Gain Margin   16.688496dB
Phase Margin   68.808011degrees

$V_{in} = 370\,\text{V dc}$
Gain Crossover Frequency   2.6985687kHz
Gain Margin   19.63565dB
Phase Margin   76.12686degrees

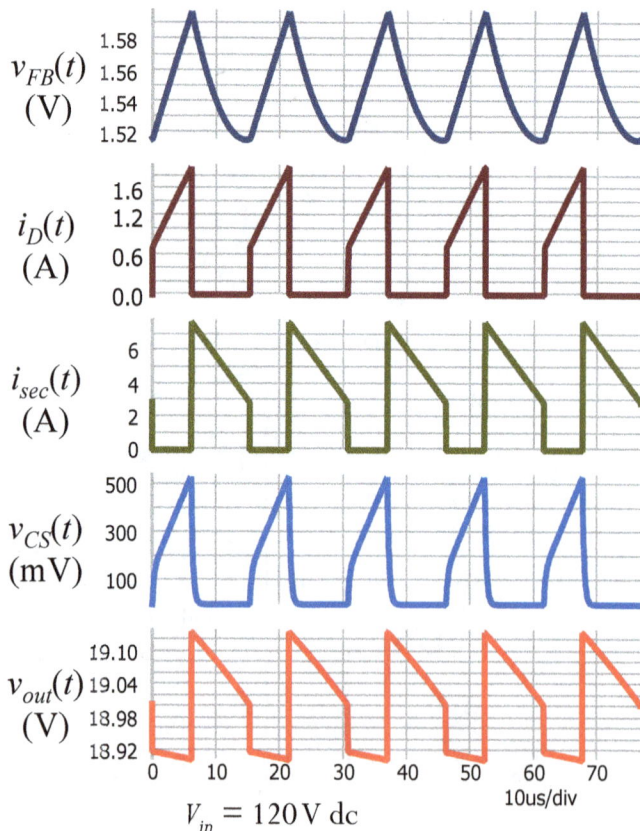

Transient response is good at both input levels.

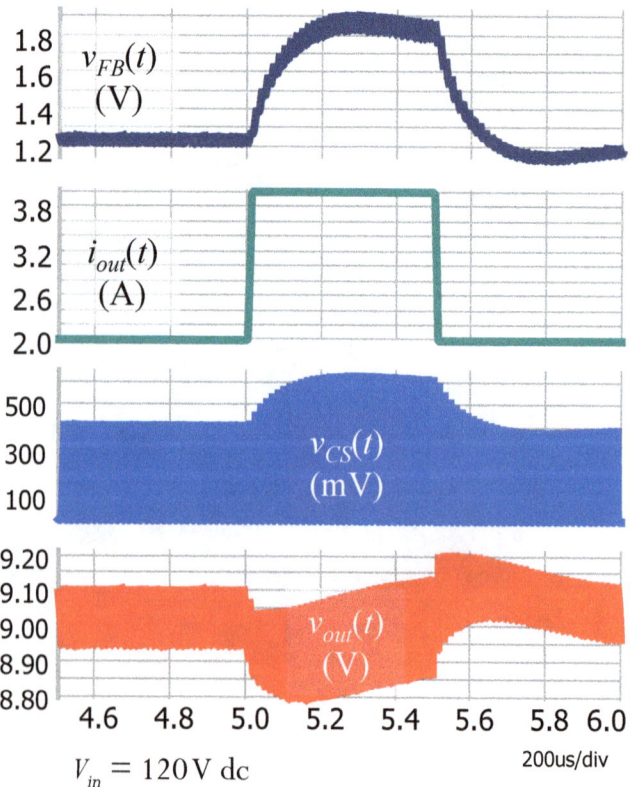

$v_{FB}(t)$ (V)

$i_D(t)$ (A)

$i_{sec}(t)$ (A)

$v_{CS}(t)$ (mV)

$v_{out}(t)$ (V)

$V_{in} = 120\,\text{V dc}$

10us/div

$v_{FB}(t)$ (V)

$i_{out}(t)$ (A)

$v_{CS}(t)$ (mV)

$v_{out}(t)$ (V)

$V_{in} = 120\,\text{V dc}$

200us/div

Flyback Converter in QR CM

THE FLYBACK CONVERTER can be operated in quasi-square-wave resonant mode, also known as quasi-resonant or QR. It is similar to the borderline operation of the boost converter used in PFC stages. The switching frequency varies with operating conditions and various schemes exist to keep it under control. In some controllers, a voltage-controlled-oscillator (VCO) can take the lead when light-load operation is entered. The VCO increases the switching period to reduce losses and improves efficiency as the load is getting lighter. If a deadtime is inserted before starting a new cycle, zero-voltage switching (ZVS) or quasi-ZVS can be obtained, further improving the overall efficiency. The below circuit shows a QR flyback operated in current mode. It relies on the auxiliary winding in the primary to detect the core demagnetization, cycle-by-cycle.

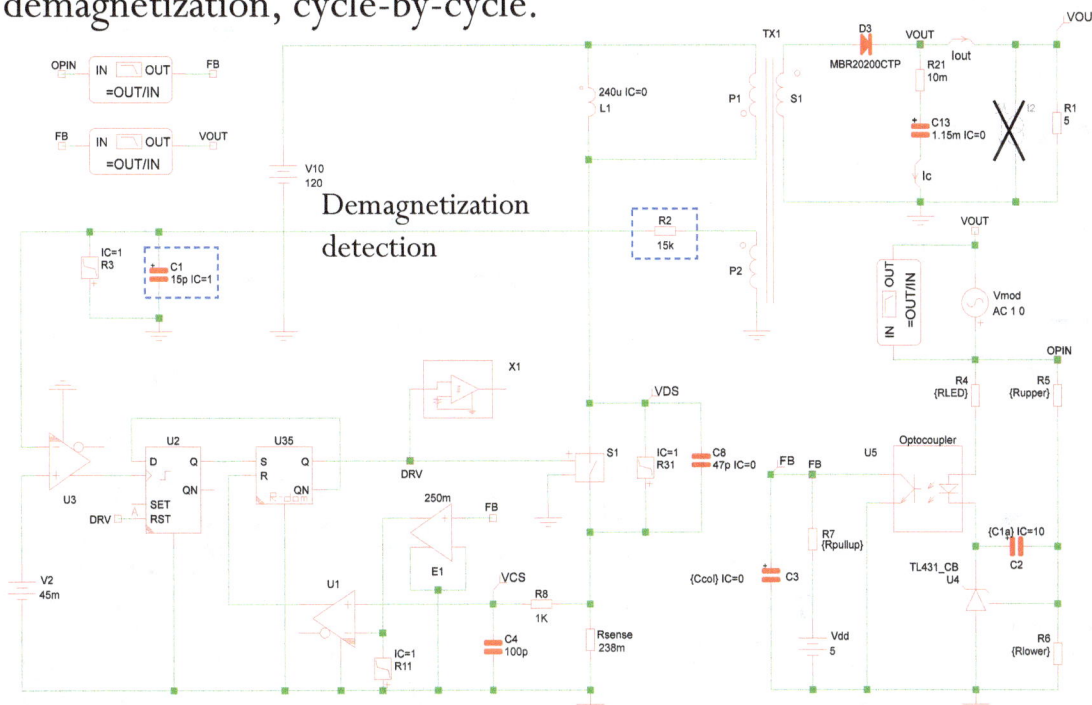

A small delay is inserted via $R_2 C_1$ which ensures ZVS in this operating point, with a switching frequency of 108 kHz.

In steady-state, when the secondary-side blocks, V_{out} is no longer reflected and the drain voltage drops. When it goes through a valley, the transistor turns back on, eliminating the turn-on switching loss when $v_{DS}(t)$ is zero.

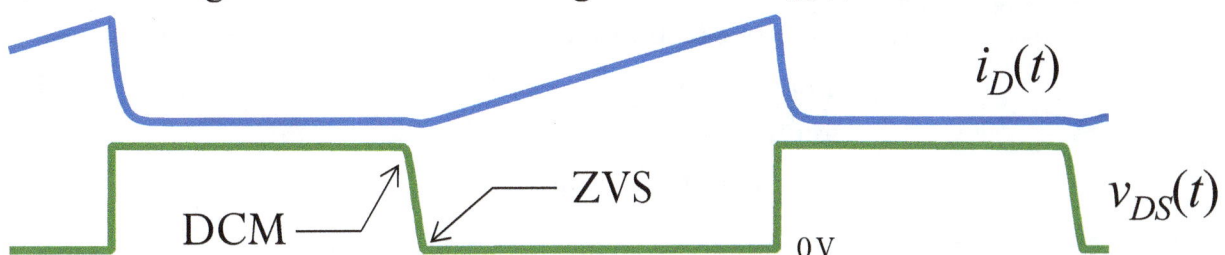

A 1st-Order Response

THE FLYBACK CONVERTER operated in QR and current-mode exhibits a first-order response with a high-frequency RHP zero. It thus naturally offers the possibility to push the crossover frequency. The ac response is shown below and is easily compensated via an automated type 2 compensator, built around a TL431 and an optocoupler. In this example, f_c is selected at 1 kHz.

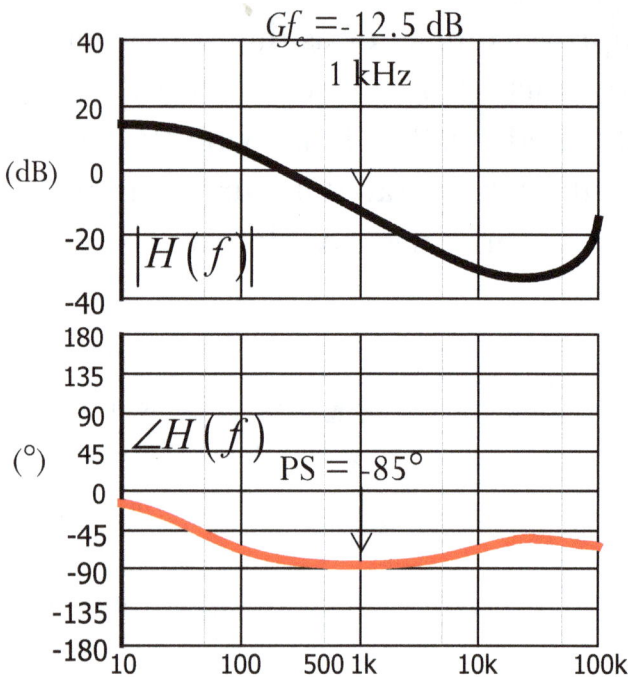

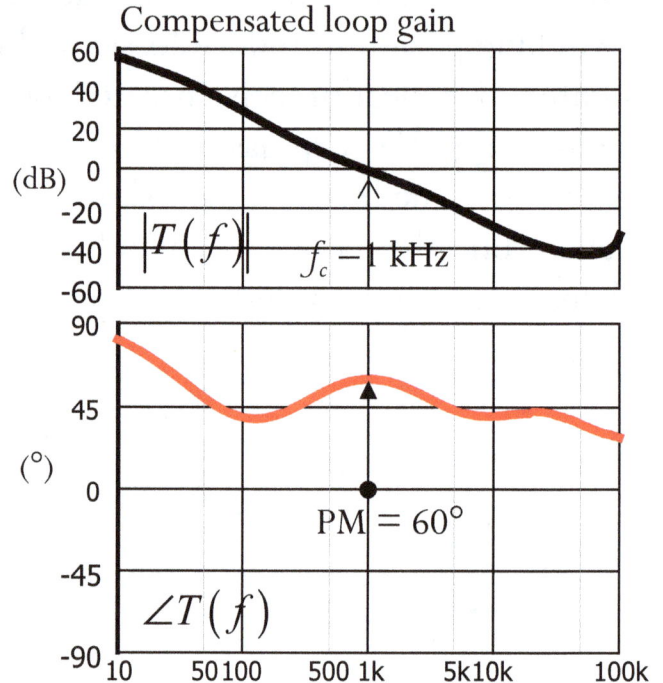

$Gf_c = -12.5$ dB

1 kHz

$|H(f)|$

$\angle H(f)$

PS = -85°

Compensated loop gain

$|T(f)|$ $f_c = 1$ kHz

PM = 60°

$\angle T(f)$

```
*
.VAR Gfc=-13.5 * magnitude at crossover *
.VAR PS=-85 * phase lag at crossover *
*
* Enter Design Goals Information Here *
*
.VAR fc=1k * targetted crossover *
.VAR PM=60 * choose phase margin at crossover *
*
* Enter the Values for Vout and Bridge Bias Current *
*
.VAR Vout=19
.VAR Ibias=250u
.VAR Vref1=2.5
.VAR Rlower=Vref/Ibias
.VAR Rupper=(Vout-Vref)/Ibias
*
* Optocoupler specifications *
*
.GLOBALVAR Rpullup=20k
.GLOBALVAR Fopto=6k
.GLOBALVAR Copto=1/(2*pi*Fopto*Rpullup)
.GLOBALVAR CTR=0.33
*
* Rupper  = 66000
* Rlower  = 10000
* C2 = 2.50906804004891e-09
* C1 = 7.64810591334346e-09
* Ccol = 1.18277684761645e-09
* Boost = 55
* Fz = 315.298788878984
* Fp = 3171.59480236321
* Rmax = 24995.7446808511
* RLED = 1394.90276629219   ← Ok
```

The compensator places a pole at 3.1 kHz and a zero at 315 Hz. The capacitor at the feedback pin is 1.2 nF while the LED series resistance is well below the maximum authorized. The transient response is good but the frequency variation is wide with this circuit.

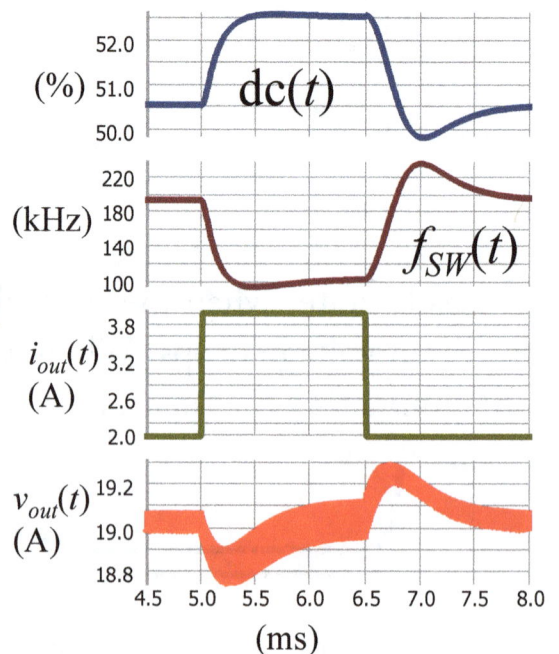

$dc(t)$

$f_{SW}(t)$

$i_{out}(t)$ (A)

$v_{out}(t)$ (A)

(ms)

Multi-Output QR Flyback Converter

THE SIMULATION CIRCUIT for a multi-output version does not really change. You just need to specify more windings on the transformer and adjust the turns ratios. The below example shows a typical QR flyback featuring an auxiliary winding and two outputs, 12 V/500 mA and 5 V/1 A. Regulation is done with the auxiliary winding (V_{cc} = 15 V) while the secondary windings are cross-regulated via the dc-stack of the 12-V over the 5-V output.

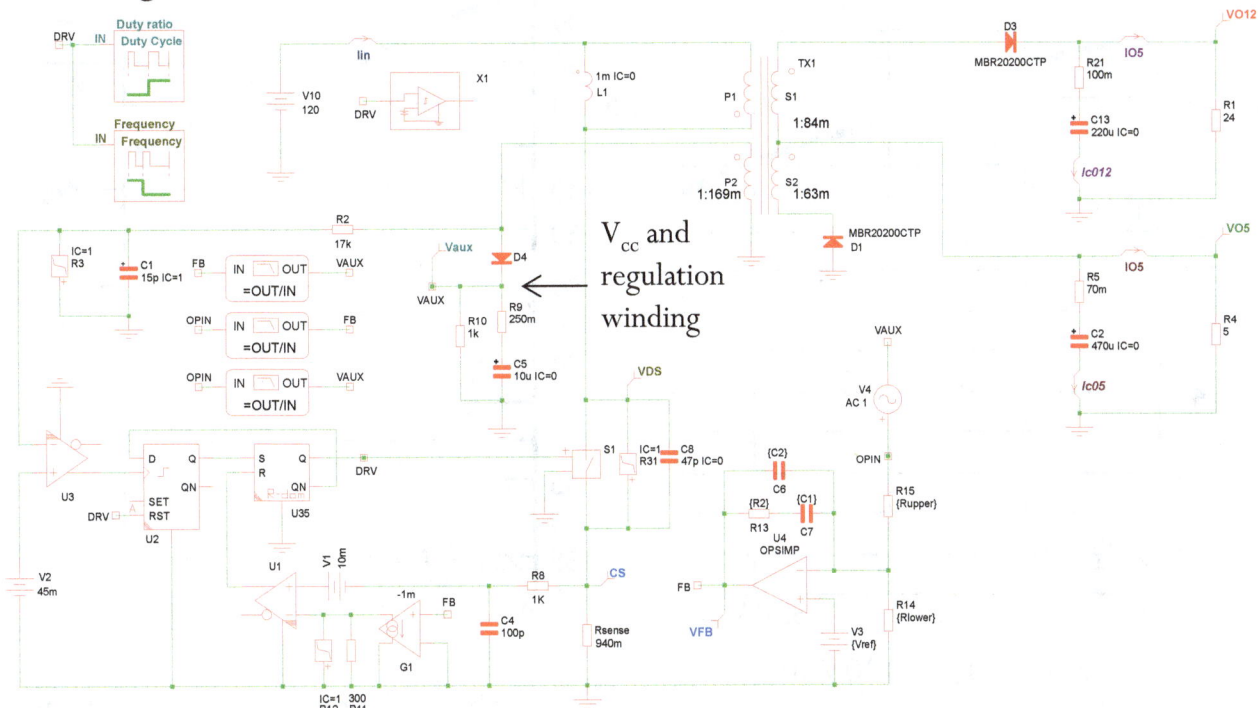

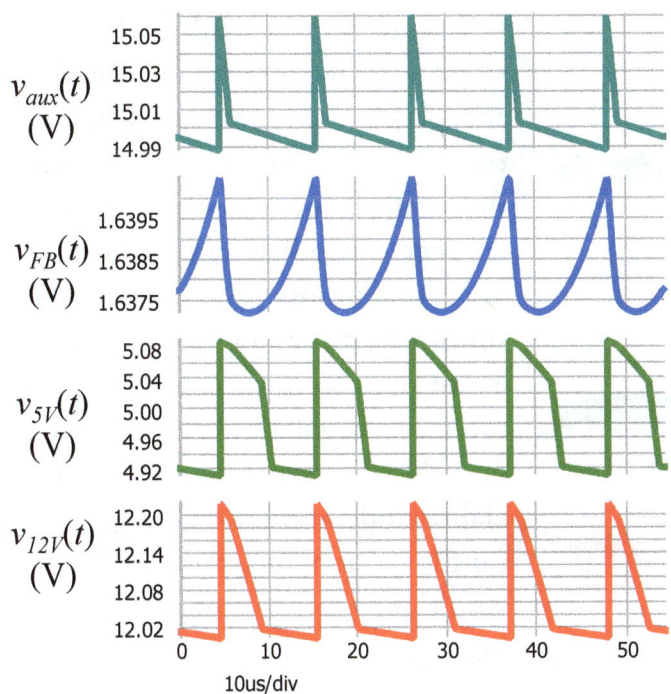

Define Ideal DC Transformer

Configuration

Primaries 2   # Secondaries 2

Winding	Number of turns	Polarity
1 Primary 1	1	-
2 Primary 2	169m	+
3 Secondary 1	84m	+
4 Secondary 2	63m	+

The multi-output transformer is modeled in the simplest way and does not reflect all the inter-winding leakage inductances, crucial to predict cross-regulation for the coupled windings. Ac injection takes place in series with the auxiliary V_{aux}.

Compensated QR Multi-Output

THE COMPENSATION EXERCISE does not differ from the previous examples and the automated macro featuring the k-factor method remains identical. A 1-kHz crossover frequency is adopted here, with a 60° phase margin goal.

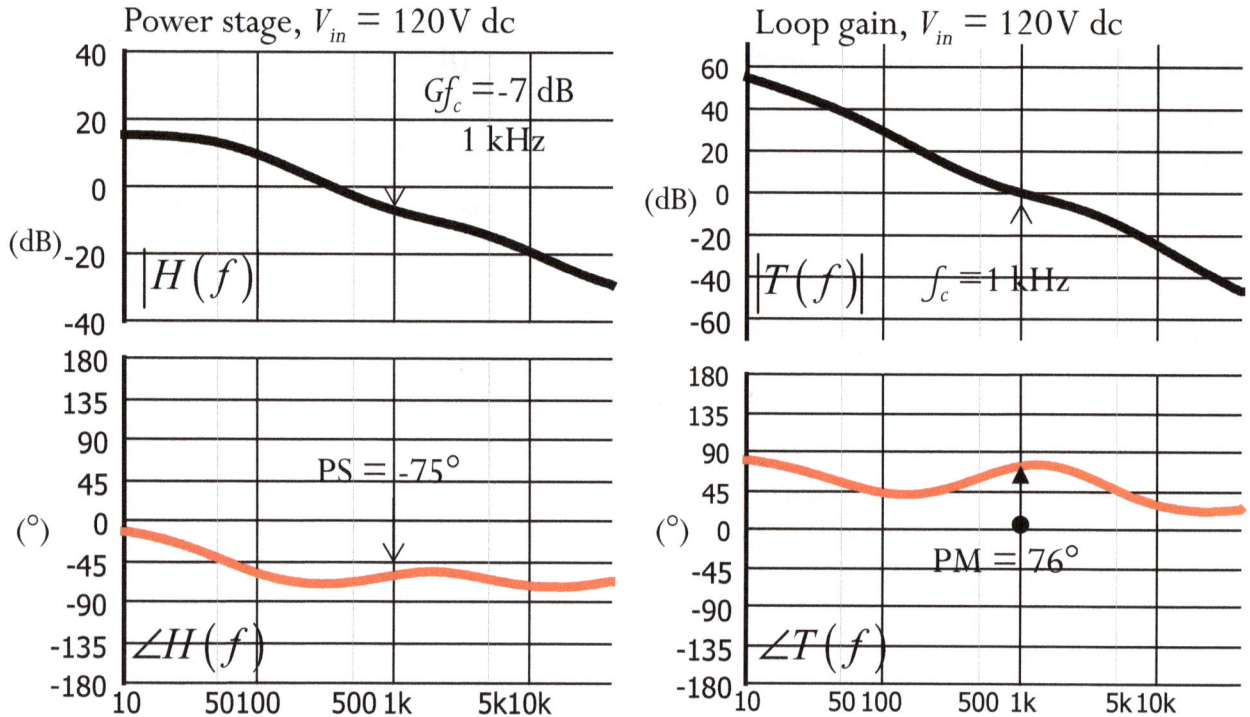

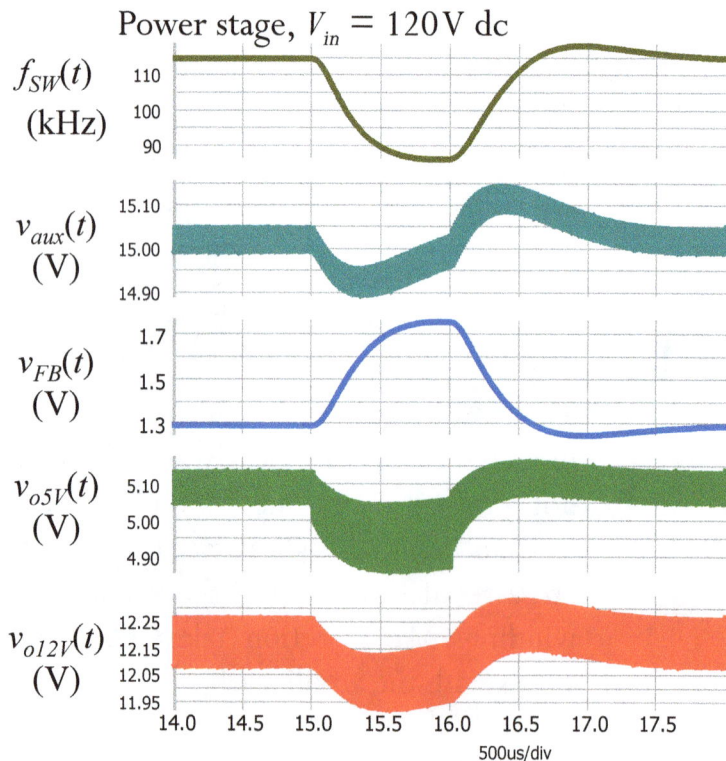

Power stage, $V_{in} = 120\,\text{V}$ dc

$Gf_c = -7$ dB
1 kHz

$|H(f)|$

(dB)

$PS = -75°$

$\angle H(f)$

(°)

Loop gain, $V_{in} = 120\,\text{V}$ dc

(dB)

$|T(f)|$ $f_c = 1$ kHz

$PM = 76°$

$\angle T(f)$

(°)

Power stage, $V_{in} = 120\,\text{V}$ dc

$f_{SW}(t)$ (kHz)

$v_{aux}(t)$ (V)

$v_{FB}(t)$ (V)

$v_{o5V}(t)$ (V)

$v_{o12V}(t)$ (V)

500us/div

The 5-V output is stepped from 0.5 A to 1.1 A with a 1-A/µs slope. The drop remains acceptable but, as underlined, this simulation is purely theoretical to check for the correct stability in different operating points. Considering the simple transformer subcircuit, it is not possible to precisely model the cross-regulation performance of this converter. A more elaborated model – e.g. the cantilever model from CoPEC – would be better suited for this exercise.

Weighted Feedback Arrangement

FOR MULTI-OUTPUT APPLICATIONS in which a better cross-regulation is required, so-called *weighted feedback* may be implemented. The principle is to assign a weight to each output, depending on its importance in terms of regulation. In our two-output QR example, assume the 5-V is the most important output. We thus assign it a weight of 75% while the 12-V will be at 25%. The assignment of these weights is made by adding more sensing resistances:

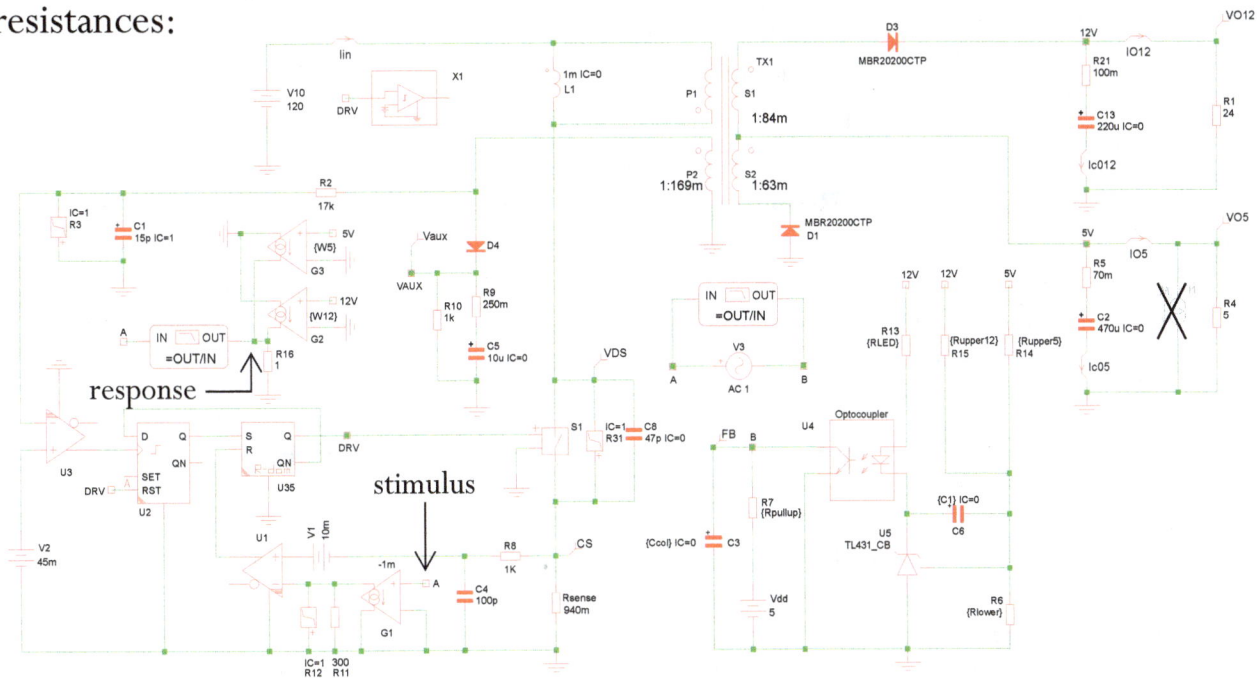

The small-signal analysis of this circuit would be extremely complicated, now considering the two weighted outputs driving the compensator and the LED current contributed by the 12-V output. In this circuit, I have used an approximate reconstruction of the power stage response by combining the two outputs, weighted accordingly via sources G_2 and G_3. Regarding the compensator, the resistance used for determining the capacitor value C_1 is actually made by paralleling the 12- and 5-V resistances. It is an approximation, but good enough for this exercise.

```
.VAR Vout1=5
.VAR Vout2=12
.VAR W5=0.75        75% weight for the 5 V
.VAR W12=0.25       25% weight for the 12 V
.VAR Ibias=250u
.VAR Vref=2.5
.VAR Rlower=Vref/Ibias
.VAR Rupper5=((Vout1-Vref)/Ibias)/W5
.VAR Rupper12=((Vout2-Vref)/Ibias)/W12
.VAR Rupper=Rupper5*Rupper12/(Rupper5+Rupper12)
*
```

```
* Rupper5 = 13333.3333333333
* Rupper12 = 152000
* Rupper = 12258.064516129
* Rlower = 10000
* C2 = 3.29620679736906e-09
* C1 = 3.13454070685956e-08
* Ccol = 1.9699156049366e-09
* Boost = 45
* Fz = 414.213562373095
* Fp = 2414.2135623731
* Rmax = 15165.9574468085
* RLED = 1657.84504479632
```

Compensated Loop and Transient Step

THE RECONSTRUCTION of the response with the two weighted controlled sources gives a first indication of the power stage control-to-output transfer function. It allows us to run a first analysis to feed the macro with gain/attenuation and phase values at the selected 1-kHz crossover frequency. Once the compensated loop gain is obtained, the compensator input data are manually adjusted until the right crossover is obtained. A 3-dB reduction was necessary after the 1st pass to obtain exactly 1 kHz.

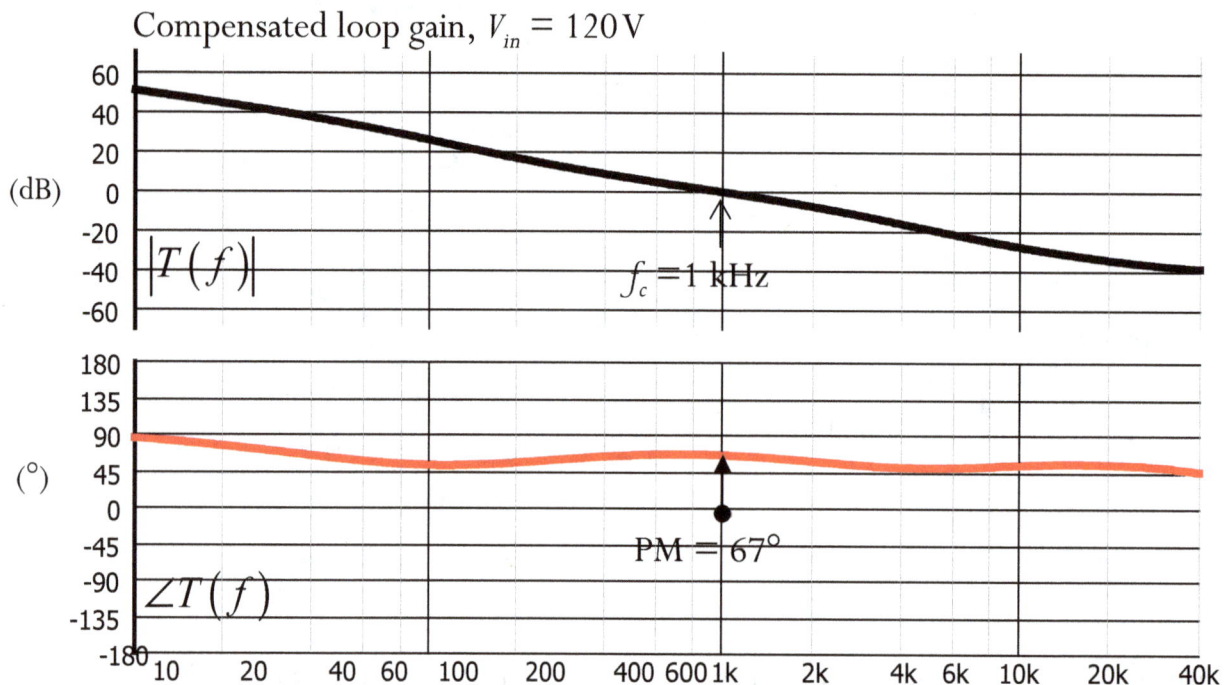

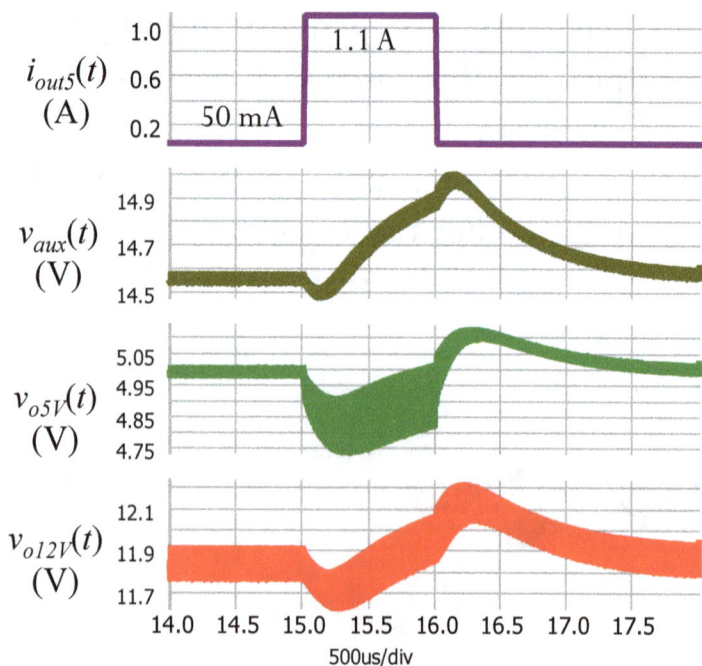

Compensated loop gain, $V_{in} = 120\,\text{V}$

The 5-V output is stepped from 0.5 A to 1.1 A with a 1-A/µs slope. The drop remains acceptable but, as underlined, this simulation is purely theoretical to check for stability at different operating points. Considering the simple transformer subcircuit, it is not possible to precisely model the cross-regulation performance of this converter. A more elaborated model – e.g. the cantilever model from CoPEC – would be better suited for this exercise.

Second-Stage *LC* Filter

YOU OFTEN SEE it in ac-dc adapters: a small inductor is installed after the main capacitor and combines with the last capacitor to form a low-pass filter. It reduces the differential ripple voltage on the output, before supplying the downstream circuit. As you can imagine, the ac response of this extra filter can affect the loop stability if no precautions are taken. In particular, the LED current must be provided *before* the *LC* filter, considering the zero it introduces. On the contrary, the output voltage is still sensed after the filter.

A rule of thumb for this *LC* filter is to keep its resonant frequency at least 10 times above the *zero* frequency [10] (300 Hz in this example).

The filter is tuned at ≈ 11 kHz

Insert the ac source after the compensator

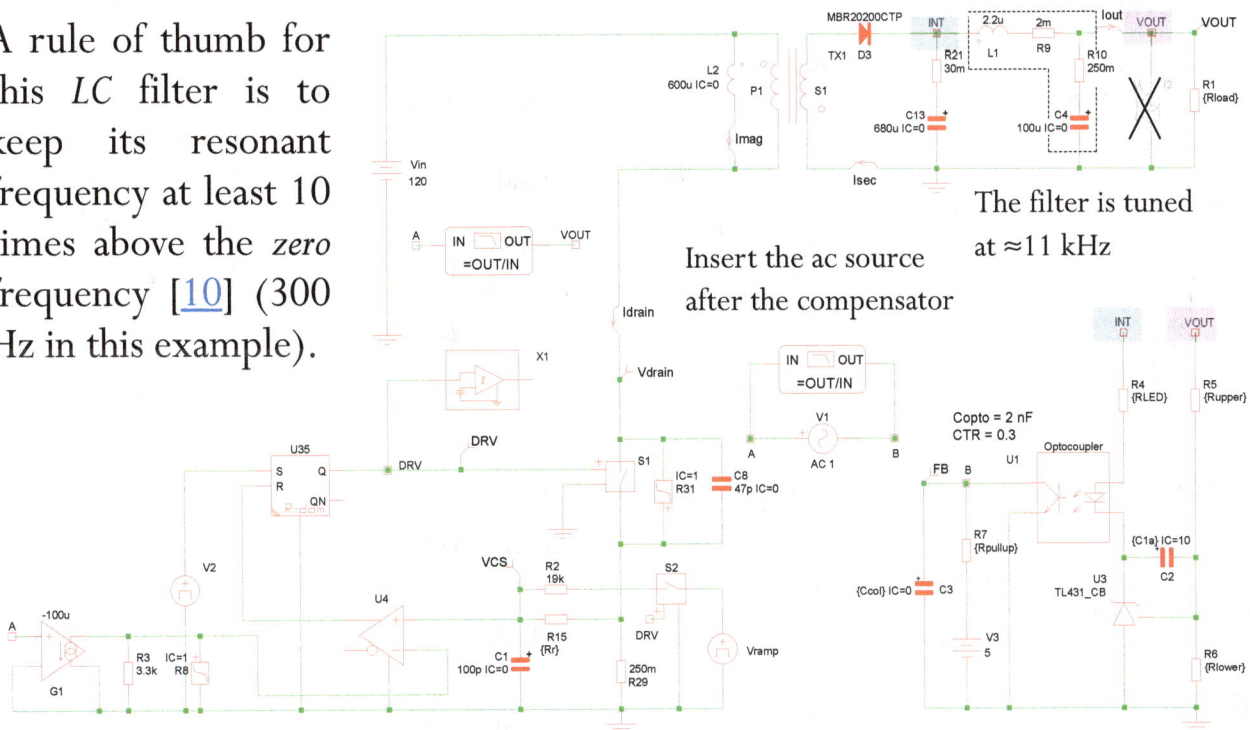

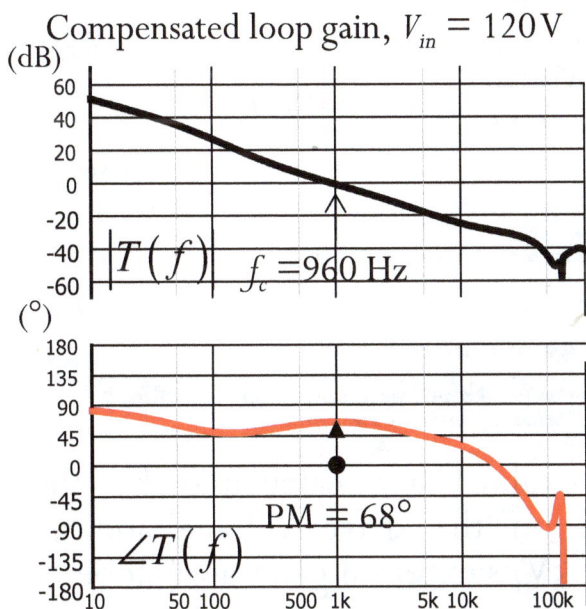

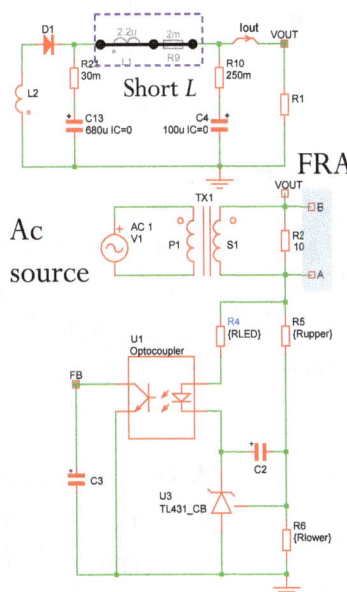

Compensated loop gain, $V_{in} = 120$ V

$|T(f)|$ $f_c = 960$ Hz

$\angle T(f)$ PM $= 68°$

Ac source

Short *L*

FRA

For the loop measurement with a frequency-response analyzer (FRA), you will need to short the inductor and connect both INT and VOUT nodes together for proper injection. Sweeping either one alone would lead to a wrong plot.

Slope Compensation with a UC384x

THE UC384x PWM CONTROLLER is a popular integrated circuit released in the 80's by Unitrode (later acquired by Texas Instruments) and now available from many different vendors. It used a bipolar process and some Bi-CMOS versions now exist for reduced power consumption and better performance. It does not feature leading-edge blanking (LEB) or slope compensation when then must be externally provided. For this latter, I do not recommend using Unitrode's recommendation, as this may disturb the oscillator:

Unitrode U-111, *Practical Considerations in current mode power supplies*

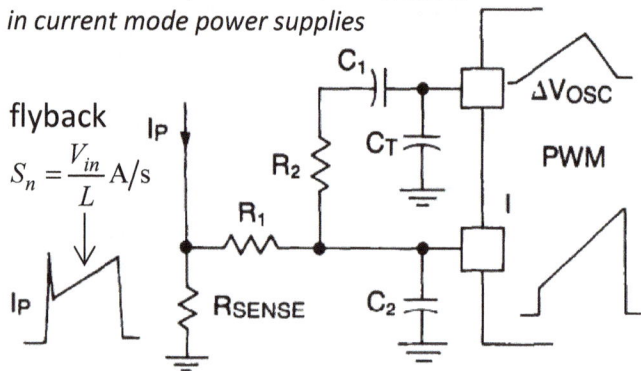

flyback

$$S_n = \frac{V_{in}}{L} \text{A/s}$$

Recommended solution from Unitrode

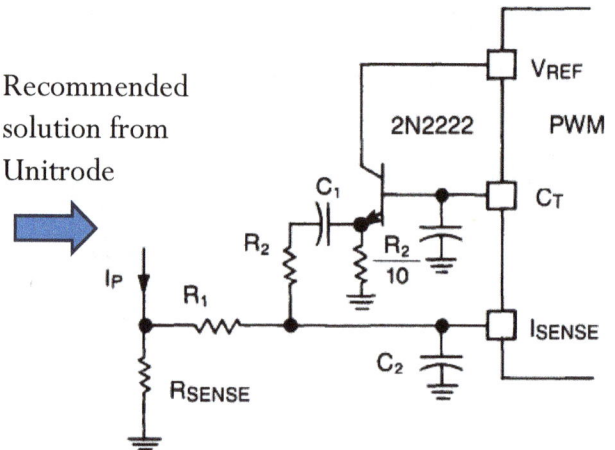

Rather, an *RCD* network supplied from the low impedance drive pin can provide a raw sawtooth, suitable for slope compensation purposes [6].

R_1 connects to the DRV pin if V_{cc} is fixed. If not, it must go to a stable dc source, e.g. 14-16 V for a 2-V typical linear ramp.

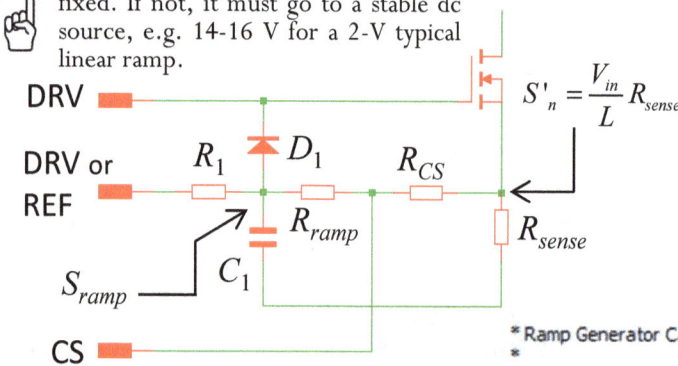

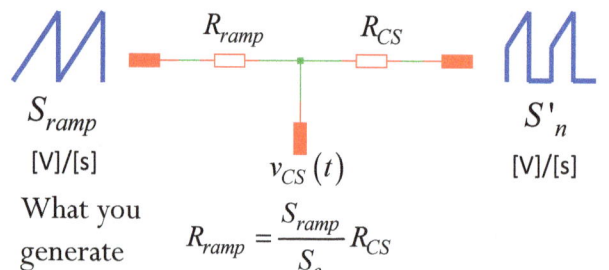

$$S'_n = \frac{V_{in}}{L} R_{sense}$$

S_{ramp} [V]/[s]

What you generate

$$R_{ramp} = \frac{S_{ramp}}{S_e} R_{CS}$$

$v_{CS}(t)$

S'_n [V]/[s]

What you need to damp poles.

Considering $R_{CS} \ll R_{ramp}$:

$$S_{ramp} \approx \frac{V_{drv} \dfrac{R_{ramp}}{R_{ramp} + R_1}\left(1 - \dfrac{t_{on}}{C_1\left(R_1 \parallel R_{ramp}\right)}\right)}{C_1\left(R_1 \parallel R_{ramp}\right)}$$

$$C_1 = V_{drv} \frac{1 + \sqrt{1 - \dfrac{4 S_{ramp} t_{on}\left(R_1 + R_{ramp}\right)}{R_{ramp} V_{drv}}}}{2 R_1 S_{ramp}}$$

* Ramp Generator Calculation
*
.VAR Vcc=20 * The Vcc is the supply voltage and the peak DRV
.VAR Rramp=19k * Ramp injection resistance
.VAR Ichg=3m * Ramp charge current
.VAR Rchg=Vcc/Ichg
.VAR Sr=300k * create a slope of 300 mV/µs
.VAR ton=D*Tsw
.VAR Cramp=(1+sqrt(1-(4*Sr*ton*(Rchg+Rramp)/(Rramp*Vcc))))*Vcc/(2*Rchg*Sr) *
.VAR RCS=Rramp*Se/Sr * Sense resistance to CS pin resistance
*

Components values are calculated by the macro based on the drive voltage and an arbitrary 300 kV/s external generator slope.

A Flyback DC-DC with a UC384x

A NON-ISOLATED FLYBACK converter is shown below. It uses the internal op-amp of the integrated circuit whose non-inverting pin is referenced to a 2.5-V bias voltage. The input voltage is 120 V and it delivers 19 V / 3 A.

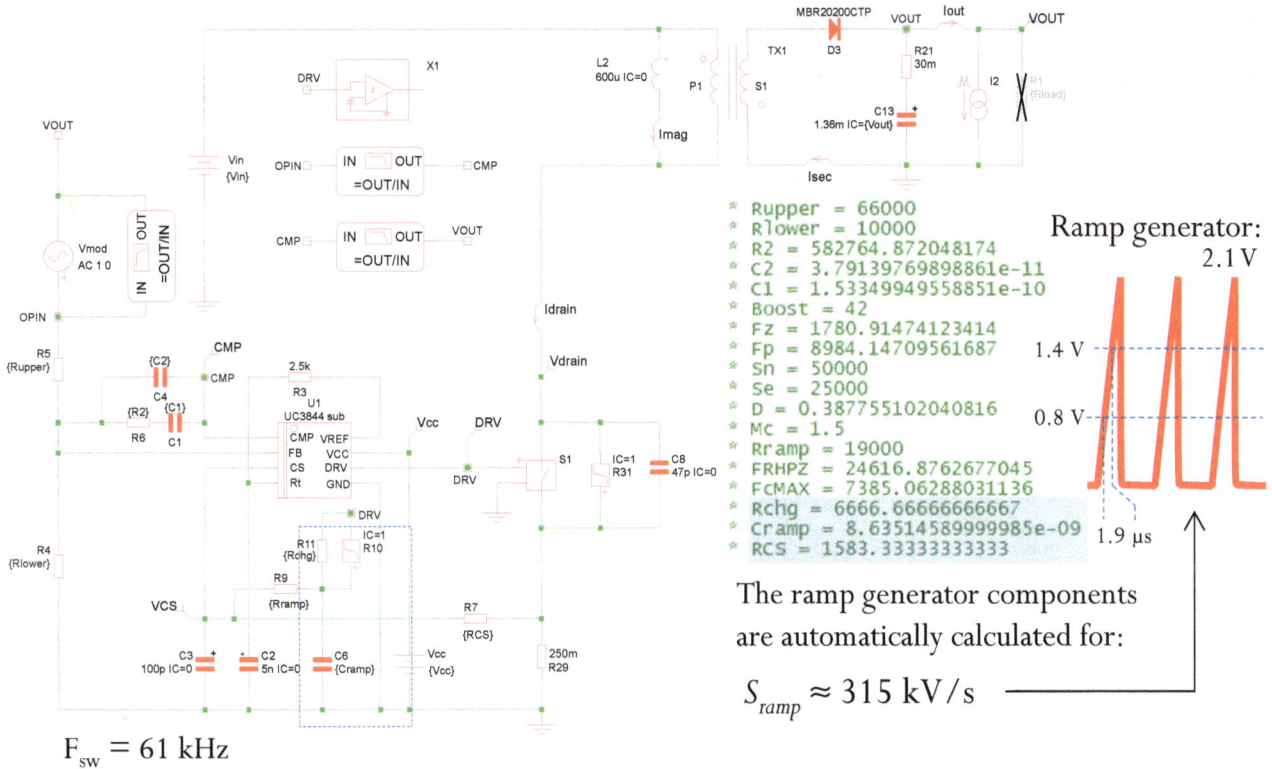

```
*  Rupper = 66000
*  Rlower = 10000
*  R2 = 582764.872048174
*  C2 = 3.79139769898861e-11
*  C1 = 1.53349949558851e-10
*  Boost = 42
*  Fz = 1780.91474123414
*  Fp = 8984.14709561687
*  Sn = 50000
*  Se = 25000
*  D = 0.387755102040816
*  Mc = 1.5
*  Rramp = 19000
*  FRHPZ = 24616.8762677045
*  FcMAX = 7385.06288031136
*  Rchg = 6666.66666666667
*  Cramp = 8.63514589999985e-09
*  RCS = 1583.33333333333
```

Ramp generator: 2.1 V
1.4 V
0.8 V
1.9 μs

$F_{sw} = 61$ kHz

The ramp generator components are automatically calculated for:

$$S_{ramp} \approx 315 \text{ kV} / \text{s}$$

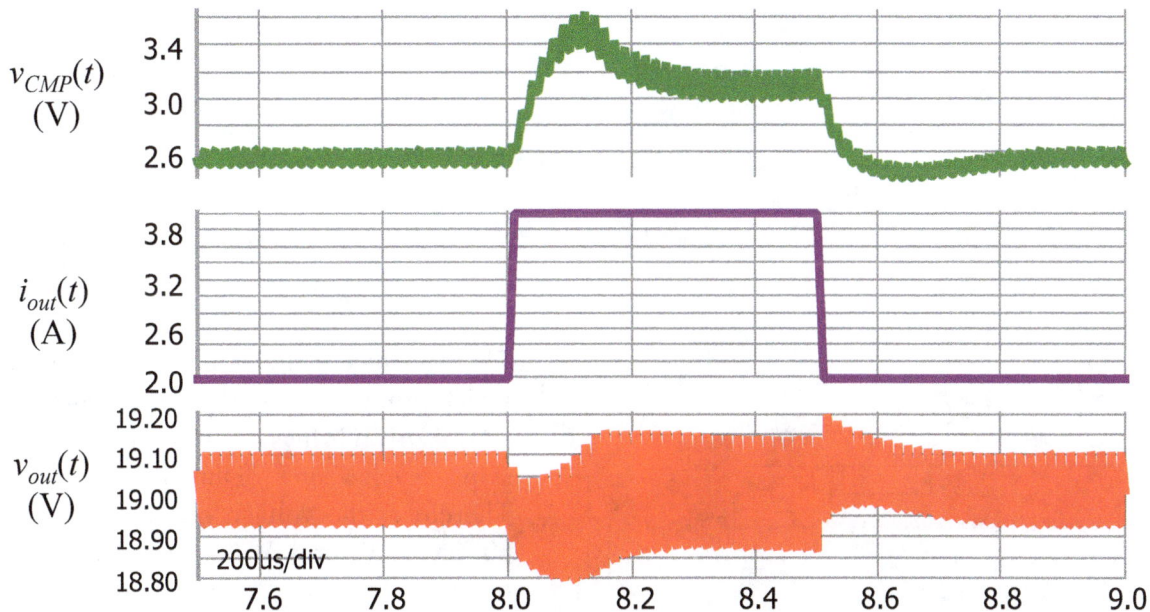

The transient response is stable and zooming-in does not reveal any signs of subharmonic instabilities.

Isolated Feedback with UC384x

THE PREVIOUS CONVERTER is a non-isolated flyback power supply. However, the UC controller lends itself well to building isolated circuits, for instance with a TL431 and an optocoupler. In that case, I recommend to completely disable the internal op-amp by grounding the FB pin and connecting the CMP pin to the 5-V reference voltage via a pull-up resistance. This resistance will affect the optocoupler parasitic pole. If you want to push it sufficiently high, a 4.7- to 10-kΩ resistance is a possible choice but to the detriment of standby power.

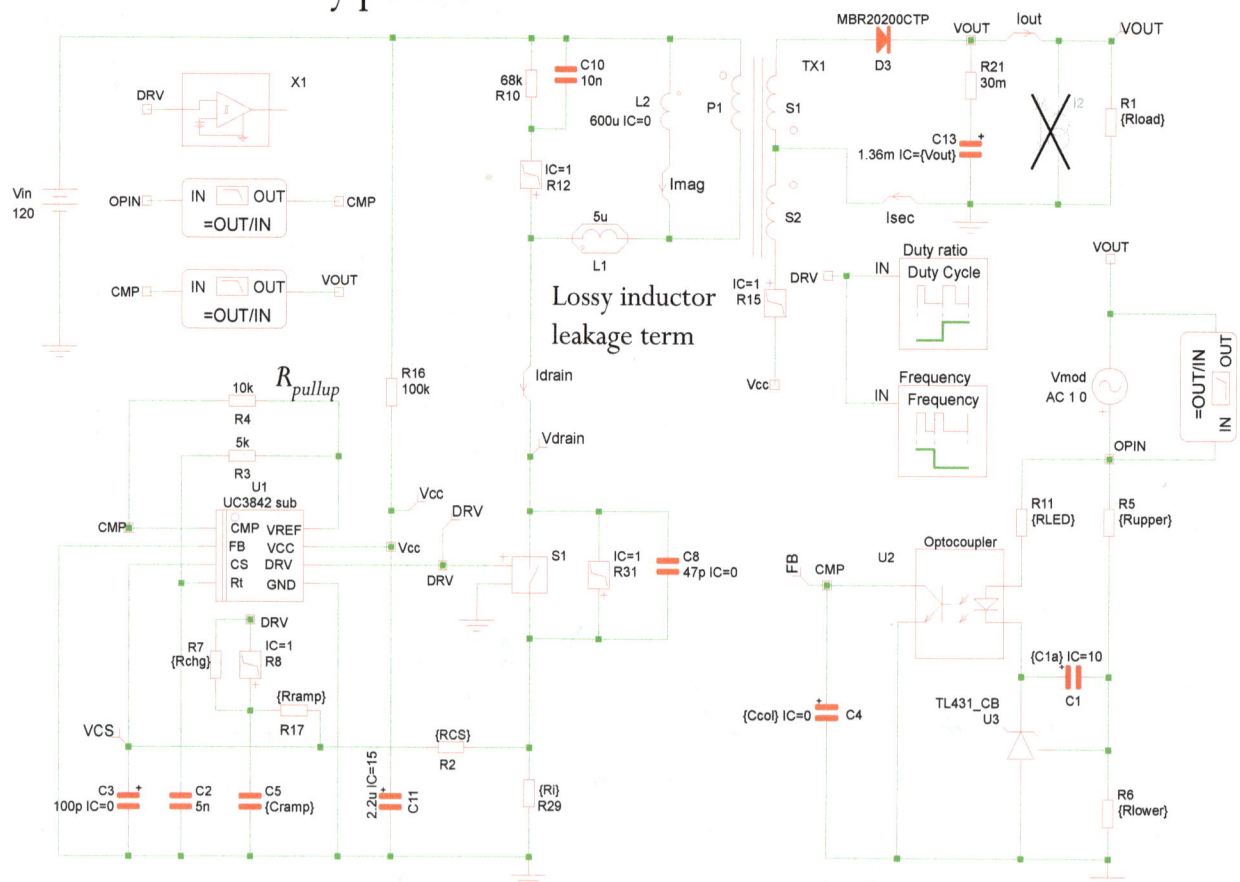

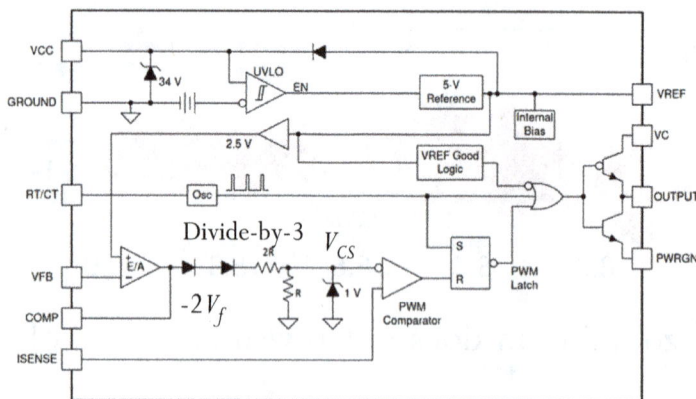

The UC384x includes a divide-by-3 circuit which improves the op-amp dynamics. A $\approx$1.2-V drop brings a 0-V setpoint and offers skip cycle in light-load conditions. Therefore, the voltage on CMP or COMP pin is 4.2 V for 1-V V_{CS} level. In light load, the IC will enter 0% duty ratio for V_{COMP} below 1.2 V. The op-amp max. current is 1 mA hence the safe control of CMP or COMP via the optocoupler directly.

Single-Stage Flyback Converter

A SINGLE-STAGE CONVERTER combines in one structure the power factor correction and the dc-dc stages. The flyback converter is well suited for this operation and is extremely popular in lighting circuits. The structure is fairly similar to that of a classical flyback converter except that there is no front-end bulk capacitor. The high-voltage rail is actually a rectified sine wave and a large electrolytic can is present on the isolated dc output to minimize the 100- or 120-Hz ripple. In the below circuit, a multiplier is implemented to build the peak current setpoint whose value is adjusted by a low-frequency feedback loop. For the sake of simplicity, there is no isolation.

The ac response is extracted with a dc bias whose value is the rms line voltage.

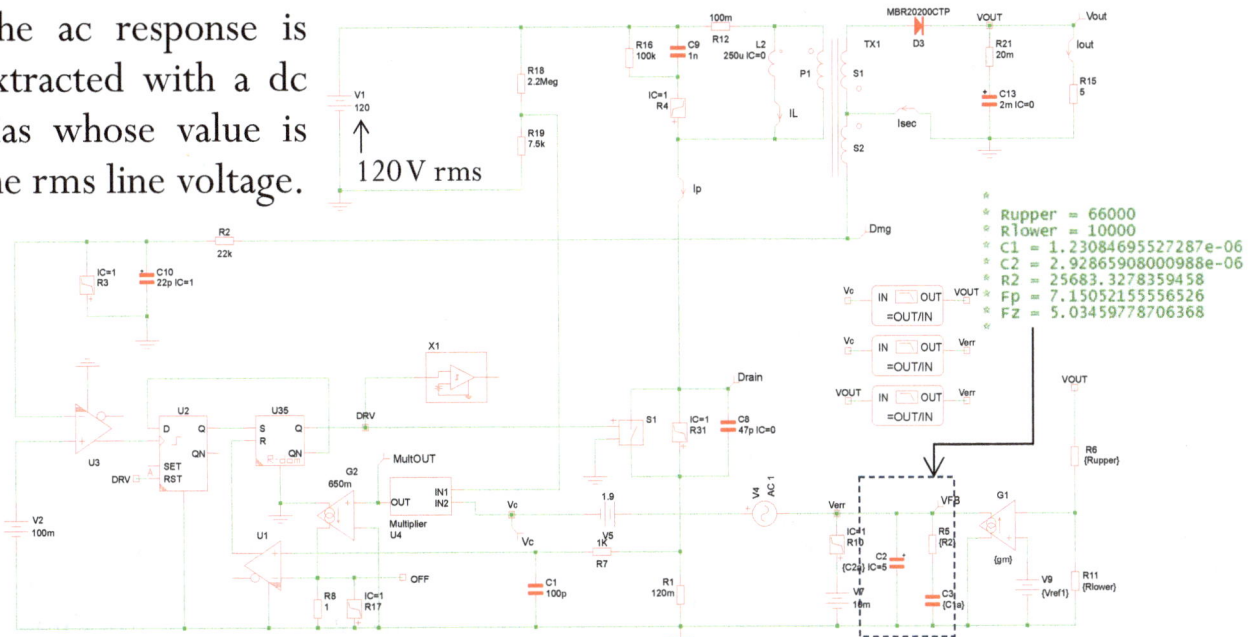

The compensation with an OTA is straightforward and leads to a f_c of 6.5-Hz.

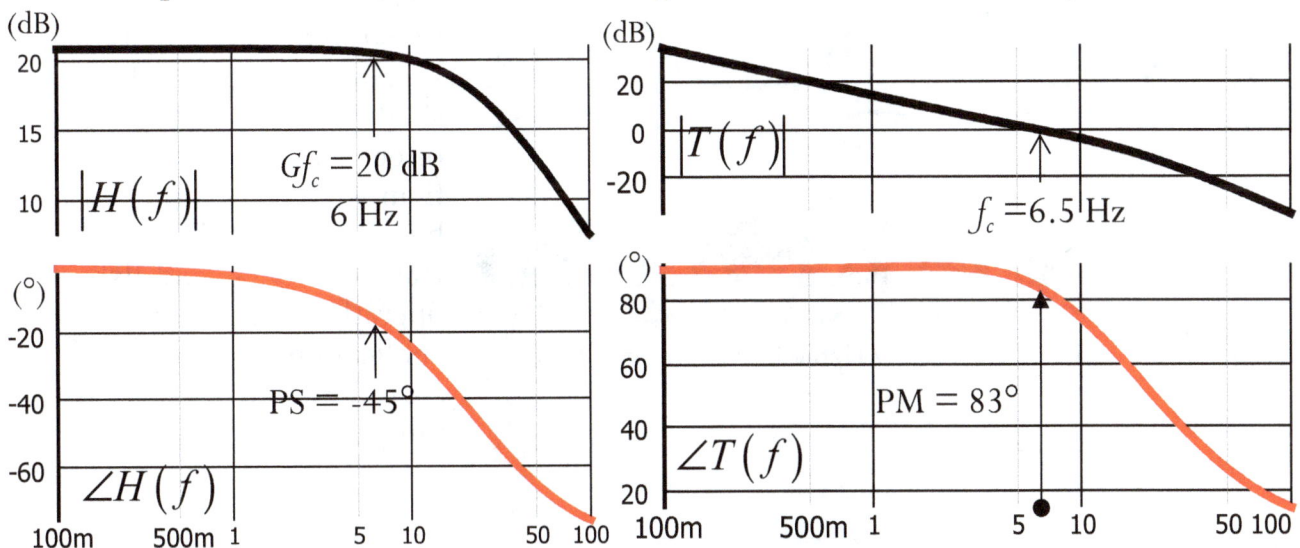

Operating Waveforms

ONCE THE CONVERTER is stabilized with the dc-biased circuit, we can run the full circuit with a sinusoidal source and check the startup sequence followed by the steady-state signals. The current-mode version requires the sensing of the input voltage to feed the multiplier but a voltage-mode option is also doable and won't require a high-voltage network.

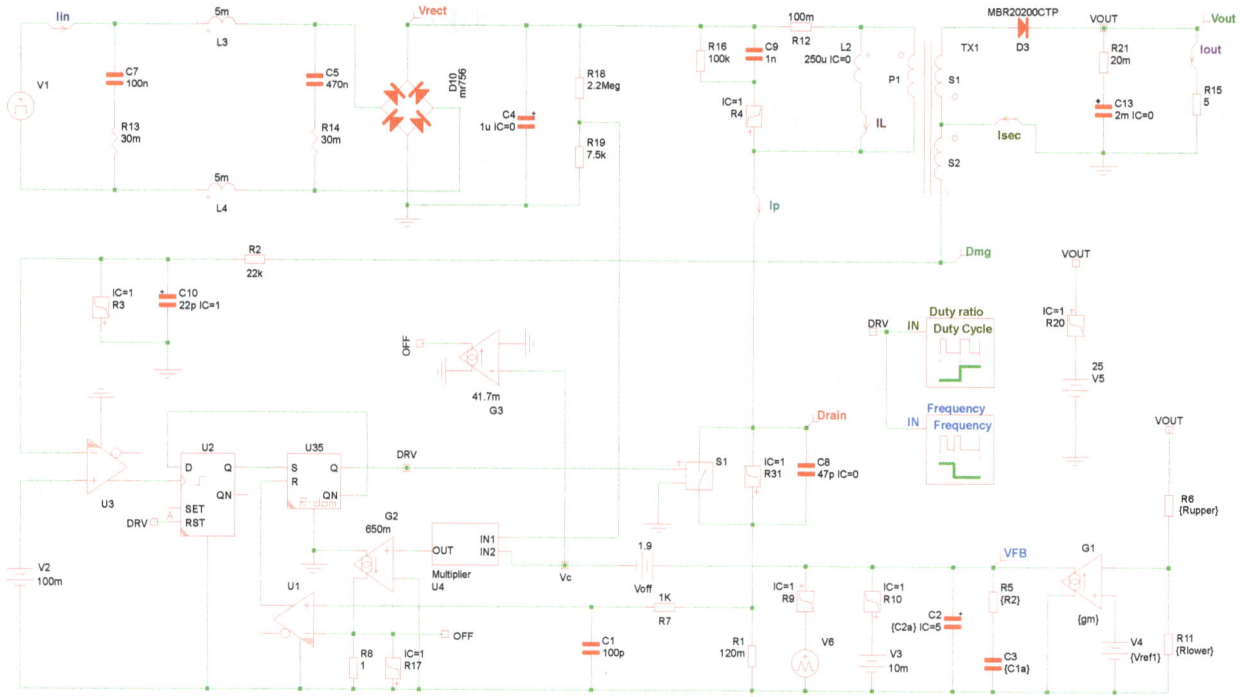

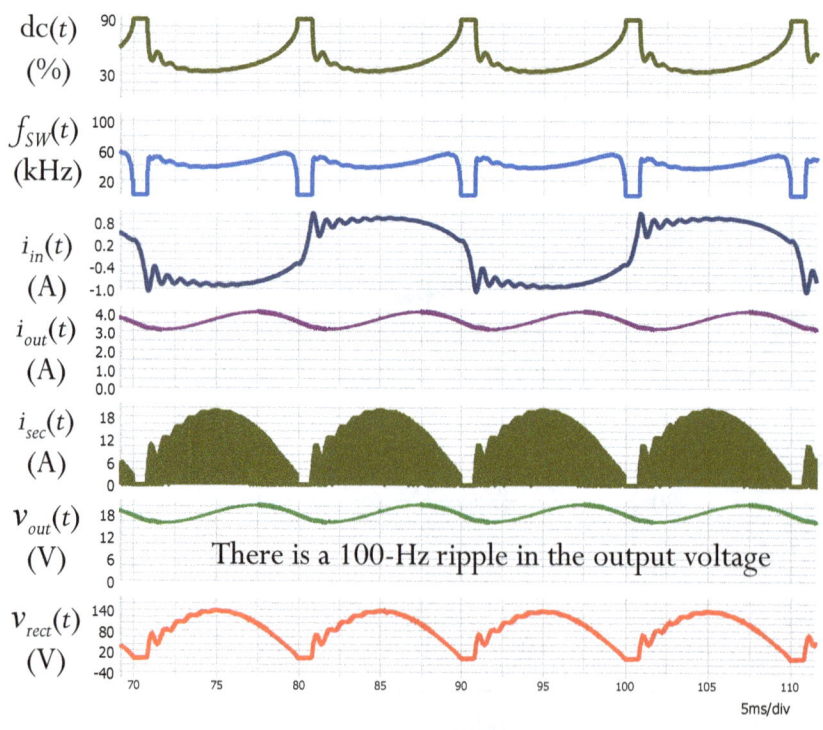

There is a 100-Hz ripple in the output voltage

The input current is approaching a sinusoidal waveform but still exhibits a high distortion level of 27%. Newly-released integrated circuits, specifically tailored for single-stage converters, offer better distortion figures, but the extra burden on the power transistor (PFC and flyback operations) limits the application of these single-stage PFCs to a low-power range.

SEPIC CM with Coupled Inductors

THE SINGLE-ENDED PRIMARY-INDUCTOR CONVERTER can be described as a boost front-end converter, coupled to an inverted buck-boost. It can be assembled in different ways, with coupled or un-coupled inductors and operated in voltage- or current-mode control. Even in CM, it remains a 5th-order system and requires care in closing the loop around it. Coupled inductors in SIMPLIS$^{®}$ must be modeled without resorting to the coupling factor k, which is only available in SIMetrix$^{®}$. You can either go through an equivalent transformer featuring a leakage inductance or use the keyword M which refers to the mutual inductance linking two inductors. We will first determine M and add its value in the control window by pressing F12:

The below SEPIC delivers 15 V/3 A from a 10-20-V dc source. The coupling coefficient of 0.96 translates into a mutual inductance of 96 μH.

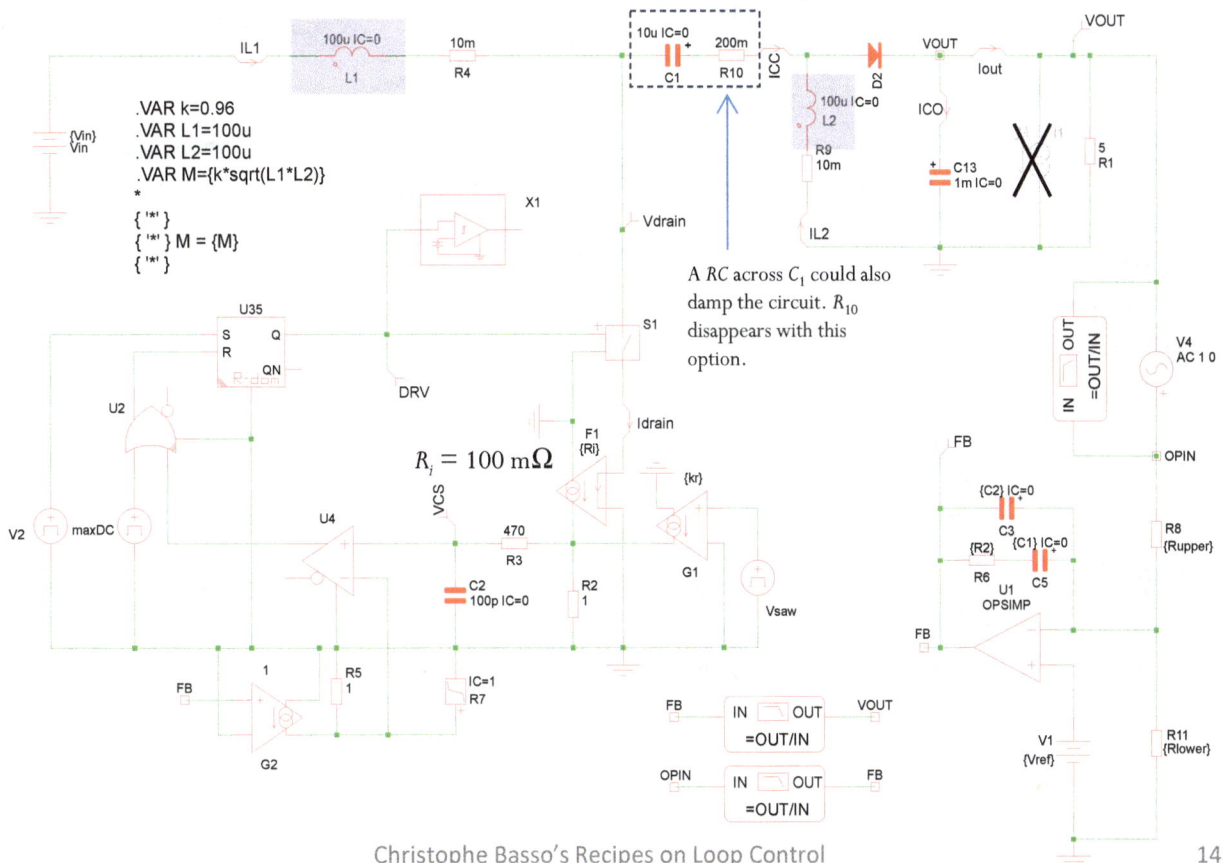

Operating Point

THE CIRCUIT delivers 3 A with a duty ratio exceeding 50% at 10 V. At this input voltage, the rms current in the coupling capacitor is high (4 A) and the series damping resistance R_{10} will dissipate power. The ripple current in the output capacitor also approaches 4 A and you will have to select it carefully.

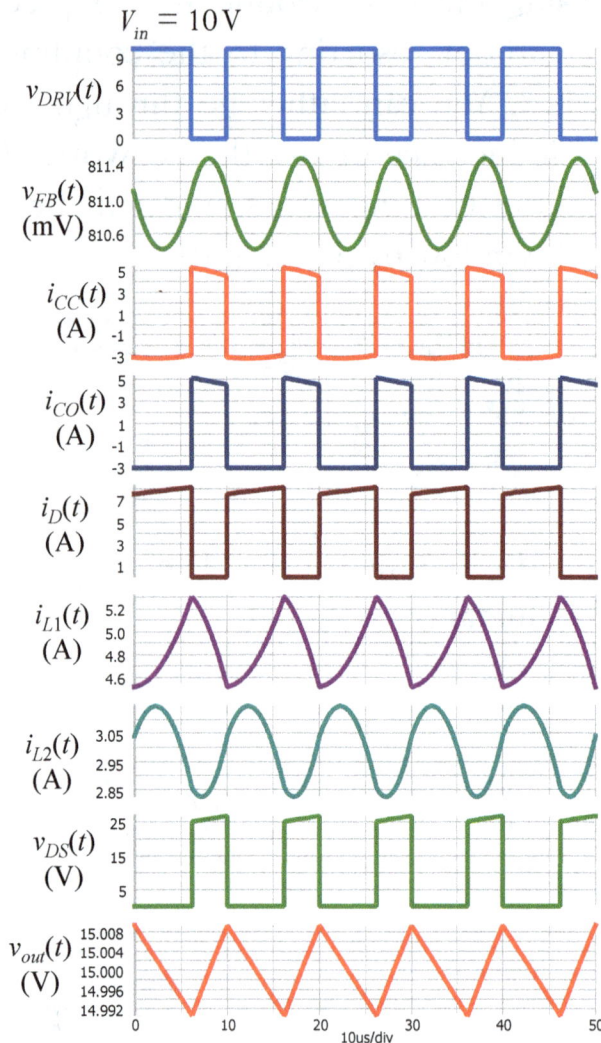

$V_{in} = 10\,\text{V}$

The upper crossover limit is linked to a RHP zero and severely limits the options here:

```
.VAR fRHPZ={((1-D)^2*Rload/(D*L))/(2*pi)}
.VAR fcMAX=0.3*fRHPZ
```

$V_{in} = 10\,\text{V}$

```
* D = 0.6
* fRHPZ = 2122.06590789194
* fcMAX = 636.619772367582
*
```

We will inject 50% of slope compensation and choose a 400-Hz crossover frequency.

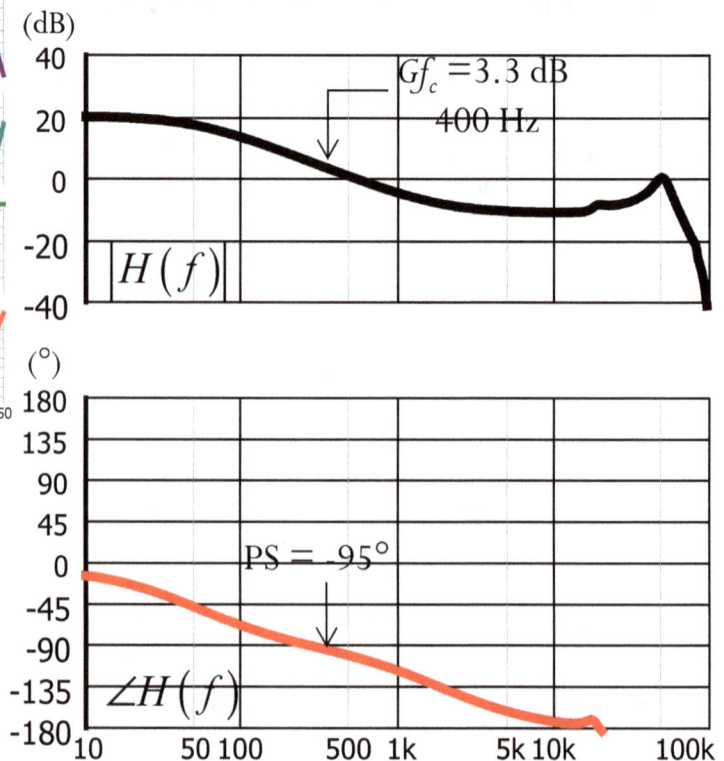

The type 2 will place a zero at 70 Hz and a pole at 2.3 kHz. The boost in phase amounts to 70°. At high-line (20 V), the power stage magnitude will slightly shift up and naturally pushes the RHPZ higher, improving phase margin.

AC Response and Transient Step

THE SEPIC IS NOW compensated and shows an acceptable phase margin at the 400-Hz crossover frequency and a 10-V input. f_c slightly increases at the 20-V input with a phase margin approaching 70°.

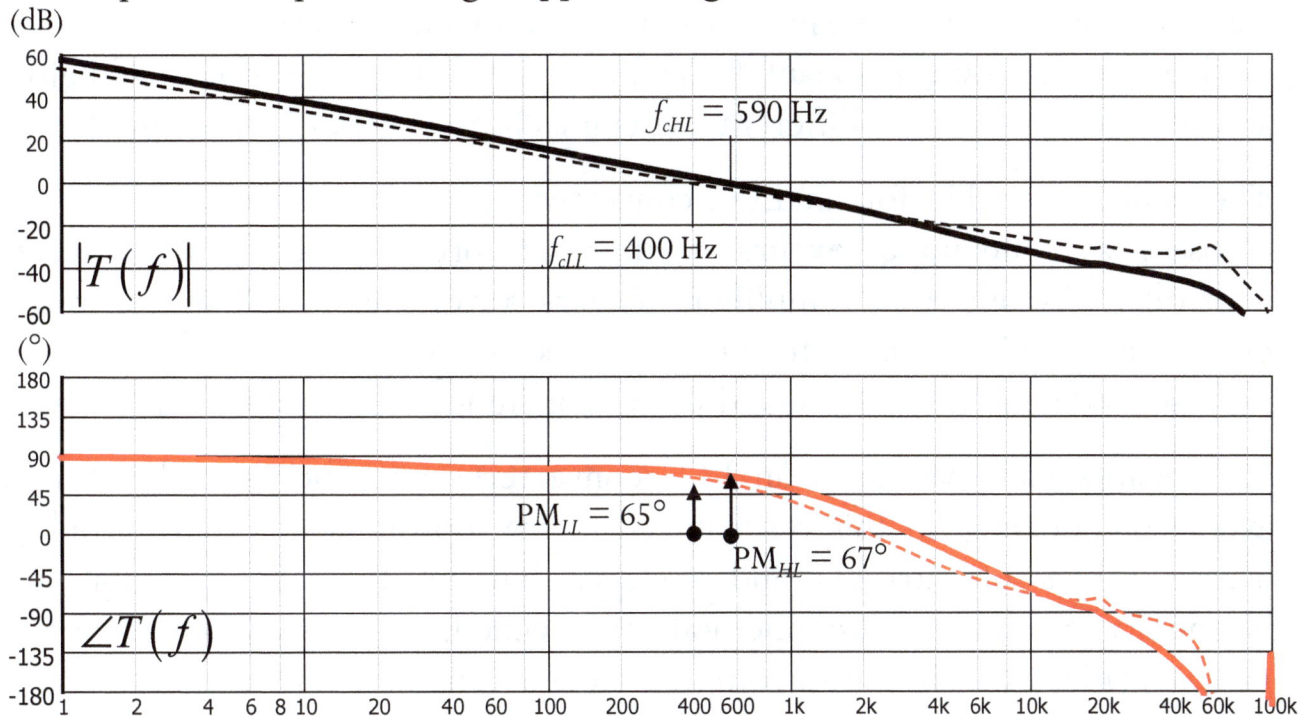

The transient response shows a 250-mV drop from a 20-V input when the output current is stepped from 2 to 3 A in 1 A/μs.

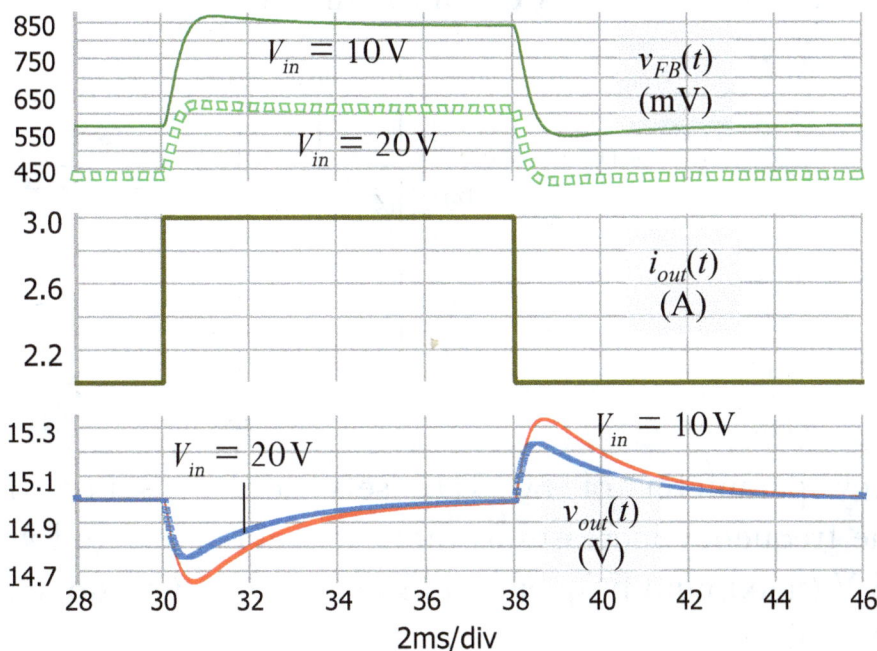

The drop increases to around 340 mV when the SEPIC is supplied from a 10-V rail. The transition from buck to boost mode is seamless with this converter.

LLC with Direct Frequency Control

LLC CONVERTERS ARE resonant power supplies based on three energy-storing elements: a capacitor C_r, the magnetizing inductance of the transformer L_m and another inductance, L_r, which is either part of the transformer – a leakage term – or externally added. This resonant converter is characterized by a resonant frequency F_0 and it can be operated below, above or at F_0, exhibiting different switching characteristics in these modes.

The control variable for an LLC converter is the switching frequency F_{sw}, whose lower and upper excursion limits are precisely set by the selected controller. As our error amplifier delivers a voltage, a voltage-controlled oscillator (VCO) is used to drive the power stage. Its slope in Hz/V represents the modulator gain and must be included in the stability analysis.

The small-signal analysis of an LLC converter cannot be carried out with averaged modeling. In this resonant structure, power is not transmitted by *average* values of currents and voltages but by their fundamental and harmonics. Specific analysis methods are needed, such as *extended describing functions* (EDF) which lead to complex models. They are difficult to use in my opinion, in particular because equations change depending on the operating point with respect to the resonant frequency. You are left with two options: hardware measurements – you need a working prototype – or SIMPLIS® simulations. We start with the VCO block, essential to our converter.

The minimum frequency parameter of 60 kHz sets current source I_1 parameter {IFMIN}. The frequency increases as more current is injected via G_2 when V_{FB} goes up to 5 V (maximum frequency 175 kHz). The VCO slope is thus $(175-60)/5 = 23$ kHz/V.

Power Stage Response

THE DIRECT CONTROL OF FREQUENCY by the VCO leads to various control-to-output transfer functions which make loop compensation a difficult exercise. Depending on the resonant tank design, loading conditions and the input source level, the LLC converter may cross different modes, all leading to different ac responses. The design must remain stable in all of these modes. The test circuit is given below, it delivers 24 V/10 A from a PFC stage set to 400 V. However, the LLC converter must still operate with a high-voltage rail down to 340 V, for example in case of hold-up time specifications.

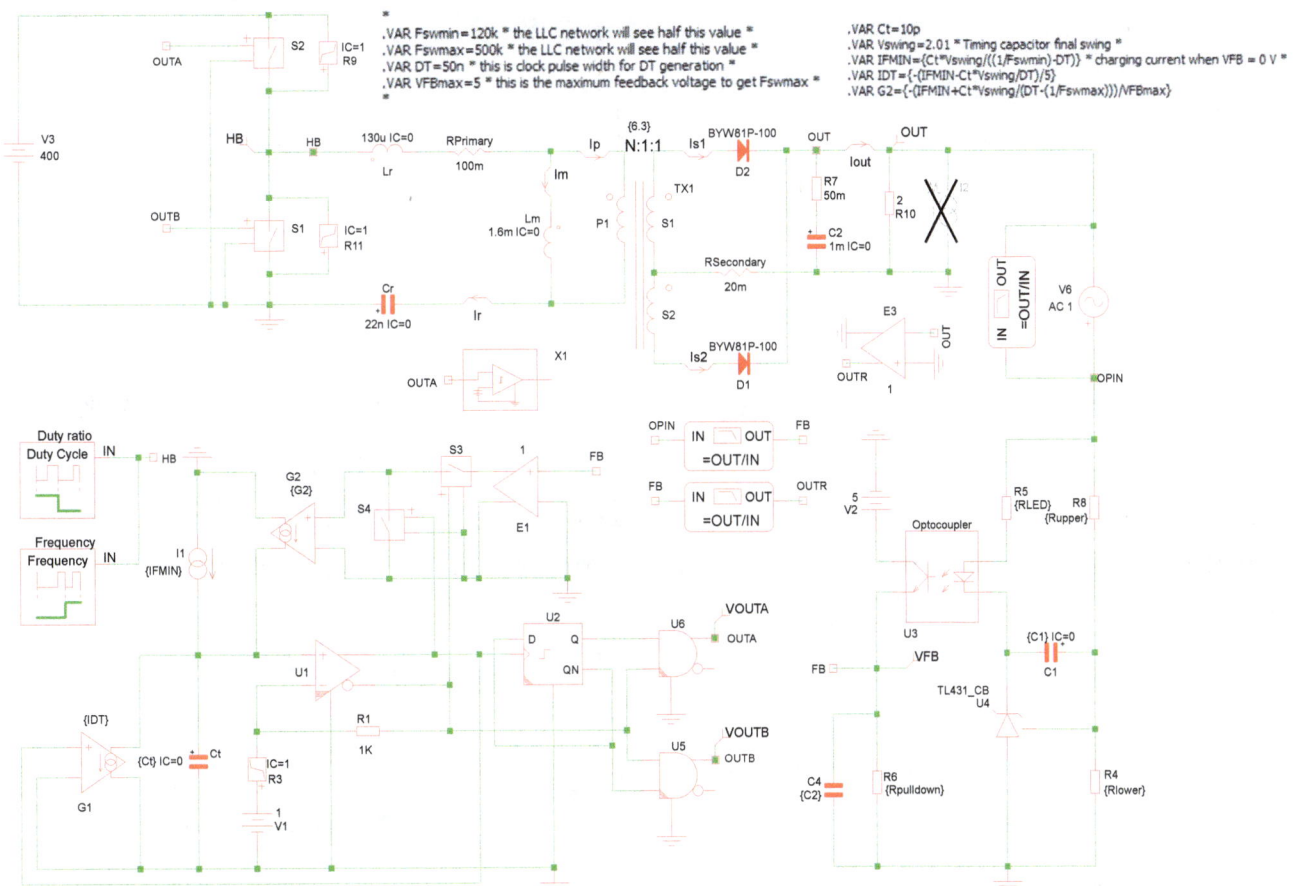

To show the impact of input and output conditions, it is possible to take advantage of multi-step analysis. I have asked SIMPLIS® to sweep the load from 2 Ω to 12 Ω in 10 steps while recording the power stage response. The input source will later be adjusted from 340 V to 400 V with a constant load.

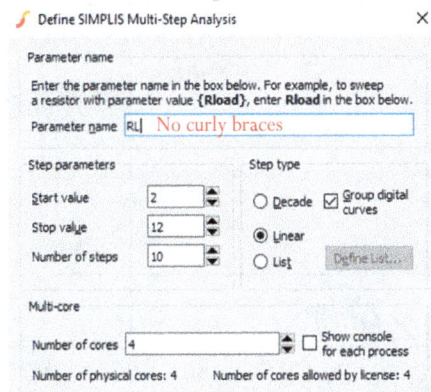

A Highly-Variable AC Response

DEPENDING ON THE OPERATING POINT, the power stage ac response will change significantly, affecting the crossover frequency once the loop is closed. The below plot illustrates the extreme variability of the structure.

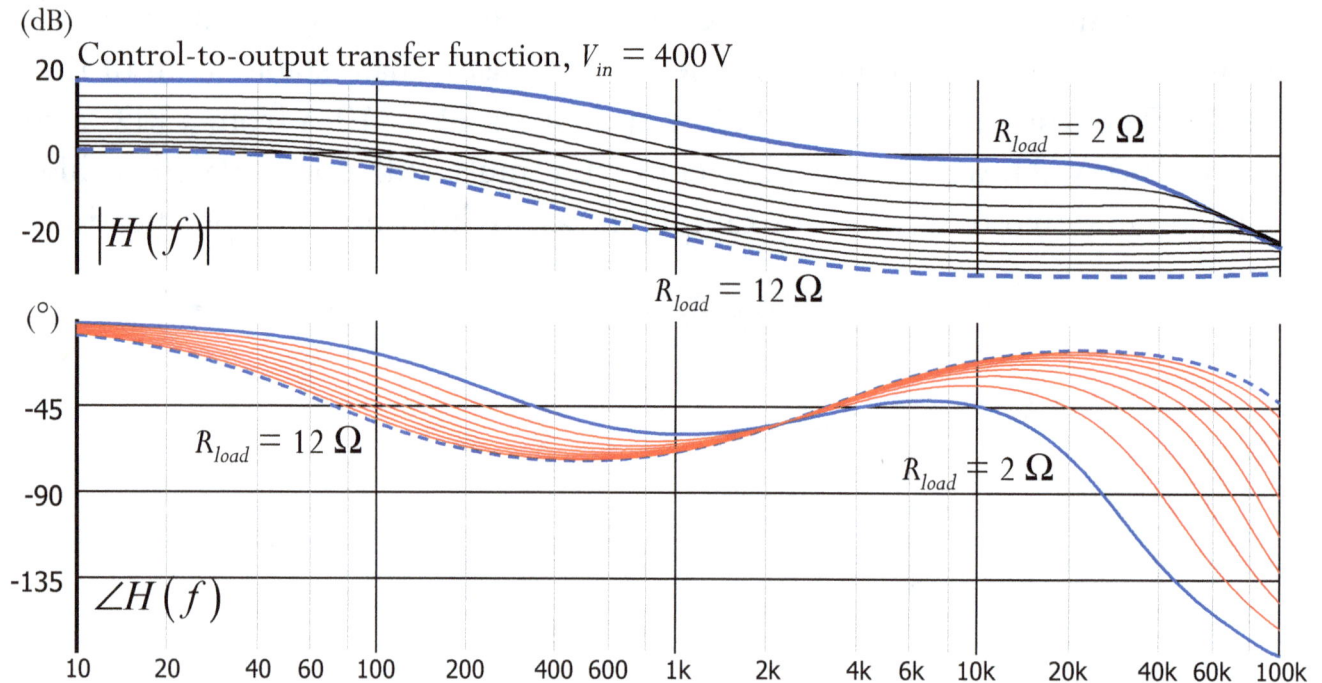

The input voltage is now swept between 360 V and 400 V. The variability is less compared to the load chart. A 1-kHz crossover frequency seems to be a good choice in light of these graphs. It is important to note that another set of curves will be obtained (peaky magnitude response, beat frequency etc.) if you design the resonant network differently (turns ratio, L_m/L_r ratio).

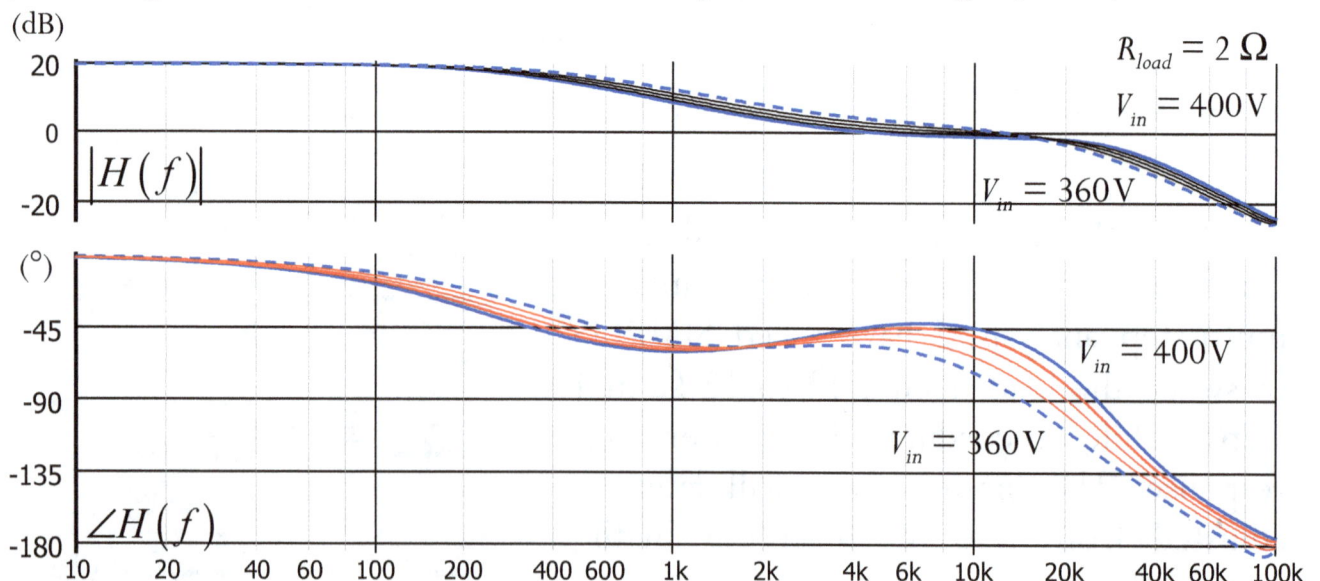

Closed-Loop Response

WE HAVE IMPLEMENTED a type 2 compensator driving an optocoupler. Considering the previous power stage magnitude variations, crossover will also be significantly impacted, in particular when the load gets lighter. It is important to verify margins remain within safe limits in all of these extreme conditions. Please note that many LLC switching controllers limit the maximum frequency excursion by programming skip cycle when the converter enters light- or no-load conditions.

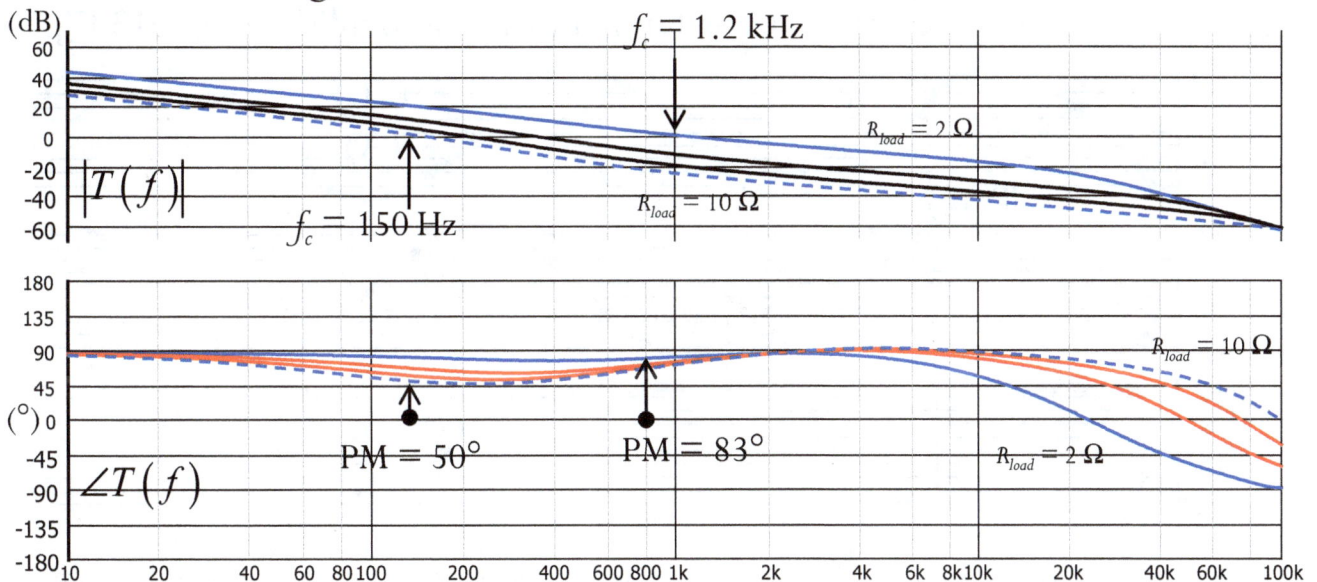

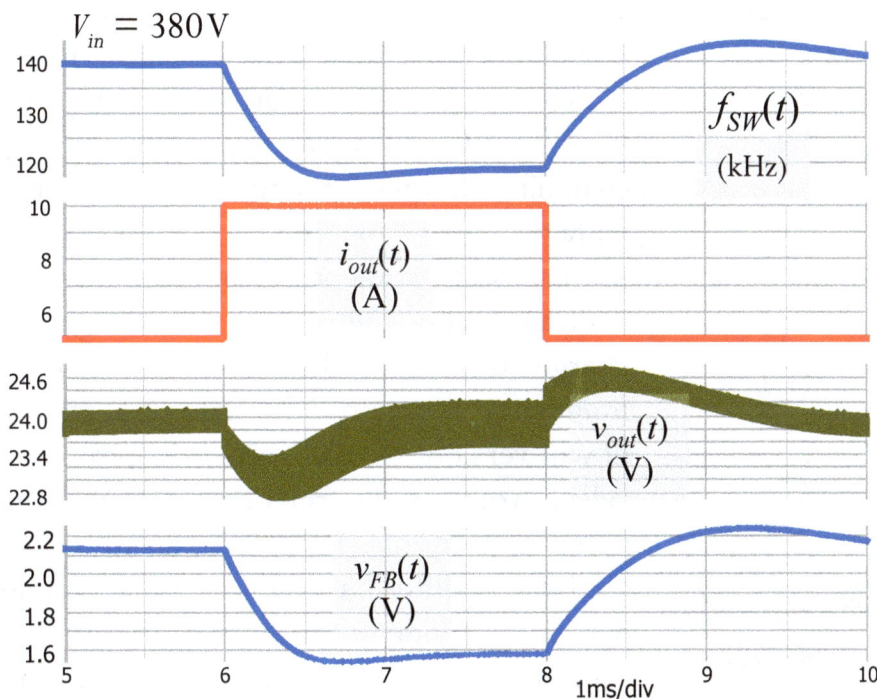

The load current is stepped from 5 to 10 A, with a 1-A/μs slope. The drop is 5% of the nominal voltage. If you can't push crossover higher, you can certainly increase the output capacitor. A 2.2-mF value with a 2-kHz crossover frequency reduces the drop to 3% of the nominal V_{out}.

LLC with Bang-Bang Charge Control

THE CONTROL-TO-OUTPUT transfer function of the LLC converter operated in direct frequency control or DFC, can make loop compensation a complicated exercise. Rather than *directly* controlling frequency, it is possible to adjust the transmitted power by monitoring the peak and valley levels of the resonant capacitor. That way, the error voltage sets the energy conveyed from the primary to the secondary, cycle-by-cycle, and *indirectly* sets the operating frequency. This technique has been described in a paper written by Queen's University (Kingston, Canada) members and published in 2013 [7].

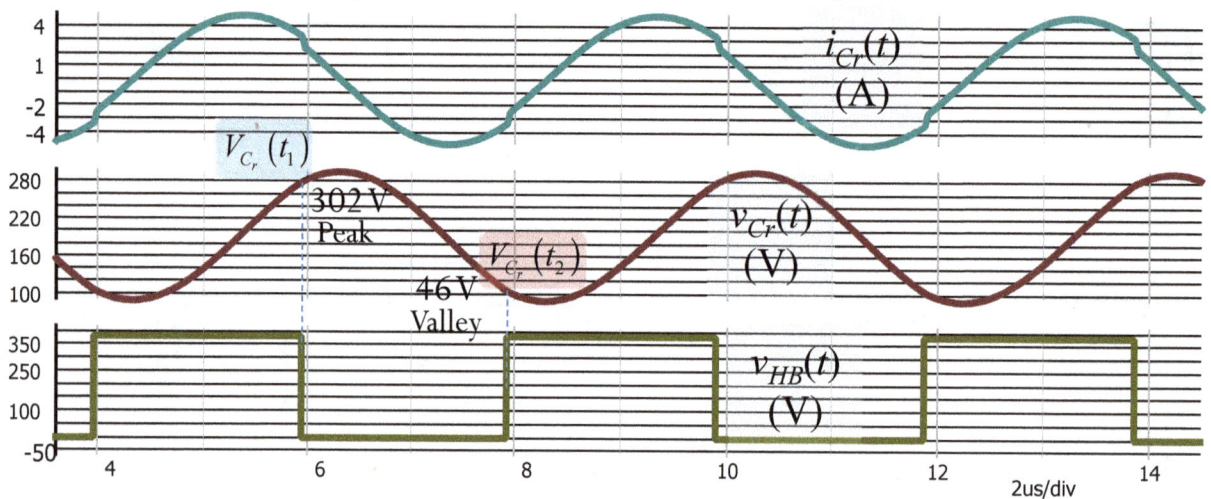

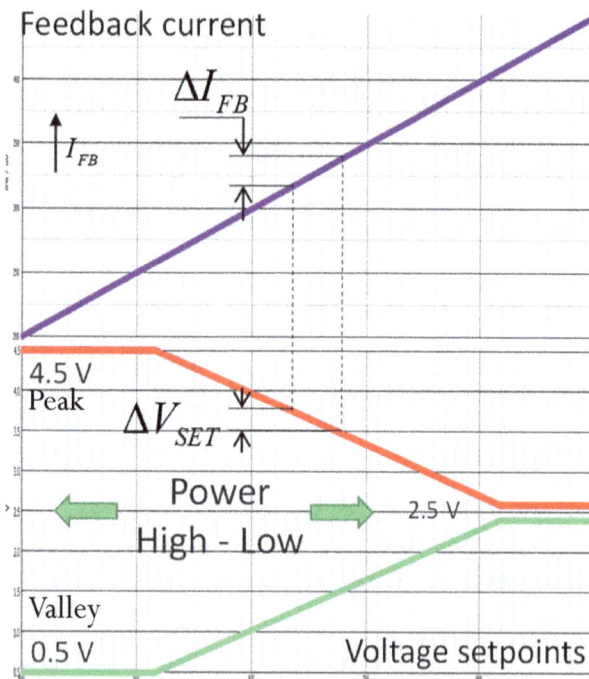

The modulator adjusts the peak and valley levels of the scaled-down capacitor voltage.

In the above waveforms, you see the voltage across the 36-nF resonant capacitor C_r going through peaks and valleys. We can write $P_{in} = \Delta Q \cdot F_{sw} \cdot V_{in}$, in which $\Delta Q = (302-46).36n = 9.22 \ \mu C$. This charge, scaled by the switching frequency and the 350-V rail voltage, leads to 268-W of processed power. NXP semiconductors has released dedicated controllers for this bang-bang charge-controlled LLC, in which a modulator symmetrically adjusts a scaled-down version of the capacitor peak and valley voltages in relationship with the power to transmit. Texas Instruments offers a slightly different version – a hybrid hysteretic mode – in which the sensed capacitor voltage is supplemented with an artificial ramp.

Power Stage Characteristics

THE GOAL OF THIS TECHNIQUE is to obtain a reduced-order control-to-output transfer function, easier to compensate. The authors have shown that this expression is of first order for the bang-bang charge-controlled LLC:

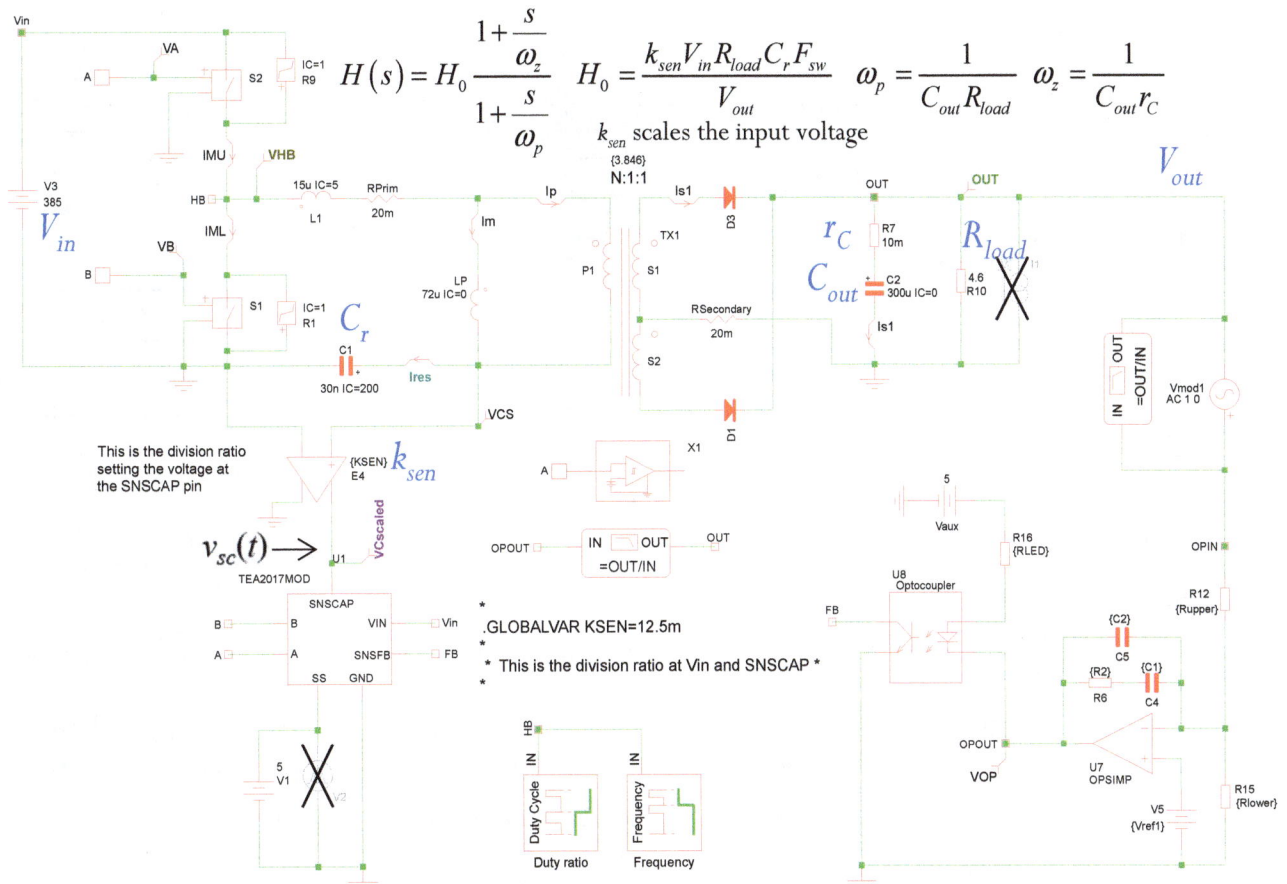

$$H(s) = H_0 \frac{1 + \dfrac{s}{\omega_z}}{1 + \dfrac{s}{\omega_p}} \quad H_0 = \frac{k_{sen} V_{in} R_{load} C_r F_{sw}}{V_{out}} \quad \omega_p = \frac{1}{C_{out} R_{load}} \quad \omega_z = \frac{1}{C_{out} r_C}$$

k_{sen} scales the input voltage

This is an application example of the TEA2017 which includes a front-end PFC and a charge-controlled power modulator. It is a 500-W converter.

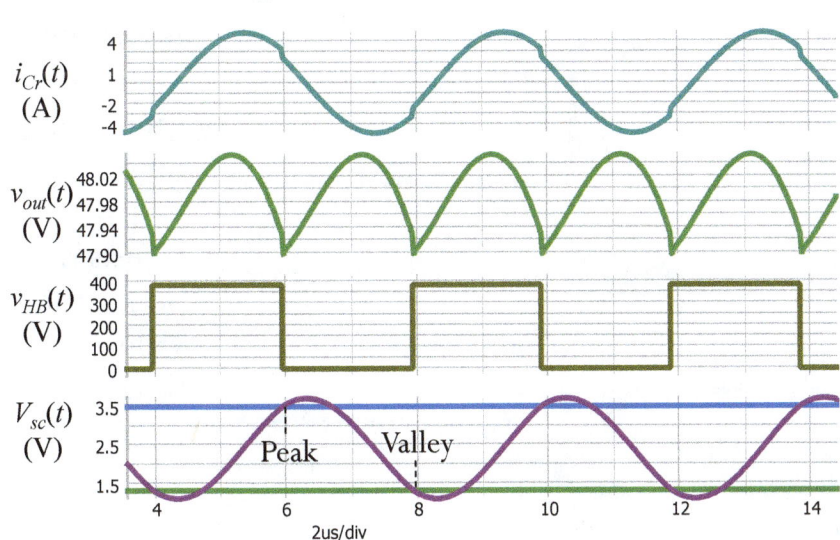

The nominal switching frequency approaches 250 kHz at 385-V dc input. A type 2 compensator is a good choice for compensating this converter. The macro places a zero at 2.2 kHz and a pole at 45 kHz for a 10-kHz nominal crossover. There is no way I could meet this number with direct frequency control.

A Well-Behaved AC Response

WE CAN IMMEDIATELY CHECK the load impact on the transfer function by using the multi-step resource of SIMPLIS®. As shown below, the crossover frequency is almost unaffected when the load is swept between 4.6 Ω (500 W) and 15 Ω (153 W):

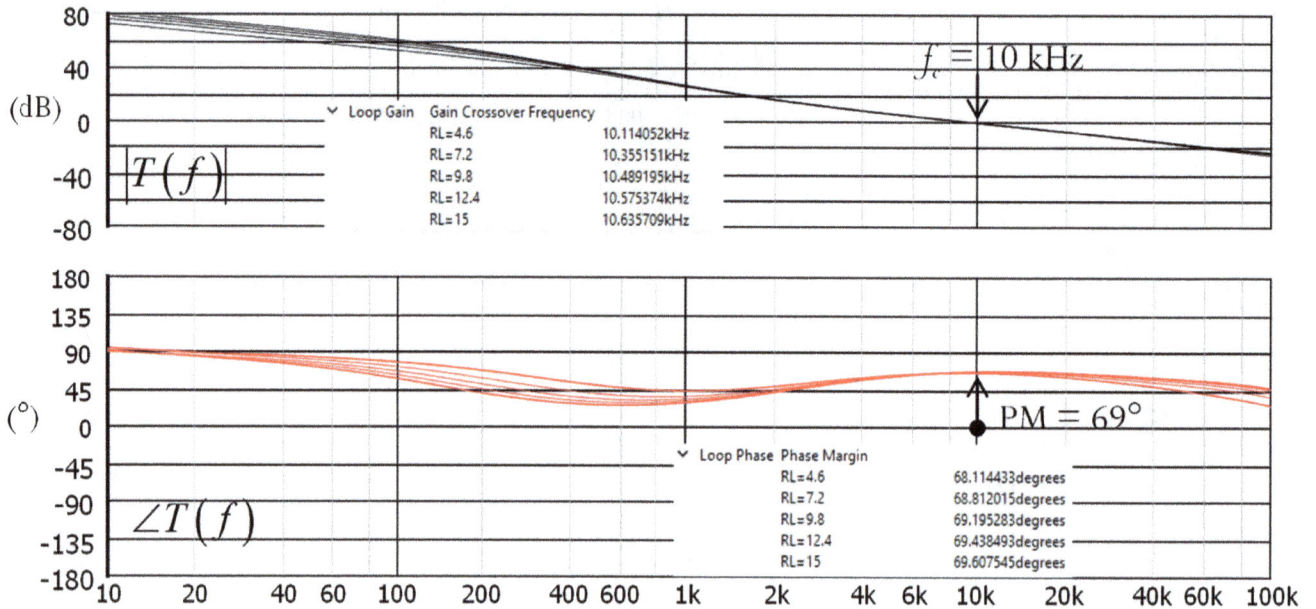

⌄ Loop Gain	Gain Crossover Frequency
RL=4.6	10.114052kHz
RL=7.2	10.355151kHz
RL=9.8	10.489195kHz
RL=12.4	10.575374kHz
RL=15	10.635709kHz

⌄ Loop Phase	Phase Margin
RL=4.6	68.114433degrees
RL=7.2	68.812015degrees
RL=9.8	69.195283degrees
RL=12.4	69.438493degrees
RL=15	69.607545degrees

Similarly, sweeping the dc input from 360 V to 410 V at 500-W P_{out} shows f_c varying from 8.5 to 11 kHz, a very narrow change. The load step from 5 to 11 A with a 1-A/µs slope confirms an excellent transient response.

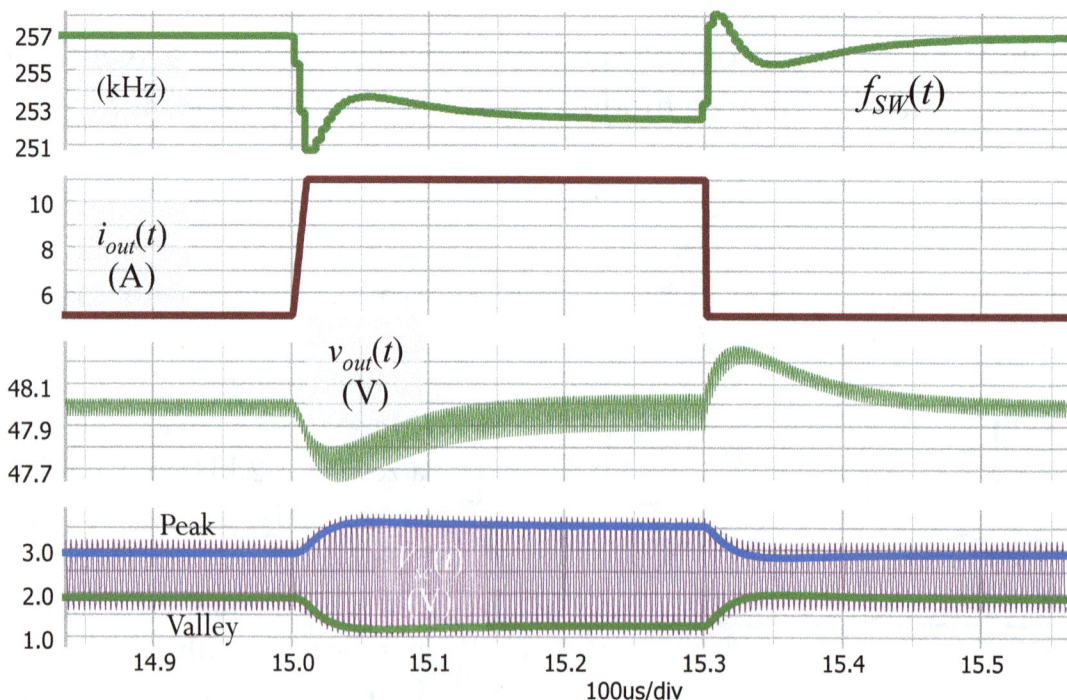

The transient response is fast and clean. You can see how the loop adjusts the peak and valley thresholds.

LLC in Current-Mode Control

THE LLC CONVERTER can also be operated in current-mode control and different embodiments are possible. For example, the NCP1399 from *onsemi* turns the high-side switch on while monitoring the resonant current. When the current reaches the setpoint, the controller turns the switch off. It precisely records this on-time and activates the low-side switch on for the exact same duration. A 50% duty ratio is obtained and the switching frequency is, again, *indirectly* set by the peak current setpoint. I built the modulator around a 10-pF timing capacitor charged and discharged from a peak to exactly 0 V via identical currents. The capacitor reaches its peak voltage at the end of the on-time. The off-time is then the time needed to bring the capacitor voltage down to 0 V. With ideal components, the duty ratio is exactly 50%.

The current-mode control of the LLC also requires the injection of some slope compensation. The below circuit shows how it can be done, by subtracting a downslope from the error voltage setting the peak current level. Dead-time is then added before driving the half-bridge transistors.

This circuit generates a voltage slope of 82 kV/s and increases to 164 kV/s with a gain of 2 in G_4.

Simulating the CM LLC

THE COMPLETE CIRCUIT appears below and shows the distinct blocks. I have added individual dead-time subcircuits to reduce the shoot-through currents during the high-voltage commutations on the bridge. The resonant current is sensed via a capacitive differentiator monitoring the voltage of capacitor C_1. This signal is then compared with the setpoint imposed by the regulation loop which ensures a 48-V output at a 10-A current.

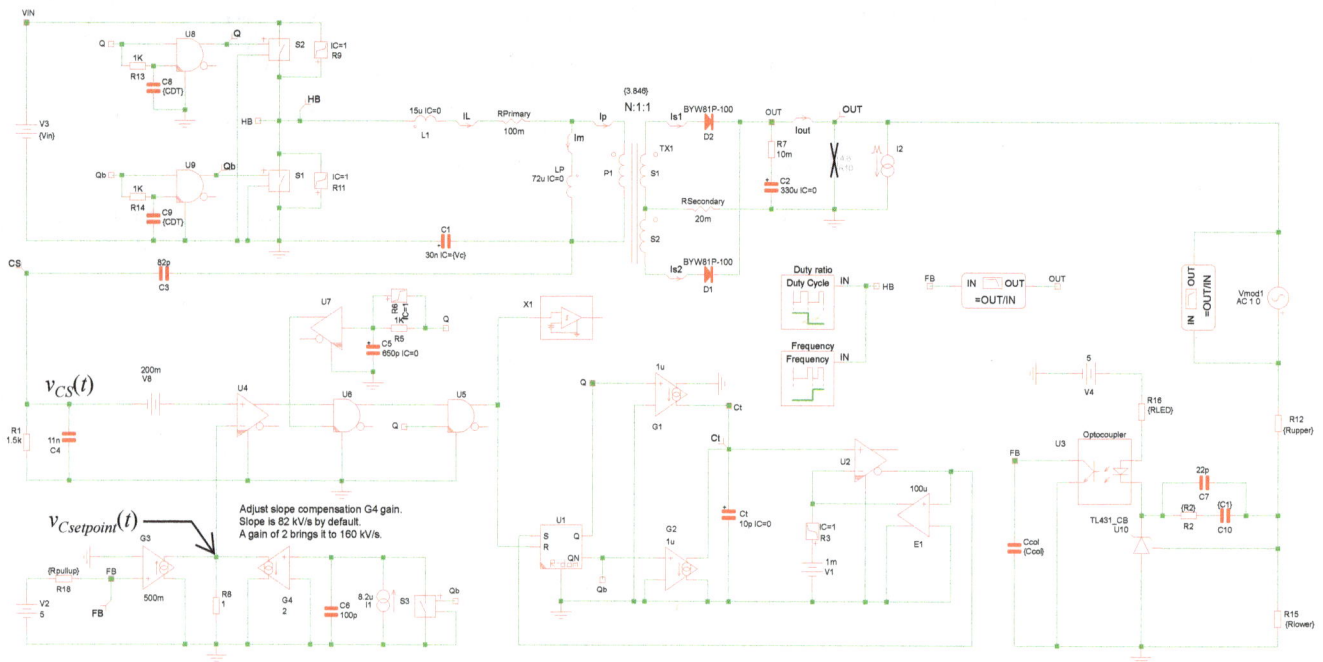

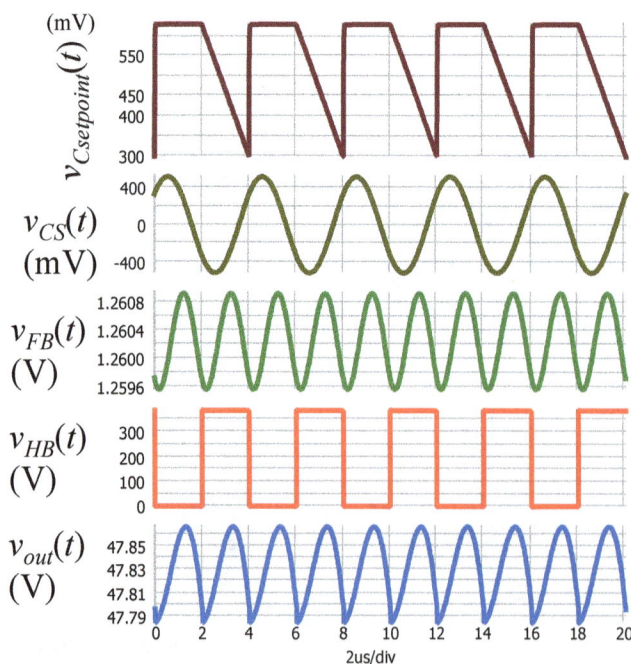

The converter delivers 48 V from the 385-V dc input. The switching frequency is 249 kHz, very close to the operating point of the bang-bang charge-controlled example. The type 2 compensator is designed without a fast lane considering the output voltage exceeding the TL431 breakdown voltage. The LED resistance is arbitrarily set to 1 kΩ. For a 10-kHz crossover, the zero is located at 2 kHz and the pole at 49 kHz. You will need to ensure the optocoupler pull-up resistance is low enough to push its pole at 20 kHz or so. If it does not work, a cascode arrangement might be the solution to favor.

Compensation of the CM LLC

The below control-to-output transfer function does not reveal any particular difficult areas and a 10-kHz crossover is a possible choice, provided the optocoupler is fast enough.

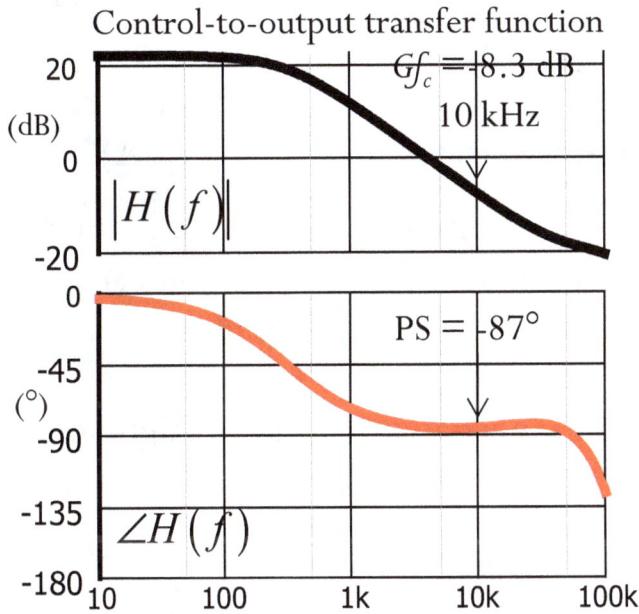

Control-to-output transfer function

$Gf_c = 8.3$ dB
10 kHz

$|H(f)|$

$PS = -87°$

$\angle H(f)$

$V_{in} = 385$ V, 48 V/10 A
Compensated loop gain

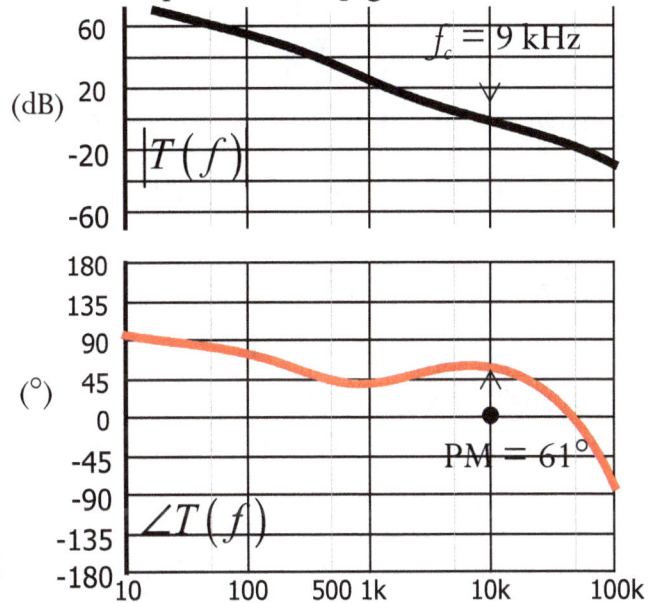

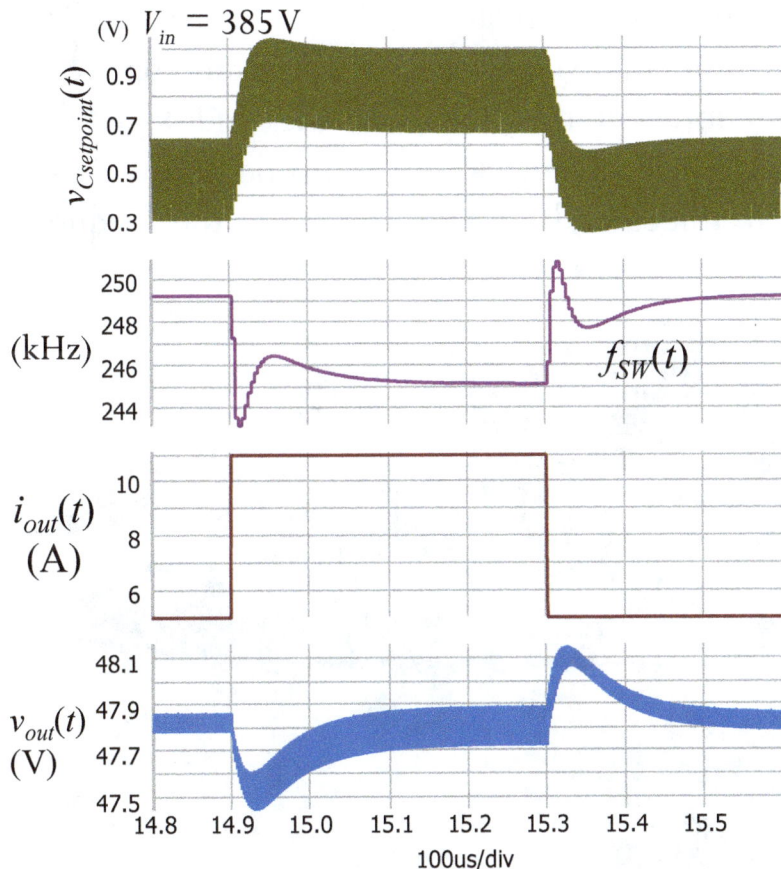

$f_c = 9$ kHz

$|T(f)|$

$PM = 61°$

$\angle T(f)$

$V_{in} = 385$ V

$v_{Csetpoint}(t)$

$f_{SW}(t)$

$i_{out}(t)$ (A)

$v_{out}(t)$ (V)

100us/div

The output current is stepped from 5 A to 11 A with a slope of 1 A/μs and you see how stable the output response is, without overshoot. I have run the same circuit from 360 V to 410 V and the response remains unchanged. Needless to say how easy to compensate this current-mode LLC is, compared to its direct frequency control counterpart.

Practical Experiments

THE BENCH MEASUREMENT represents the ultimate step to validate the theoretical analysis, whether it was conducted analytically or through simulation. If you have correctly extracted the parasitics of your components – in particular the ESR of the capacitors – then theoretical results should approach the bench results quite well, validating the model. With this confirmed model, you can then confidently extend the exploration on the computer and run Monte Carlo or worst-case circuit analyses (WCCA).

For those of you starting in the field of loop control, I recommend you start with low-power and low-voltage dc-dc converters. First of all, you can safely probe the operating waveforms or touch the board without suffering an electric shock. Second, if you miss the loop design, it won't end up in a dangerous situation as I have often experienced with unstable high-voltage circuits going straight to maximum duty ratio at the first power on.

In my [APEC 2018](#) and [APEC 2019](#) seminars, I have purposely developed some small dc-dc converters – a boost and a buck – operated in VM and CM. They are powered from a venerable UC3843 controller and the error amplifier is compensated with leaded components you insert into small connectors. Below is the voltage-mode version and the switch in the middle lets you toggle between a physically-opened loop and closed-loop operation. Pins on the board let you conveniently inject and probe signals for the frequency response analyzer (FRA).

A Buck Converter with UC3843

FOR THESE EXPERIMENTAL BOARDS, I had to modify the original application circuit of the UC3843 which is a current-mode controller. A sawtooth generator has been built for the VM version and I added an external soft-start and open-loop protection (not shown below). The high-side driver is using a simple arrangement as described by Monsieur Balogh in his popular application note on MOSFETs drivers, SLUA618.

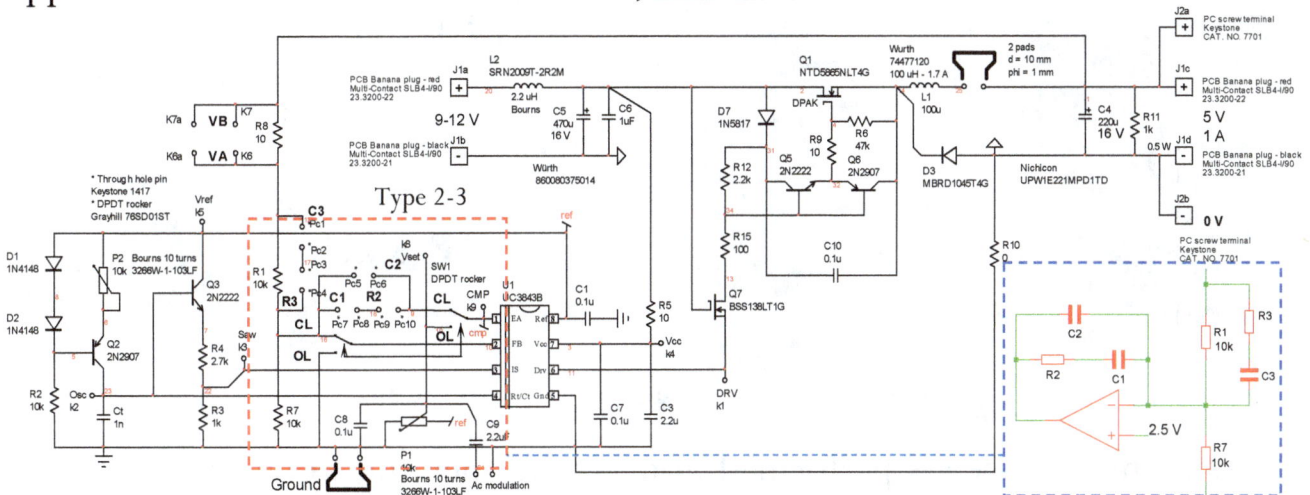

If you now run the measurement with the FRA and you did correctly extract parasitics, this is the typical plot you may obtain. Ok, maybe not at the first attempt as you need a bit of practice to set the filters and place probes on the board (far from switching nodes), but you see how close the curves can be.

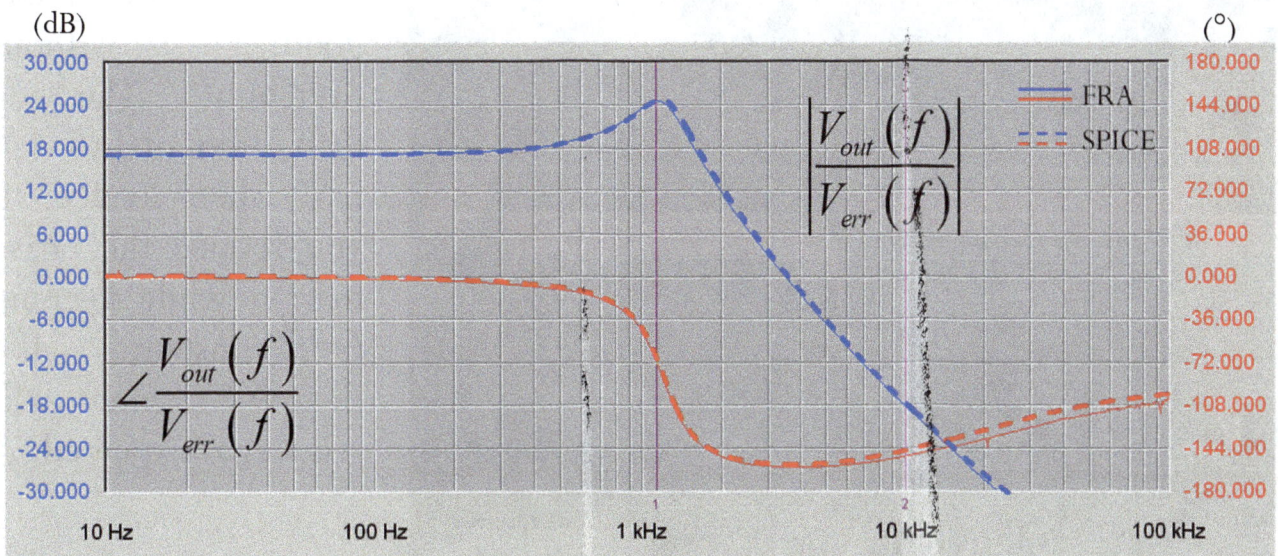

I like to have phase and magnitude centered at zero, it makes the reading easier, especially for phase margin with the loop gain plot.

Checking the Loop Gain

ONCE THE BUCK IS COMPENSATED with the methods described in this manual, you check the loop gain and plot crossover, phase and gain margins. As the graph shows, simulation and bench results match very well:

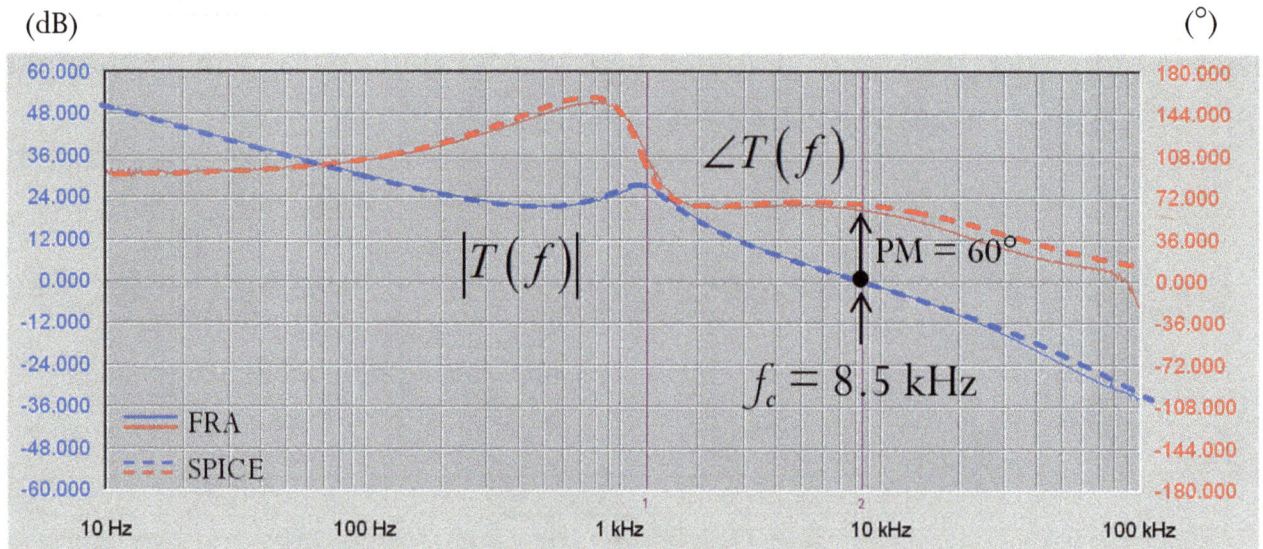

A transient load step is performed in different operating conditions. Keep in mind that if we can predict a transient response with the small-signal output impedance (or simulation), a real test may bring different results owing to the many nonlinear behaviors and stray elements not accounted for.

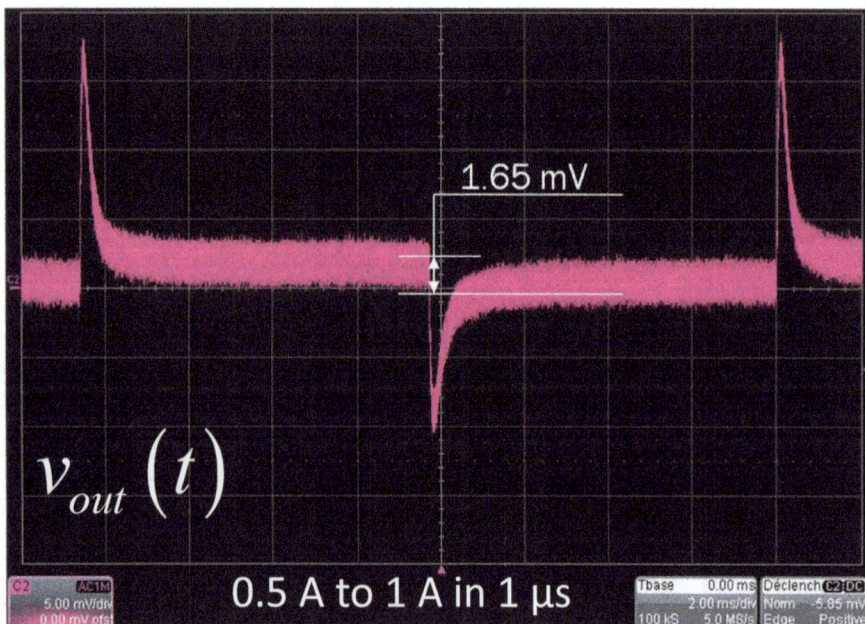

The response looks good, without overshoot. The dc drift amounts to 1.6 mV when loaded. If you reduce the phase margin while keeping f_c constant, the recovery time will reduce, as expected.

Should you like to build these boards for educational purposes, let me know by email and I will send you the schematic diagrams with the bill of materials.

A Few Books to Consider

I HAVE GATHERED some of the books I believe you could acquire for strengthening your knowledge in the field of control. There are many sources but few are specifically focused on our field specifically and, most importantly, include practical details for immediate implementation.

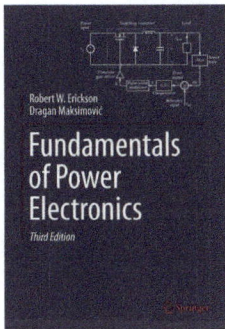

Fundamentals of Power Electronics – Robert Erickson and Dragan Maksimović.

This is an excellent and comprehensive text on power electronics. A must read for engineers willing to understand theory.

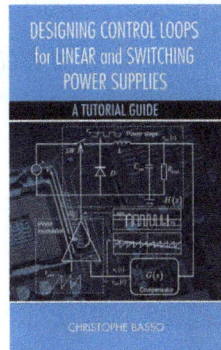

Designing Control Loops for Linear and Switching Power Supplies – Christophe Basso.

A good balance between theory you must know and design considerations with many examples.

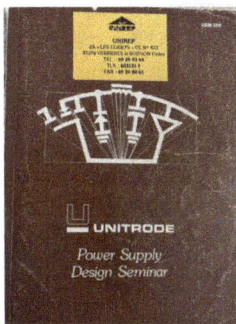

Unitrode – Power Supply Seminar series.

You just can't miss these manuals written by the best application engineers years ago, Lloyd Dixon, Bob Mammano, Bill Andreycak…

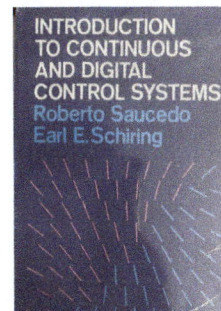

Introduction to Continuous and Digital Control Systems – Roberto Saucedo and Earl Schiring

An old book but packed with a lot of useful theoretical data, still valid of course!

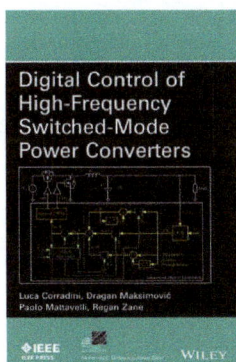

Digital Control of High-Frequency Switched-Mode Power Converters – L. Corradini, D. Maksimović, P. Mattavelli, R. Zane

If you look for a rigorous and extensive coverage of digital control, it is an excellent read.

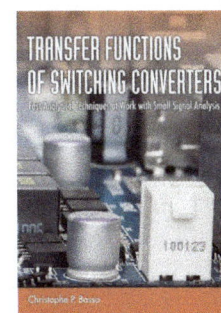

Transfer Functions of Switching Converters – Christophe Basso

If you wonder how to derive the transfer functions of switching converters, this is the book.

Papers and Articles

I HAVE TRIED TO LIST some papers I believe are relevant to the topic of loop control. As any list, it is far from being comprehensive but the references in each paper is a link to another source of knowledge.

1. V. Vorpérian, *Simplified Analysis of PWM Converters using Model of PWM Switch, parts I and II*, IEEE Transactions on Aerospace and Electronic Systems, Vol. 26, NO. 3, 1990

2. V. Vorpérian, *Analysis of Current-Controlled Converters using the Model of the Current-Controlled PWM Switch*, Power Conversion and Intelligent Motion Conference, pp. 183-195, 1990

3. K. Yao et al., *Critical Bandwidth for the Load Transient Response of Voltage Regulator Modules*, IEEE Transactions on Power Electronics, Vol. 19, NO. 6, 2004

4. C. Hymowitz, *Monte Carlo gone Wrong*, on-line EDN Magazine, 2019

5. R. D. Middlebrook, *Input Filter Considerations in Design and Application of Switching Regulators*, IEEE Proceedings, 1976

6. R. B. Ridley, *A New Small-Signal Model for Current-Mode Control*, PhD Dissertation, Virginia Polytechnic Institute and State University, November, 1990.

7. Z. Hu, Y. Liu, *Bang Bang Charge Control for LLC Resonant Converters*, IEEE Transactions on Power Electronics, Vol. 30, NO. 2, 2015

8. D. Venable, *The k Factor; A new Mathematical Tool for Stability Analysis and Synthesis*, Proceeding of Powercon 10, San-Diego, CA, march 22-24, 1983

9. Y. Panov, M. Jovanović, *Small-signal analysis and control design of isolated power supplies with optocoupler feedback*, IEEE Transactions on Power Electronics, Vol. 20, NO. 4, 2005

10. R. B. Ridley, *Designing with the TL431*, Designer Series XV, Switching Power Magazine, 2005

11. C. Basso, *The TL431 in the Control of Switching Power Supplies*, onsemi Seminar, September 2009

12. L. Dixon, *Closing the Feedback Loop*, SEM300, Unitrode Seminars, SLUP068, Texas-Instruments

13. C. Basso, *Understanding Op Amp Dynamic Response In A Type-2 Compensator*, 2-part article, How2Power.com, January 2017

14. C. Basso, *Modern Control Methods For LLC Converters Simplify Compensator Design*, How2Power.com, December 2021

15. C. Basso, *Analysis, Simulation and Experimentation Enable Successful Design of Power Supply Compensation*, How2Power.com, July 2020

16. Y. Panov, M. Jovanović, L. Huber, *Principles of Converter Control*, Internal Course, Delta Electronics, 2002

17. M. Jovanović, Introduction to Digital Control of Switch-Mode Power Converters, Internal course, Delta Electronics, 2014

18. J. Turchi, *Compensation of a PFC Stage Driven by the NCP1654*, onsemi, AND8321/D, 2009

Internet Links

BELOW IS A SHORT LIST of links where you will find interesting materials to download or videos to watch.

My webpage in which you will find books, application notes, seminars and the free ready-made templates I used in the book: http://powersimtof.com/Spice.htm

How2Power is the on-line monthly newsletter edited by David Morrison, entirely dedicated to power electronics. A must-read for power engineers: http://www.how2power.com/

Electronic StackExchange is an interesting professional forum in which users ask questions on various subjects, including power and control. The answers are often very well documented: https://electronics.stackexchange.com/questions

Plenty of YouTube channels to learn from!

Brian Douglas is an enthusiast of control systems and an excellent teacher too:

https://www.youtube.com/@BrianBDouglas/videos

Robert Bolanos is well-known in the field of power supply and posts many interesting tutorial videos on loop design:

https://www.youtube.com/@RobertBolanos

Ali Shirsavar of Biricha Digital covers many aspects of loop control, in the digital and analog domains: https://www.youtube.com/@BirichaLectures/videos

Katherine Kim teaches at the National Taiwan University and runs an instructive YouTube channel with plenty of good videos on power electronics:

https://www.youtube.com/user/katkimshow

Marcos Alonso teaches at the university of Oviedo, Spain, and posts many interesting videos on control, illustrated with LTspice® or QSPICE®:

https://www.youtube.com/@MarcosAlonsoElectronics/featured

Sam Ben-Yaakov offers many excellent videos on power electronics, ranging from simple to complicated structures, a must-visit site: https://www.youtube.com/user/sambenyaakov

Eddie Aho presents many interesting things on power circuits, instruments but also books he reviews: https://www.youtube.com/c/kissanalog

www.ingramcontent.com/pod-product-compliance
Lightning Source LLC
Chambersburg PA
CBHW061817210326

41599CB00034B/7031